XIV All-Russian School-Conference of Young Scientists with International Participation "Actual Problems of Thermal Physics and Physical Hydrodynamics" 2016

EPJ Web of Conferences Volume 159 (2017)

Novosibirsk, Russia
22 - 25 November 2016

Editors:

Dmitry Markovich
Dmitry Zaitsev
Andrey Semenov

ISBN: 978-1-5108-5054-5

Printed from e-media with permission by:

Curran Associates, Inc.
57 Morehouse Lane
Red Hook, NY 12571

Some format issues inherent in the e-media version may also appear in this print version.

Printed by Curran Associates, Inc. (2017)

For additional information, please contact EDP Sciences – Web of Conferences at the address below.

EDP Sciences – Web of Conferences
17, Avenue du Hoggar
Parc d'Activité de Courtabœuf
BP 112
F-91944 Les Ulis Cedex A
France

Phone: +33 (0) 1 69 18 75 75
Fax: +33 (0) 1 69 28 84 91

contact-edps@webofconferences.org

Additional copies of this publication are available from:

Curran Associates, Inc.
57 Morehouse Lane
Red Hook, NY 12571 USA
Phone: 845-758-0400
Fax: 845-758-2633
Email: curran@proceedings.com
Web: www.proceedings.com

TABLE OF CONTENTS

edp sciences EPJ Web of Conferences

Statement of Peer review

In submitting conference proceedings to EPJ Web of Conferences, I certify to the Publisher that I adhere to the **Policy on Publishing Integrity** of the journal in order to safeguard good scientific practice in publishing.

1. All articles have been subjected to peer review administered by the proceedings editors.
2. Reviews have been conducted by expert referees, who have been requested to provide unbiased and constructive comments aimed, whenever possible, at improving the work.
3. Proceedings editors have taken all reasonable steps to ensure the quality of the materials they publish and their decision to accept or reject a paper for publication has been based only on the merits of the work and the relevance to the journal.

Title, date and place of the conference

XIV All-Russian school-conference of young scientists with international participation «Actual problems of thermal physics and physical hydrodynamics» (AVTFG2016), November 22–25, 2016, Novosibirsk, Russia

Proceedings editor(s):

Dmitry Markovich, Dmitry Zaitsev and Andrey Semenov

Date and editor's signature 6.10.2017

Development of advanced air-blown entrained-flow two-stage bituminous coal IGCC gasifier

Nikolay A. Abaimov*, and *Alexander F.* Ryzhkov

Federal State Autonomous Educational Institution of Higher Education «Ural Federal University named after the first President of Russia B.N.Yeltsin», 620002, 19 Mira street, Ekaterinburg, Russia

Abstract. Integrated gasification combined cycle (IGCC) technology has two main advantages: high efficiency, and low levels of harmful emissions. Key element of IGCC is gasifier, which converts solid fuel into a combustible synthesis gas. One of the most promising gasifiers is air-blown entrained-flow two-stage bituminous coal gasifier developed by Mitsubishi Heavy Industries (MHI). The most obvious way to develop advanced gasifier is improvement of commercial-scale 1700 t/d MHI gasifier using the computational fluid dynamics (CFD) method. Modernization of commercial-scale 1700 t/d MHI gasifier is made by changing the regime parameters in order to improve its cold gas efficiency (CGE) and environmental performance, namely H_2/CO ratio. The first change is supply of high temperature (900°C) steam in gasifier second stage. And the second change is additional heating of blast air to 900°C.

1 Introduction

Coal is the most widespread kind of fossil fuel in the world. Issues of highly efficient and environmentally friendly usage of coal fuel will be relevant of at least several hundred years. One of the promising options for solving these problems is the integrated gasification combined cycle (IGCC) technology. Coal or other solid fuel is converted (gasified) into a combustible synthesis gas by special reactors called gasifiers. The large capacity (over 250 MW) IGCC structure includes entrained-flow type gasifier, because of their high unit performance. The IGCC efficiency depends largely on the gasifier operation parameters.

The paper aim is development of advanced air-blown entrained-flow two-stage bituminous coal IGCC gasifier. To achieve the aim it is necessary to accomplish the following tasks:

1) examine the gasifiers development international experience and choose the original design of commercially successful gasifier, based on which advanced gasifier will be developed;

2) create and verify the model to investigate the features of entrained-flow gasifiers and develop new designs;

3) modify the original gasifier to increase its efficiency and improve environmental performance using the verified model.

* Corresponding author: nick.sum41@mail.ru

2 MHI gasification technologies

One of the most perspective and cost-effective entrained-flow gasification technology is the air-blown two-stage gasification. Using air blast instead of oxygen blast allows refusing from expensive air separation plant, and two-stage principle provides the efficiency of the air-blown gasifier on oxygen-blown unit level. A world leader in the air-blown entrained-flow two-stage gasification technology is Mitsubishi Heavy Industries (MHI) power engineering company. The company built the first Nakoso 250 MW demonstration IGCC plant with air-blown entrained-flow gasifier in September 2007 [1]. The plant was owned by the Joban Joint Power Company and start commercial exploitation in April 2013 [2]. MHI needed to create experimental facilities fleet (table 1) of different scales and operating parameters gasifiers for development of commercial-scale air-blown entrained-flow two-stage bituminous coal IGCC 1500-2000 t/d gasifier.

Table 1. Experimental air-blown entrained-flow two-stage MHI gasifiers

Parameters	Center Research Institute of Electric Power Industry [3, 4]		MHI, Ltd. [5]	MHI Nagasaki R&D Center [6]	IGC Research Association [7]	Clean Coal Power R&D Company, Ltd. [1]
Coal mass flow, t/d	2	3	8	24	200	1500-2000
Operation period, yr	1983-1995	2004-NA	NA	1998-2002	1991-1996	2007-2012
Operating time (total/ continuous), hr	2220/ NA	NA	NA	201/NA	4770/789	18788/2238
Number of investigated coals	22	1	NA	NA	3	8
Gasifier dimensions (height/ diameter), m	5.85/ 0.3	NA	NA	NA	11/1.2	3.5/18 (estimated)
Pressure, MPa	2	2	0.1	2	2.7	2.7
Features	A lot of exp. data	Possibility of a steam supply	Atmospheric pressure	Testing of fuel supply and automatics	Resolves issue of diffuser slagging	Converted into commercial operation

The MHI gasifier operation principle is constructive separation of combustion and gasification regions. The gasifier is divided into two chambers (stages): combustion chamber (first stage) and gasification chamber (second stage). The chambers are connected by confuser-diffuser pinch to prevent ingress of the fuel mineral part from the first stage to the second one. The primary coal and char are tangentially fed into the combustion chamber to create a cyclonic effect which allows intensifying of the combustion process. Besides this effect favors to create a molten slag film on the first stage walls, the slag gradually flows down into the slag bath. The coal particle carbon on the chamber walls burn out much more rapidly than in the air flow as it airspeed increases substantially. Further, the conversion gas products pass into the gasification chamber, where they are mixed with a secondary coal. There occurs a so-called chemical quenching, i.e. hot gas products of complete combustion are cooled as a result of the endothermic reactions between CO_2, H_2O and coal carbon. The secondary coal char is separated from the synthesis gas after the second stage by the cyclone-separator. Then, the char is fed into the gasifier first stage for combustion that improves the carbon conversion rate of the gasifier.

3 Model verification

The development of such gasifiers requires costly experimental studies and high precision mathematical modeling. There is very limited number of the experimental installations, which are capable to obtain sufficient results for modeling and development. However, there are many approaches to the entrained-flow gasifiers modeling. Over recent decades Computational fluid dynamics (CFD) method has become increasingly popular. The CFD approach has broad functionality and high accuracy.

To confirm the applicability of the our CFD model (described in [8]) for the MHI-type gasifiers calculations we carried out model verification using experimental data which was obtained at MHI 2 t/d gasifier [9]. Verification results are shown in Figure 1 and Table 2.

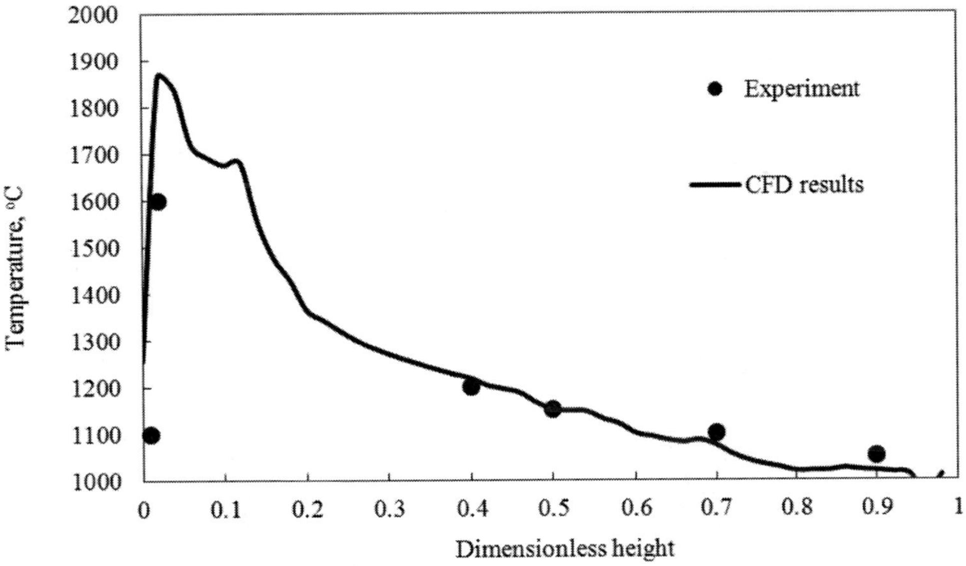

Fig. 1. Comparison of experimental [5] and CFD calculation temperature along gasifier height.

Table 2. Comparison of experimental [5] and CFD calculation syngas components concentrations.

Syngas component concentrations, % vol.	Experiment	CFD results
H_2	7.3	6.9
CO	20.2	21.5
CO_2	6.3	5.1
H_2O	2.0	3.7

Verification results show that used CFD model satisfactorily meets the requirements of adequate MHI-type gasifiers modeling. Thus, this model will continue to be used in the development of the advanced gasifier.

4 Gasifier development

When creating high-efficiency advanced air-blown entrained-flow two-stage IGCC gasifier the most obvious way is to upgrade the existing commercial MHI gasifier. The modernization consists in several fundamental technical solutions:

1) steam supply into gasifier second stage;

2) blast air extra heating;

3) optimization of the gasifier design, in particular the location of the steam supply points into the second stage taking into account the control mechanisms of the gasification process, used earlier in the paper [8].

The first two points are considered in this paper. Implementation of the third point requires additional experimental and computational data, so it will be resolved in future studies.

With increasing CO content relative to hydrogen in the synthesis gas the fuel nitrogen oxides quantity produced by combustion is reduced [10]. To increase the synthesis gas hydrogen concentration the steam is supplied into the gasifier second stage, and the heat required for the steam gasification is compensated by additional heating of the blast air. This improves the cold gas efficiency (CGE) of the gasifier, the carbon conversion rate and the synthesis gas hydrogen concentration to the required level.

Three-dimensional view and two-dimensional sections of the modernized MHI gasifier are presented in Figure 2.

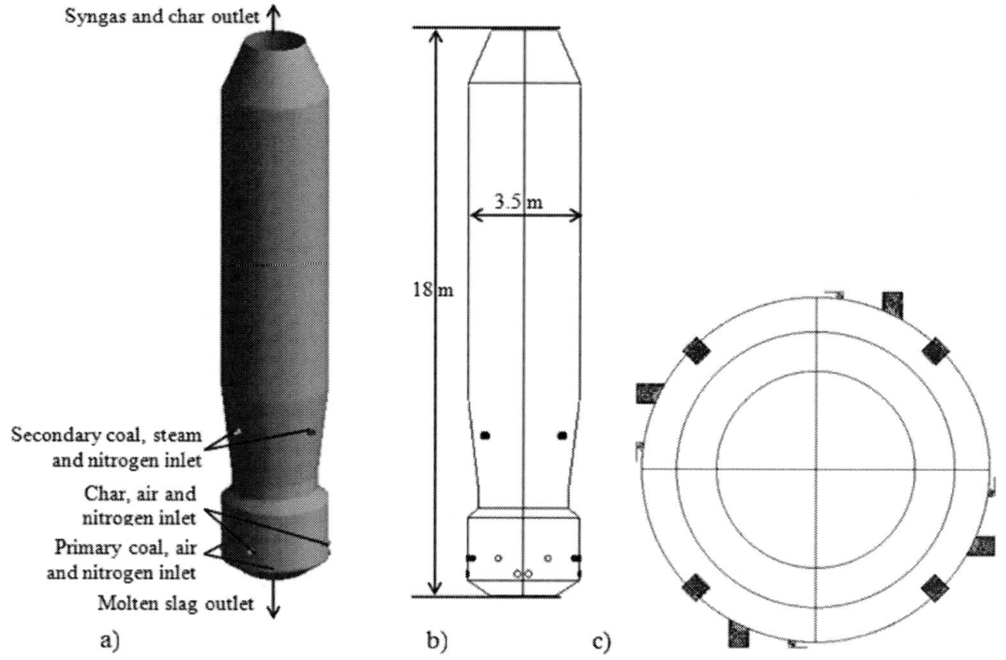

Fig. 2. Modernized gasifier: a) 3-D view; b) longitudinal section; c) transverse section.

Used air stoichiometric factor ensures the process flow near the point of thermodynamic equilibrium taking into account the amount of supplied steam. The blast air is enriched with oxygen to 25 %. Transport of coal and char is carried out by nitrogen. The second stage steam is fed at a rate sufficient to ensure that the synthesis gas temperature at the gasifier outlet is approximately 1100-1200°C. The blast air and steam have a temperature of 900°C. The compositions of the synthesis gas produced in the original (MHI) and modernized (UrFU) gasifier designs are given in Table 3.

Table 3. Comparison of MHI [1] and Ural Federal University (UrFU) gasifier operation parameters.

Gasifier	Dry syngas composition, % vol.					CGE, %
	CO	H_2	CO_2	CH_4	N_2	
MHI	30,5	10,5	2,8	0,7	55,5	77,2
UrFU	27,8	16,8	4,78	2,4	46,3	84,9

Modernization of the MHI gasifier improves the heating value of the produced synthesis gas, increases the H_2 concentration, reduces the CO concentration and increases gasifier CGE.

5 Conclusion

The proposed modernization (steam supply into gasifier second stage and blast air extra heating) of the MHI-type gasifier, conducted with the use of verified CFD model, improves the CGE and the ratio H_2/CO of the produced synthesis gas. A further way of the gasifier improving is optimization of location and directions of coal and steam feed into the gasifier second stage.

This work was carried out at the Ural Federal University and financially supported by the Russian Science Foundation (project no. 14-19-00524).

References

1. M. Susaki, Y. Takashima, H. Ishii, Y. Kitagawa, O. Shinada, T. Hashimoto, Mitsubishi Heavy Industries Technical Review, **46**, 5-8 (2009)
2. K. Sakamoto, Gasification and Syngas Technologies Conf. (2014) http://www.gasification-syngas.org/uploads/eventLibrary/2014_13.3_Mitsubishi_KS_UPDATED.pdf
3. S. Hara, J. Inumaru, M. Ashizawa, K. Ichikawa, JSME Int. J. Series B Fluids and Thermal Engineering, **45**, 518-522 (2002)
4. S. Hara, Second Int. Freiberg Conf. on IGCC and XtL Technologies (2007) http://tu-freiberg.de/sites/default/files/media/institut-fuer-energieverfahrenstechnik-143/publikationen/2007-5-4.pdf
5. C. Chen, T. Miyoshi, H. Kamiya, M. Horio, T. Kojima, The Canadian J. of chem. engineering, **77**, 745-750 (1999)
6. O. Shinada, A. Yamada, Y. Koyama, Energy Conversion and Management, **43**, 1221–1233 (2002)
7. S. Araki, Y. Hanai, J. of the Japan Institute of Energy, **75**, 839-850 (1996)
8. N. A. Abaimov, A. F. Ryzhkov, Thermal Engineering, **62**, 767–772 (2015)
9. H. Watanabe, M. Otaka, Fuel, **85**, 1935–1943 (2006)
10. T. Hasegawa, Energies, **3**, 335–449 (2010)

EPJ Web of Conferences **159**, 00002 (2017) DOI: 10.1051/epjconf/201715900002
AVTFG2016

Numerical simulation of gas flow and droplet motion in a wave-plate eliminator of the separator-steam-generator system in the waste-heat-utilisation complex

Valerij Artemov[1], *Konstantin* Minko[1,*], *Georgij* Yankov[1], *Anton* Ptakhin[2], *Anton* Kondratev[2] and *Oleg* Milman[2]

[1]National Research University "Moscow Power Engineering Institute", 111250, Moscow, ul. Krasnokazarmennaya 14

[2]Scientific Production Company "Turbocon", 248010, Kaluga, ul. Komsomol'skaya Roshcha 43

Abstract. This paper studied the droplet transport and deposition in the turbulent flow inside a wave-plate eliminator of the waste-heat utilisation complex (WHUC). The Lagrangian discrete particle approach was used to simulate the process of liquid separation from wet steam flow. Two different models for droplet-eddy interaction were tested using data from the available literature. The tested numerical model was used to predict the WHUC performance.

1 Introduction

The droplet transport and deposition in the turbulent flow inside a wave-plate eliminator of the WHUC (Fig. 1) [1] were studied using an Eulerian-Lagrangian computational approach. Flash boiling occurs in the nozzle because of a sharp pressure drop which results in a metastable liquid state. The steam-liquid mixture is sent to the expansion unit and then to the wave-plate eliminator. The temperature and pressure of the water entering the nozzle are t_{in} = 257°C and p_{in} = 10 MPa. The pressure of the steam-liquid mixture leaving the expansion unit is approximately p_{out} = 1 MPa. This paper presents the result of numerical simulation of liquid droplet behaviour in the separator. The information about processes in the nozzle and the operating condition at the inlet of the wave-plate eliminator were obtained using the model presented in [2].

2 Mathematical model

There is presently significant number of works devoted to the theoretical and experimental study of two-phase dispersed flow [3]. The Lagrangian discrete particle approach (LDP-approach) was used in this paper to simulate the process of liquid separation from wet-steam flow in the wave-plate eliminator of the WHUC.

* Corresponding author: minkokb@gmail.com

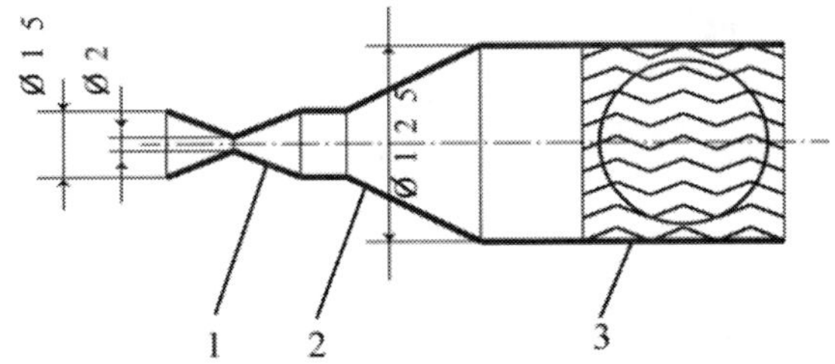

Fig. 1. The separator-steam-generator system of a prototype WHUC (sizes in mm): (1) nozzle; (2) expansion unit; (3) wave-plate eliminator.

In the LDP-approach, the continuous-phase flow field is computed using the Reynolds averaged Navier-Stokes equations and discrete-phase droplet motion is obtained from the time integration of Newton's law. Different modifications of the LDP-approach are available in commercial CFD codes. In this work, the approach realised in the author's CFD code ANES [4] was used. The main features are as follows: droplets have no feedback effect on the carrier gas; there is no heat and mass transfer between the dispersed and continuous phases; there is no droplet-droplet interaction (and, therefore, coalescence); droplets behave as hard spheres; drag is the only force acting on the droplets; there is no droplet-film interaction at the wall; only one droplet-wall interaction is considered—the liquid droplets trapped by the wall.

The dispersion of particles due to turbulence in the continuous phase is simulated using a stochastic tracking model [5]. The instantaneous continuous phase velocity is given by

$$\mathbf{u}(x,y,z,t)=\bar{\mathbf{u}}(x,y,z)+\mathbf{u}'(t) \quad , \tag{1}$$

where $\bar{\mathbf{u}}(x,y,z)$ – the mean fluid phase velocity which was obtained using one of the models to describe the turbulent transport of momentum (the k-ω model [6] was used in this paper), $\mathbf{u}'(t)$ – the fluctuating continuously phase velocity. Many droplets should be injected in the domain until stationary statistics are achieved.

Components of the fluctuating velocity $\mathbf{u}'(t)$ are step functions of time. The basic parameters of the function are amplitude u_t and step width Δt_t. Two different models were tested. In the first model, all the values calculated at the start of the droplet-eddy interaction used characteristics of the flow at the particle location and remained constant during the droplet-eddy interaction (model №1). In the second model, all the necessary parameters of the droplet-eddy interaction (amplitude of the function and the eddy interaction time) were "modified" by the local flow characteristics at the particle location (model №2).

Amplitude of the fluctuating continuous-phase velocity is calculated by the equation

$$u_{t,i}=\Lambda_i\cdot u_*, \qquad u_*=\sqrt{\frac{2\cdot k}{3}}, i=1,2,3\,, \tag{2}$$

where k – turbulent kinetic energy, Λ_i – a random number drawn from a normal probability distribution with zero mean and standard deviation equal to unit.

In the model №1, the characteristic scale of fluctuation u_* and Λ_i was calculated at the start of the particle-eddy interaction and remained constant. In the model №2, Λ_i remained constant, but u_* was updated using k at the droplet location.

The droplet is assumed to interact with the eddy over the smaller (Δt_i) of the two characteristic times:

1) Δt_{life} – the turbulent eddy lifetime,

2) Δt_{cross} – the eddy crossing time.

The eddy lifetime is given by equation

$$\Delta t_{eddy} = 2C_{TE}\frac{k}{\varepsilon}, \quad (3)$$

The different value of C_{TE} is used in the literature. The most commonly used are the following equations

$$2C_{TE} = \sqrt{\frac{3}{2}}C_{\mu}^{\ 3/4} \text{ or } 2C_{TE} = \sqrt{\frac{3}{2}}C_{\mu}^{\ 1/2}, \quad (4)$$

which correspond to C_{TE} from 0.1 to 0.18 (C_{μ} equals 0.09). In this work, constant C_{TE} was equal to 0.15. For the *k-ω* models, substitute $\varepsilon = C_{\mu}k\omega$ is used.

In the model №1, the interaction time Δt_i is calculated at the start of the droplet-eddy interaction and time before the end of droplet interaction with the current eddy is set at $t_{end} = \Delta t_i$. This time is reduced by time step $t_{end} = t_{end} - \Delta t$ at each integration time step. At $\Delta t_{end} = 0$, characteristics of the new eddy are calculated. In our model (model №2), time before the end of particle interaction with the current eddy is modified at each time step:

$$t_{end,new} = t_{end,old}\frac{\Delta t_{t,new}}{\Delta t_{t,old}}, \quad (5)$$

where $\Delta t_{t,new}, \Delta t_{t,old}$ – droplet-eddy interaction time, which are calculated using characteristics of the flow at the new and old droplet locations.

3 Validation of mathematical model

The validation analyses are performed through comparison with available experimental data on removal efficiencies of the wave-plate eliminator [7]. Fig. 2 shows comparison results of calculation and experimental data [7].

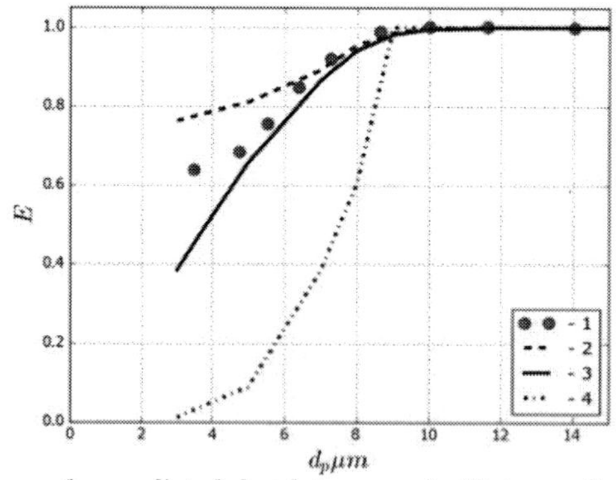

Fig. 2. Comparison between the predicted droplet-removal efficiency *E* and the experimental data of Ghetti [7] at gas velocities equal to 3 m/s. 1 – experimental data [7], 2 – model №1, 3 – model №2, 4 – without turbulence dispersion.

The calculation results show that the experimental dependence of the efficiency of the separator (the relation of liquid weight at the outlet and at the inlet) from the particle diameter is well reproduced with model №2. However, such a parameter as the cut-off diameter (a particle diameter at which 95% of particles were trapped at the wall) can be determined without using a turbulent dispersion-particles model. For example, the experimental value of the cut-off diameter for the bulk velocity of continuous-phase flow equals 3 m/s is 8.5 μm, and the value of this parameter is calculated without considering the turbulent dispersion equals to 9 μm.

4 Problem statements and results

The wave-plate eliminator of the WHUC is a cylindrical channel with an inner diameter of $D_{sep} = 85$ mm, in which several zig-zag plates are installed. At the inlet, wet steam has a flow rate of $G_0 = 0.317$ kg/s and a vapor mass fraction of $x = 0.139$ at $p_{out} = 1044$ kPa (it corresponds to $t_{in} = 257$ °C and $p_{in} = 10$ MPa at the inlet of WHUC. For details see [2]). The geometry of the typical element is shown in Fig.3 and the dimensions of the channel are given in Table 1. Two designs are considered: a separator without hooks and a separator with hooks. The mean velocity at the inlet of the typical element of the wave-plate eliminator equals 1.5 m/s.

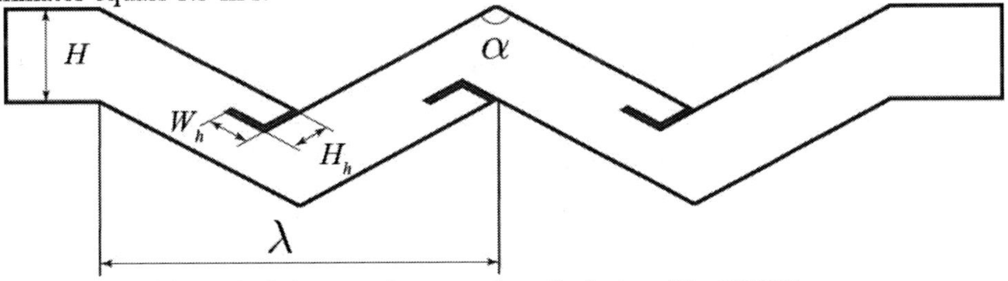

Fig. 3. Scheme of the typical element of a wave-plate eliminator of the WHUC.

Table 1. The dimensions of the channel.

H, mm	α, °	λ, mm	H_h, mm	W_h, mm
9	117	19	0/4	0/4

The first series of calculations has been devoted to assessing the impact of droplet-size-distribution parameters on the separation efficiency of the separator without hooks. A Rosin–Rammler model [8] for the particle-size-distribution analysis was used. Three cases were tested: the constants in the Rosi-Rammler models were equal to 2.25 and 4.52 and the case of monodisperse particles. Fig. 4 shows the calculation results. It can be noted that narrowing of the droplet-diameter-distribution function results in an increasing separation efficiency.

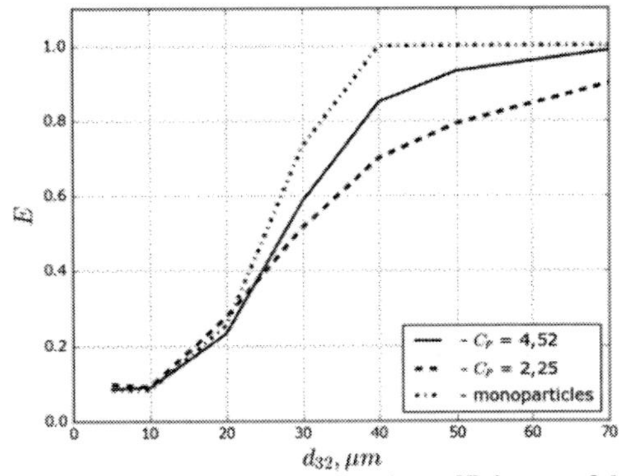

Fig. 4. Effect of particle-size distribution on the separation efficiency of the separation without hooks.

Fig. 5 shows the influence of the mean velocity on the separation efficiency at a fixed value of a mean particle diameter of d_{32} = 40 μm. Its value was estimated using nozzle parameters and equations from [9]. The constant in the Rosi-Rammler model was equal to 4.52.

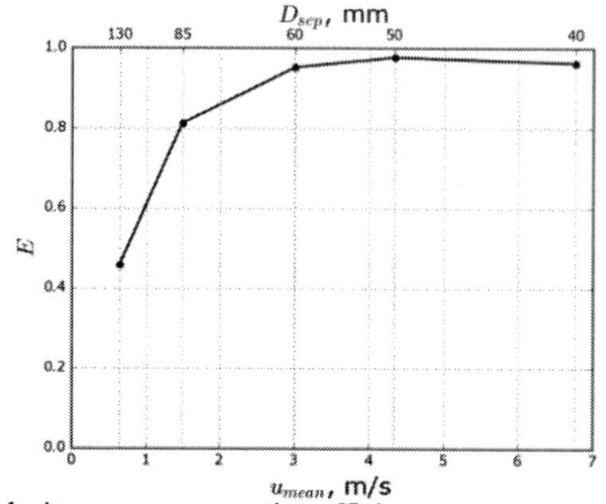

Fig. 5. Effect of mean velocity u_{mean} on separation efficiency.

The next series of calculations was performed for various parameters at the inlet of the WHUC nozzle. Table 2 shows the results of the calculation.

Table 2. Influence of inlet parameters of the WHUC.

p_{in}, MPa	t_{in}, °C	G_0, g/s	x	d_{32}, μm	E
10	257	317	0.139	40	0.81
10	220	372	0.034	58	0.72
10	300	214	0.259	37	0.81
15	257	442	0.059	47	0.77

For the basic WHUC parameters (p_{in} = 10 MPa, t_{in} = 257 °C), the effectiveness of particle separation for a separator with hooks increases to 99.6% from 81.3%.

Financial support was provided by the Russian Ministry of Education and Science (contract 14.579.21.0031 with ZAO NPVP Turbokon, identifier RFMEFI57914X0031, 2014) and the Council for Grants of the President of the Russian Federation (MK-6236.2016.8 grant).

References

1. V. A. Fedorov, O. O. Mil'man, D. V. Fedorov, A. M. Trinoga, *Energy Saving Technology of Electrical Energy Production at Natural Gas Swapping through Pipeline System* (Mos. Gos. Tekhn. Univ., Moscow, 2011)
2. V. I. Artemov, K. B.Minko, G. G. Yan'kov, Therm. Eng., **62**, 897 (2015)
3. E.P. Volkov, *Modelling of processes of combustion and pyrolysis of solid fuel* (ID MEI, Moscow, 2014)
4. URL: http://www.anes.ch12655.tmweb.ru/
5. A. D. Gosman, E. Ioannides, J. Energy, **7**, 482 (1981)
6. D. C. Wilcox, AIAA J., **46**, 2823 (2008)
7. S. Ghetti, *Investigation of Entrainment Phenomena In Inertial Separators* (University of Pisa, Pisa, Italy, 2003)
8. P. Gonzalez-Tello, F. Camacho, J. Vicaria, P. Gonzalez, Powder Technol, **186**, 278 (2008)
9. T. Ueda, Bull. JSME, **45**, 127 (1979)

EPJ Web of Conferences **159**, 00003 (2017) DOI: 10.1051/epjconf/201715900003
AVTFG2016

Plasmonic properties of gold nanoparticles covered by silicon suboxide thin film

Evgeniy Baranov[1,*], *Alexandr* Zamchiy[1], *Aleksey* Safonov[1], *Sergey* Starinskiy[1], *Sergey* Khmel[1]

[1]Kutateladze Institute of Thermophysics SB RAS, 630090 Novosibirsk, Russia

Abstract. The optical properties of nanocomposite material consisting of gold nanoparticles without/with silicon suboxide thin film were obtained. The gold film was deposited by thermal vacuum evaporation and then it was annealed in a vacuum chamber to form gold nanoparticles. The silicon suboxide thin films were deposited by the gas-jet electron beam plasma chemical vapor deposition method. The intensity of the localized surface plasmon resonance increased and the plasmon maximum peak shifted from 520 nm to 537 nm.

1 Introduction

In recent years, the light absorption enhancement in solar cells becomes the actual problem. One of the possible routes to solve this problem is using plasmonic nanoparticles for scattering of the light inside thin-film solar cell. Gold is a widely used metal with plasmonic properties. Localized surface plasmons of gold nanoparticles provide the light scattering at large angles without increasing the surface area of the solar cell. In addition, the using of plasmonic particles shifts the absorption maximum of the long-wave region of the spectrum [1]. It is possible to control the position of the absorption maximum by varying the parameters of the nanoparticles, such as size shape, concentration, as well as by changing the dielectric properties of the material around the particles.

Pulsed laser ablation method is simple and widely used technique that allows to obtain the gold nanoparticles in a wide range of sizes and concentrations [2]. The advantage of this method is that the material composition of the target fully reflects the composition of the synthesized nanoparticles.

In this paper, amorphous silicon suboxide (a-SiO_x:H) thin films were used as a protective coatings. In addition, the changing of the stoichiometric coefficient of the a-SiO_x:H thin films allows to control the plasmon resonance peak position. The aim of the work was to show that gold nanoparticles in a matrix of the a-SiO_x:H increase the intensity of its local plasmon resonance and the absorption maximum is shifted to long-wave region.

2 Experiment

* Corresponding author: itpbaranov@gmail.com

The substrates were a silicon wafer (for measurement of particles size by a scanning electron microscope) and a borosilicate glass (for measuring transmittance spectra).

2.1 Synthesis of the gold nanoparticles

The gold films of about 4.8 nm thick were deposited by thermal vacuum evaporation onto the monocrystalline silicon and the borosilicate glass substrates. The pressure was $7*10^{-5}$ Pa during the deposition. The substrate temperature was 100°C. Figure 1 shows the SEM image of as-deposited gold film with thickness about 4.8 nm. It is evident from Fig. 1 that the film is not continuous and consists of individual particles of irregular shape with average size about 10 nm.

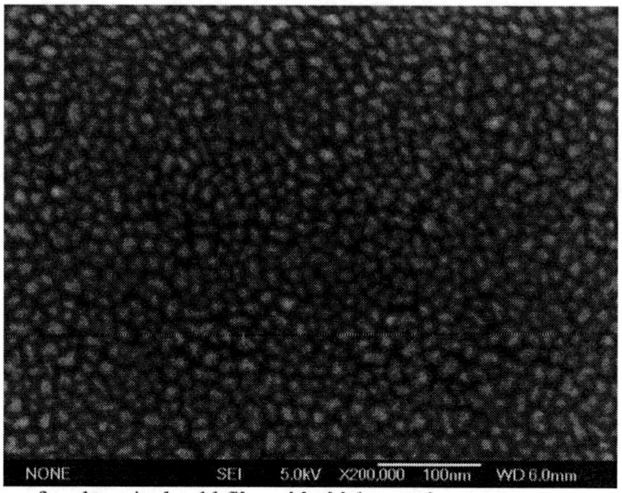

Fig. 1. SEM image of as-deposited gold film with thickness about 4.8 nm.

Then the samples were annealed in a vacuum chamber at a temperature of 400°C for 1 hour to form gold nanoparticles [3]. Fig. 2a shows the SEM image of the gold nanoparticles on the silicon substrate obtained by a scanning electron microscope JEOL JSM-6700F. The Fig. 2b shows a histogram of the size distribution of nanoparticles. The particles have a nearly spherical shape, average size is 13 nm and the filling density is low.

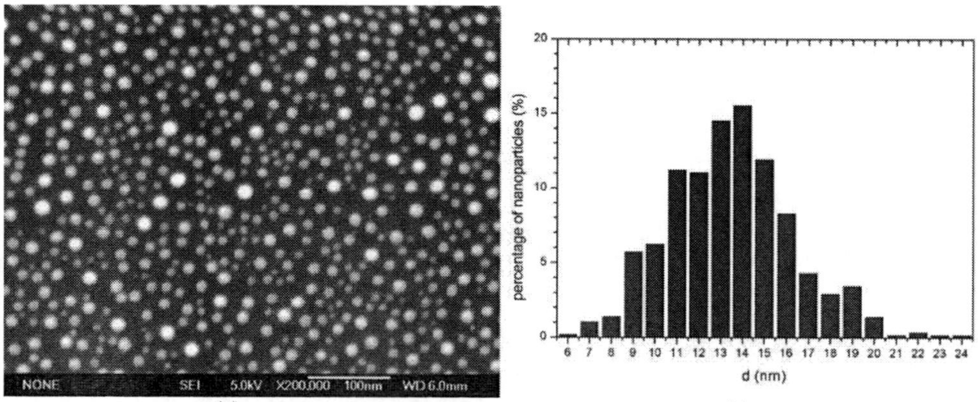

(a) (b)

Fig. 2. a - SEM image of the gold nanoparticles on silicon substrate after annealing; b - nanoparticles size distribution histogram obtained from SEM picture analysis.

2.2 Silicon suboxide thin films deposition

In the present work, a-SiO_x:H thin films were deposited by the gas-jet electron beam plasma chemical vapor deposition method [4]. Also, this method is used for the deposition of thin films of hydrogenated amorphous silicon [5] and the synthesis of silicon oxide nanowires [6, 7]. The experiments were carried out in a vacuum chamber evacuated with a forevacuum pump. Gas pressure was not exceeding 16 Pa after introduction of processing gases (argon, a mixture of 5% monosilane–argon, and oxygen). The gas flow rate was 280 sccm for argon 18 sccm for the monosilane–argon mixture, and 6 sccm for oxygen. Argon and the mixture were supplied into the chamber via a nozzle 2 mm in diameter, and oxygen was introduced directly into the vacuum chamber via a leak valve. For reagents activation, electron beam with an energy of 1600 eV and a current of 70 mA was used generated by electron gun with a plasma cathode [8]. The substrates for films synthesis were c-Si <100> wafers and Corning Eagle XG glass for infrared and optical measurements, respectively. The deposition time was 2 min that provided the samples thickness of about 130 nm. Films were synthesized at substrate temperatures of 335°C. Optical properties of silicon suboxide determined by using the software package PUMA.

3 Results and discussion

Figure 1 shows the extinction spectra of the gold nanoparticles without/with silicon suboxide thin film on the borosilicate glass obtained from transmission spectra. Oxygen concentration in the silicon suboxide film founded by FTIR measurement was 26% and hydrogen concentration was about 1%.

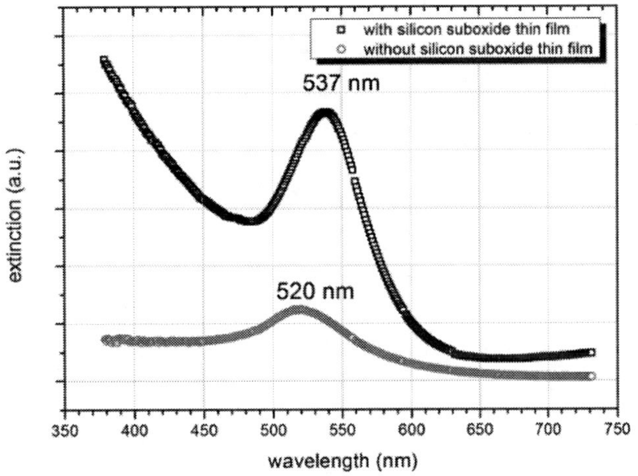

Fig. 3. Extinction spectra of the gold nanoparticles on the borosilicate glass: ○ - without silicon suboxide thin film; □ - with silicon suboxide thin film.

Localized surface plasmon resonance of the gold nanoparticles on the borosilicate glass is observed at a wavelength of 520 nm. This is characteristic of spherical nanoparticles, as shown previously in the literature [9]. When the gold nanoparticles were covered by the silicon suboxide film, plasmon peak maximum shifts to red wavelengths area. This is due to the refractive index change in the surrounding environment. The increase on absorption for wavelengths below 480 nm is due to the optical band gap of the material.

4 Conclusion

The gold nanoparticles with an average size of 13 nm were covered by silicon suboxide film with 130 nm thickness. In the issue, the intensity of the localized surface plasmon resonance increased and the plasmon maximum peak shifted from 520 nm to 537 nm.

Research is supported by the Russian Science Foundation grant №16-10-10506.

References

1. E. Moulin, P. Luo, B. Pieters, J. Sukmanowski, J. Kirchhoff, W. Reetz, T. Müller, R. Carius, F.X. Royer, H. Stiebig, Appl. Phys. Lett. **95**, 033505 (2009)
2. S.V. Starinskiy, Yu.G. Shukhov, A.V. Bulgakov, Technical Physics Letters **42**, 411 (2016)
3. A. Safonov, V. Sulyaeva, N. Timoshenko, S. Starinskiy, Thin Solid Films **603**, 313 (2016)
4. E. Baranov, S. Khmel, A. Zamchiy, M. Buyko, Physica Status Solidi (A) Applications and Materials Science **213**, 1783 (2016)
5. E. Baranov, S. Khmel, A. Zamchiy, I. Cheskovskaya, M. Sharafutdinov, Can. J. Phys. **92**, 723 (2014)
6. S. Khmel, E. Baranov, A. Zamchiy, E. Maximovskiy, D. Gulyaev, K. Zhuravlev, Physica Status Solidi (A) Applications and Materials Science **213**, 1790 (2016)
7. S. Khmel, E. Baranov, A. Zaikovskii, A. Zamchiy, E. Maximovskiy, D. Gulyaev, K. Zhuravlev, Physica Status Solidi (A) Applications and Materials Science **213**, 1774 (2016)
8. E. Baranov, S. Khmel, A. Zamchiy, IEEE Trans. Plasma Sci. **42**, 2794 (2014)
9. A. Safonov, V. Sulyaeva, N. Timoshenko, K. Kubrak, S. Starinskiy, Physics Letters, Section A: General, Atomic and Solid State Physics **380**, 3919 (2016)

EPJ Web of Conferences **159**, 00004 (2017) DOI: 10.1051/epjconf/201715900004
AVTFG2016

Experimental investigation of two-phase gas-liquid flow in microchannel with T-junction

German Bartkus[1,2*], *Igor* Kozulin[1,2], *Vladimir* Kuznetsov[1,2]

[1]SB RAS, Kutateladze Institute of Thermophysics, 630090 Ac. Lavrentieva ave. 1, Novosibirsk, Russia
[2]Novosibirsk State University, 630090 st. Pirogova 1, Novosibirsk, Russia

Abstract. Using high-speed video recording and the method of dual laser scanning the gas-liquid flow was investigated in rectangular microchannels with an aspect ratio of 2.35 and 1.26. Experiments were carried out for the vertical flow of ethanol-nitrogen mixture in a microchannel with a cross section of 553x235 μm and for the horizontal flow of water-nitrogen mixture in a microchannel with a cross section of 315x250 μm. The T-mixer was used at the channel's inlet for gas-liquid flow formation. It was observed that elongated bubble, transition, and annular flows are the main regimes for a microchannel with a hydraulic diameter substantially less than the capillary constant. Using laser scanning, the maps of flow regimes for ethanol-nitrogen and water–nitrogen mixtures were obtained and discussed.

1 Introduction

Multiphase microsystems based on microchannels are perspective for a variety of applications. Typical dimensions of the channels in these microsystems are in the range from hundreds of nanometers to hundreds of micrometers, which ensures the excellent heat and mass transfer properties. Research of gas-liquid flows in microchannels show that it is possible to distinguish 5 basic modes: isolated bubbles, the Taylor bubbles, slug-annular flow, annular and dispersion flow, which properties have been well studied [1, 2]. Important parameters for optimizing the heat and mass transfer processes in microchannels are the bubble length, the liquid film thickness, and their distribution. In the paper [3], the modes of gas-liquid flow in a microchannel with a gap of 300 μm and width from 10 to 30 mm were presented. Laser scanning method was used in [4, 5] for determination of the statistical characteristics of the gas-liquid flows.

The objective of the present work is experimental studying the mechanism of multi-scale upward vertical and horizontal gas-liquid flow formation in rectangular microchannels of the microfluidic devices. Using voltage signals of flow from the photodiodes, flow types were defined in microchannels with cross sections of 553x235 μm and 315x250 μm by dual laser scanning. Deionized water, ethanol, and nitrogen were used as working fluids.

*Corresponding author: germanbartkus@gmail.com

2 Experimental equipment and methods

The experimental setup for studying the upward vertical and horizontal gas-liquid flow in a rectangular microchannel is shown in Figure 1 (a). Nitrogen, which is fed from a tank (5) via the control valve and gas flow meter (2) to the test section, is used as working gas. Water flows from the tank (4) through a flow meter (3) and further into the T-shaped mixer (1) at the inlet of the experimental section. Experimental section is a microchannel, assembled from glued glass and aluminum plates with length L = 30 cm. The hydraulic diameter is much smaller than the capillary constant. From the exit of experimental section, the gas-liquid mixture is moved to the liquid tank (10), where the gas escapes to the atmosphere.

The study of gas-liquid flow regimes was conducted using the method of dual laser scanning, see Figure 1 (b). Two lasers (7) are positioned so that the laser beams illuminate one side of the rectangular channel. A lens (8) is used to change the diameter of laser spot; an aperture is used to reduce the diameter of the laser beam. The lasers are powered using voltage power source. The intensity of transmitted light is measured with photodiodes (6) arranged at the opposite side of the channel. The electric voltage from the photodiodes is analyzed using high-speed ADC and processed on the computer over the period from one to three minutes. Flow patterns are also registered using the high-speed camera (9). The first laser is located at a distance of 8.5 cm from the inlet, camera – 19 cm from the inlet, second laser – 24 cm for observing the stable gas-liquid flow.

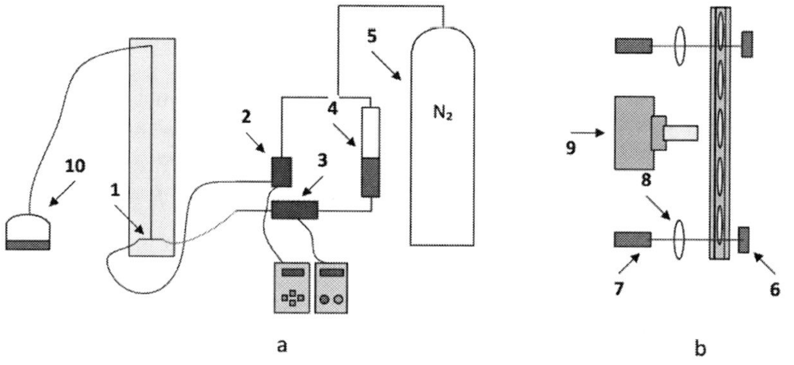

Fig. 1. Experimental setup (a), scheme of the dual laser scanning method (b).

3 Two-phase gas-liquid flow in microchannels

The wide range of phenomena, which are not typical on a large scale and are poorly understood, arises by the reason of reducing typical size of the flow in microchannel systems. The high-speed video shows that the periodic flow with elongated gas bubbles (Taylor bubbles), the transition flow and the annular flow are the main flow regimes for microchannels with the gap substantially less than the capillary constant. In the microchannel with a cross section of 553x235 μm and of 315x250 μm, the flow with elongated bubbles (slug), the transition flow and the annular one are presented. The slug flow is characterized by the periodicity of the voltage signal from the photodiode. Increasing the superficial gas velocity above the specified values leads to the destruction of the elongated bubbles, and the flow loses its periodicity.

The elongated-bubble flow and the transition flow are presented in Fig. 2 for an upward ethanol-nitrogen experiment in a microchannel 553x235 μm. The elongated bubble flow is observed in the range of superficial gas velocities J_{gas}=0.126-0.772 m/s. A fragment of the signal from the photodiode for the present flow is shown in Fig. 2 (a). The highest amplitude level from photodiode corresponds to a full filling cross section of the channel by liquid (liquid bridge), the minimum signal corresponds to gas bubble. The visualization of the flow is presented in Fig. 2 (b). The elongated bubble flow is characterized by the periodicity of gas bubbles and liquid bridges.

The transition flow is observed in the range of superficial gas velocities J_{gas}=0.617-3.012 m/s. The definition of the boundary from the transition to annular flow is required experiments with higher superficial gas velocities. A fragment of the signal from the photodiode for transition flow is shown in Fig. 2 (c) The visualization of the flow is presented in Fig. 2 (d). Such flow is unstable; liquid bridges appear by combining waves of disturbances on short opposite sides of the microchannel and they are destroyed by combining of gas bubble-slugs.

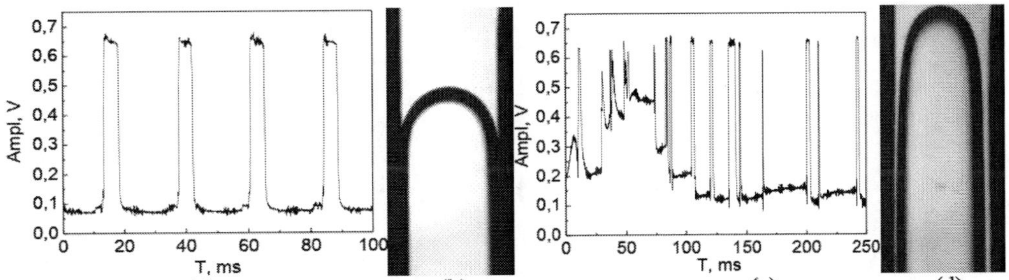

(a) (b) (c) (d)

Fig. 2. Vertical flow in microchannel 553x235 μm (ethanol-nitrogen). The dynamics of voltage on the photodiode for elongated-bubble flow at superficial velocities of liquid and gas: J_{liq}=0.042 m/s, J_{gas}=0.369 m/s (a), visualization (b), the dynamics of voltage on the photodiode for transition flow at velocities of liquid and gas: J_{liq}=0.042 m/s, J_{gas}=1.814 m/s (c), visualization (d).

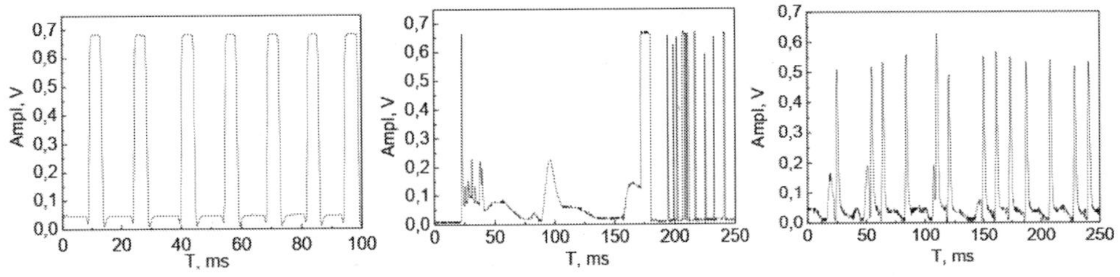

(a) (b) (c)

Fig. 3. Horizontal flow in microchannel of 315x250 μm (water-nitrogen). The dynamics of voltage on the photodiode for elongated-bubble flow at superficial velocities of liquid and gas: J_{liq}=0.035 m/s, J_{gas}=0.14 m/s (a), for transition flow at velocities of liquid and gas: J_{liq}=0.035 m/s, J_{gas}=3.73 m/s (b), for annular flow at velocities of liquid and gas: J_{liq}=0.035 m/s, J_{gas}=7.26 m/s (b).

The elongated-bubble flow and the transition flow are presented in Fig. 3 for the water-nitrogen experiment in a horizontal microchannel 315x250 μm. The elongated bubble flow is observed in the range of superficial gas velocities J_{gas}=0.074-0.786 m/s. A fragment of the signal from the photodiode for the transition flow is shown in Fig. 3 (a).

The transition flow is observed in the range of gas velocities J_{gas}=0.641-4.232 m/s. A fragment of the signal from the photodiode for the transition flow is shown in Fig. 3 (b).

The annular flow is observed in the range of gas velocities J_{gas}=4.67-21.056 m/s. A fragment of the signal from the photodiode for the transition flow is shown in Fig. 3 (c). The highest amplitude level from photodiode does not correspond to a full filling cross section of the channel by liquid (liquid bridge); waves of liquid are observed on short opposite sides of the microchannel.

DOI: 10.1051/epjconf/201715900004

4 Flow pattern maps

Experiments for horizontal water-nitrogen and vertical upward ethanol-nitrogen types' flows were carried out. Maps were obtained using visualization and signals from photodiodes. 3 regimes for a microchannel with hydraulic diameters D_h=279 μm (315x250 μm) and 2 regimes for a microchannel with hydraulic diameter D_h=326 μm were observed. As you can see for the ethanol-nitrogen experiment and water-nitrogen experiment, the boundaries for beginning transition flow are not very different.

For the water-nitrogen experiment, flow pattern map confirms the data obtained by Zhao [6] for the channel with hydraulic diameter D_h=400 μm. In Fig. 4 (b), filled area (1) corresponds to unstable slug (transition flow), filled area (2) corresponds to slug-annular flow.

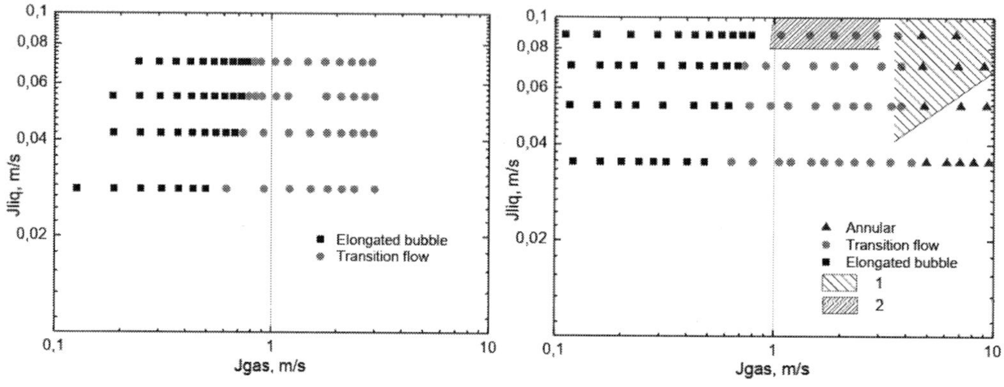

(a) (b)

Fig. 4. Maps of regimes for ethanol-nitrogen mixture in microchannel with hydraulic diameter D_h=326 μm (a), and for water-nitrogen mixture in microchannel with hydraulic diameter D_h=279 μm (b).

References

Laser scanning method and high-speed visualization are an informative technique to obtain the data of two-phase flow. It was observed that the regimes with elongated gas bubbles, the transition flow, and annular flow are the main patterns for a microchannel with a hydraulic diameter substantially less than the capillary constant. Maps of flow regimes in a rectangular microchannel with hydraulic diameter D_h=326 μm (553x235 μm) for the ethanol-nitrogen mixture, D_h=279 μm (315x250 μm) for water-nitrogen were obtained and are shown in Fig. 4.

This work performed in the Kutateladze Institute of Thermophysics SB RAS by a grant from the Russian Science Foundation (project No 16-19-10519).

References

1. J.W. Coleman, S. Garimella, Int. J. Heat Mass Trans. **42** (1999)
2. F. A. Kawahara, A., P.M.Y. Chung, M. Kawaji, Int. J. Multiphase Flow **28** (2002)
3. E. A. Chinnov, V. Ronshin, O. A. Kabov, Thermophys. Aeromech. **21** (2015)
4. R. Revellin, J. R. Thome, J. Micromech. Microeng. **17** (2007)
5. I.A. Kozulin, V.V. Kuznetsov, J. Appl. Mech. Tech. Phys. **52** (2011)
6. Yuchao Zhao, Guangwen Chen, Chunbo Ye, Quan Yuan, Chem. Eng. Sci. **87** (2013)

Pressure gradient effect at distributed excitation of 3D TS waves by freestream and wall disturbances

Vladimir Borodulin, *Andrey* Ivanov, *Yury* Kachanov, *Dmitry* Mischenko*, *and* *Anna* Fedenkova

Khristianovich Institute of Theoretical and Applied Mechanics SB RAS, 630090 Novosibirsk, Russia

Abstract. The present work is a continuation of previous experiments (carried out in the Blasius boundary layer) and devoted to quantitative investigation of influence of an adverse pressure gradient on two efficient mechanisms of excitation of 3D TS instability waves due to a distributed boundary layer receptivity to free-stream vortices. These mechanisms are associated with distributed scattering of 3D amplified free-stream vortices both on the natural boundary-layer nonuniformity (on smooth surface) and on 2D surface nonuniformities (waviness). The corresponding detailed hot-wire measurements were carried out in a self-similar boundary layer with Hartree parameter $\beta_H = -0.115$ in a wide range of the problem parameters. Complex values of quantitative characteristics of the physical phenomenon under study (the distributed receptivity coefficients) are evaluated by based on the obtained experimental data. It is found that the adverse pressure gradient leads to reduction of efficiency of the investigated vortex-roughness receptivity mechanism.

1 Introduction

The present experiments are devoted to quantitative investigation of adverse-pressure-gradient effects on the mechanisms of excitation of three-dimensional Tollmien-Schlichting (TS) waves due to distributed (in the streamwise direction) boundary layer receptivity to free-stream vortices. These mechanisms are associated with distributed scattering of three-dimensional unsteady free-stream vortices both on the natural boundary layer nonuniformity (smooth surface) and on 2D surface nonuniformities (waviness). The receptivity and the boundary-layer instability mechanisms affect simultaneously on downstream evolution of TS-waves and provide the possibility of much faster growth of the latter compared to the growth rates associated with the linear instability mechanism.

The majority of previous studies were theoretical and they dealt with 2D problems. The first experimental estimations of the distributed vortex receptivity coefficients were carried out in [1] for a two-dimensional problem. Experiments [2] were devoted to investigation of 3D vortex receptivity problem in the Blasius boundary layer. These experiments gave for the first time systematic information on the distributed excitation of 3D TS-waves by

* Corresponding author: d.a.mischenko@itam.nsc.ru

freestream vortices in presence of 2D surface roughness. The values of the distributed receptivity coefficients were obtained there in a broad range of the problem parameters. The present work is a continuation of experiments [2] and devoted to a systematic quantitative investigation of influence of an adverse pressure gradient on the distributed vortical receptivity mechanisms mentioned above. The present experiments were performed in a self-similar boundary layer with Hartree parameter $\beta_H = -0.115$.

2 Experiment setup and evolution of exited TS-modes

The measurements were carried out in a low-turbulence wind tunnel T-324 of ITAM SB RAS in a boundary layer of a high-precision experimental model consisted of a flat plate (1485×1000×10 mm) and a wall bump of a special adjustable shape mounted on the wind-tunnel ceiling just above the plate. The shape of the wall bump provided formation on the flat plate in the region of main measurements of a self-similar flow with Hartree parameter $\beta_H = -0.115$. The boundary-layer edge velocity U_e varied from 8.9 to 8.6 m/s in the region of the streamwise coordinate $x = 350$ to 620 mm having its origin at the plate leading edge. The corresponding Reynolds numbers $Re^* = U_e\delta_1/\nu$ varied between 837 and 1174 (here δ_1 is the boundary layer displacement thickness and ν is the air kinematic viscosity). The 3D freestream vortices were generated by a vibrating wire mounted normally to the flat-plate surface upstream of its leading edge. The 2D surface nonuniformities of sinusoidal shape were created by special thin-film patches applied onto the plate surface. These patches were manufactured by a special high-precision technology and had rated parameters: the streamwise wavelength λ_{sx} and the amplitude h_s. Thorough measurements of the boundary layer and freestream disturbances were carried out by a single hot-wire probe in several regimes (see the Table 1). In some of these regimes, the problem parameters (such as the disturbance frequency f, the streamwise surface-waviness wavelength, and the propagation angle θ of the excited TS-waves) were chosen in a way to provide the satisfaction of conditions of the resonance of streamwise wavenumbers of the excited TS-modes, freestream vortices, and surface nonuniformities [2].

Table 1. Main regimes of measurements.

Regime name	Roughness #	f [Hz]	λ_{sx} [mm]	h_s [μm]
R1f1	1	53.7	85.7	177
R3f2	3	81.4	63.5	199
R5f2	5	81.4	69.8	215
R6f3	6	109.1	51.8	179

The characteristics of the TS-modes' linear evolution on the wavy surfaces were measured in additional series of experiments in all main studied regimes in absence of freestream disturbances. These measurements were necessary for obtaining quantitative characteristics of the studied mechanism of excitation of the instability modes – the vortex/roughness distributed receptivity coefficients. In these additional experiments the wave-trains of controlled 3D TS-waves were excited by special point disturbance source (of blowing/suction type) mounted in the beginning of the experimental model.

The experiments have shown that a controlled antisymmetric vortex street is produced in the wake of the vibrating wire and leads to a rather efficient excitation of wave trains consisted of 3D TS-modes. The effective excitation of 3D TS-waves by the point disturbance source was also achieved in the additional (stability) experiments. In the two experiments, a number of spanwise profiles of amplitudes and phases of excited boundary layer disturbances were measured at various streamwise positions. Streamwise distributions of amplitudes and phases of the excited TS-modes were decomposed into the spanwise-wavenumber spectra. It was found that due to the action of the distributed receptivity mechanisms, the amplitudes of the excited TS-waves are able to grow much faster then those generated in the case of action of the linear instability mechanism only. Fig. 1 illustrates a typical picture of comparison of streamwise evolutions of the boundary layer disturbances (*a* – amplitudes, *b* – phases) in the main (receptivity) and in the additional (stability) experiments.

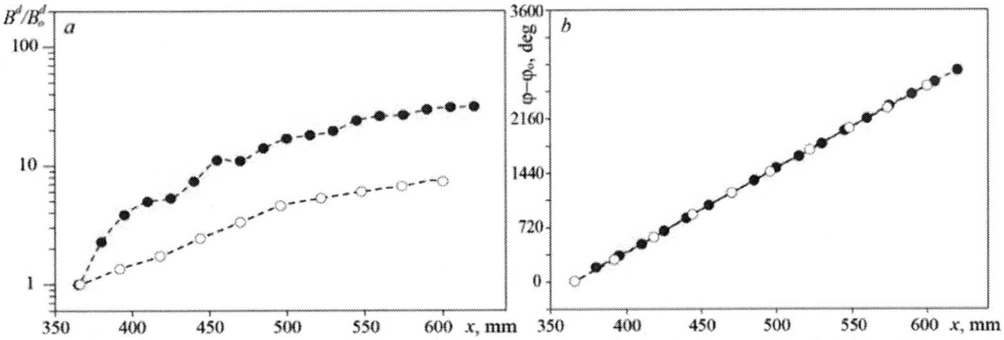

Fig. 1. Comparison of streamwise evolutions of amplitudes (*a*) and phases (*b*) of oblique TS-waves generated in receptivity experiment (dark points) and in additional stability experiment (open points). Regime *R6f3*, $\theta = 41$ deg.

Qualitatively the same results were obtained in all studied regimes in a wide range of the spanwise wavenumbers. These data were used then for subsequent processing and obtaining the amplitudes and phases of the distributed vortex-receptivity coefficients.

3 Procedure of obtaining receptivity coefficients and main results

The distributed receptivity coefficients are main quantitative characteristics of the distributed receptivity mechanism under study. They were determined in the same manner as in previous work [2] – as the corresponding coefficients of a differential equation that describes the evolution of the distributedly excited TS-modes. An analytical solution of this equation can be written in a general case as:

$$\overline{B}^d(x^*) = e^{\overline{A}(x^*)}\left\{\int_0^{x^*}\overline{B}_v(s)\overline{G}_v^d(s)e^{-\overline{A}(s)}ds + \int_0^{x^*}\overline{B}_v(s)\overline{G}_{vs}^d(s)(\overline{H}(s)+\overline{\overline{H}}(s))e^{-\overline{A}(s)}ds + \overline{B}_0^d\right\},$$

$$\overline{A}(x^*) = \int_0^{x^*}\overline{\alpha}(s)ds \qquad (1)$$

Here $\overline{B}^d$ is the complex-valued amplitude of the boundary layer disturbance; $\overline{\alpha}(x)$ is the *TS*-mode complex-valued wavenumber (obtained from processing of data of the additional, stability experiments); $\overline{B}_v$ is the free-stream vortex complex-valued amplitude (measured at the boundary layer edge); $\overline{H}$ is the surface roughness complex-valued amplitude; $\overline{\overline{H}}$ is its complex conjugate; $\overline{G}_v^d(x)$ and $\overline{G}_{vs}^d(x)$ are the complex-valued distributed vortex- (on smooth surface) and vortex-roughness receptivity functions, respectively; B_0^d is the complex-valued 'initial' amplitude of the excited boundary layer disturbances and x^* is the streamwise coordinate with its origin located (for convenience) at the first experiment point of the streamwise region of measurements.

It was assumed in the studied cases that functions $\overline{G}_v^d(x)$ and $\overline{G}_{vs}^d(x)$ are independent of the streamwise coordinate x and correspond to the receptivity coefficients. They were found by means of approximation of the experimental amplitude and phase distributions of the excited boundary layer disturbances $\overline{B}^d$ by solution (1) by means of adjustment of unknown solution parameters ($\overline{G}_v^d$, $\overline{G}_{vs}^d$, and $\overline{B}_0^d$). This fitting was carried out using optimization toolbox of MATLAB (gradient, simplex and genetic algorithms) for two different criteria of deviation of experimental and analytical complex amplitudes. As a result, the receptivity coefficients were obtained in a broad range of parameters.

Fig. 2*ab* illustrates the amplitude parts of coefficients $\overline{G}_v^d$ and $\overline{G}_{vs}^d$ versus the TS-wave propagation angle θ. Amplitudes G_{vs}^d grow with θ in a monotonous way, while amplitudes G_v^d depend on θ in a more complicated way (similar to the experiments [2]). Amplitudes of the two types of receptivity coefficients depend on the task parameters in a quite conservative manner. Amplitudes G_{vs}^d obtained in different regimes are very close to each other and sometimes even coincide. The G_v^d amplitudes also have in different regimes a quite conservative shape of distributions and similar characteristic values. Corresponding estimations showed that if amplitudes of surface roughness h_s/δ_1 is less then ~ 4%, the values of the coefficients of vortex receptivity on smooth surface are higher (in average) then those of the coefficients of the vortex-roughness receptivity. The mechanism associated with the vortex scattering on natural boundary layer nonuniformity dominates in these cases. Meanwhile, the vortex-roughness receptivity mechanism begins to become predominant when the amplitudes of surface waviness become higher than about 4%.

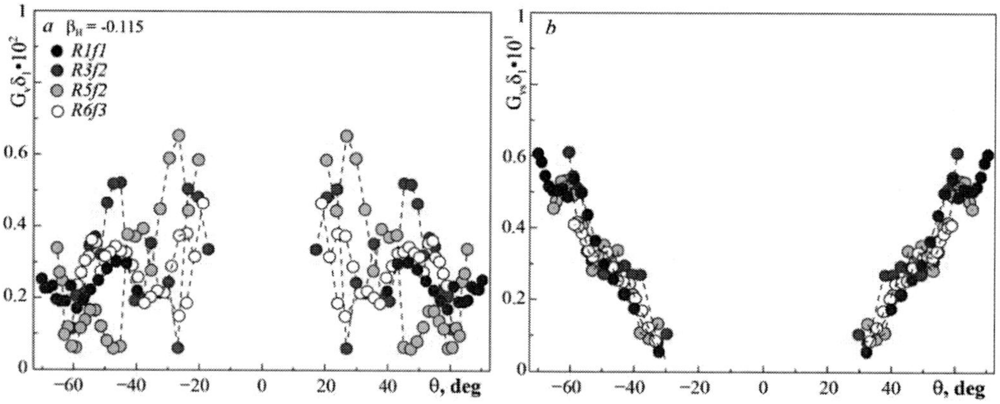

Fig 2. Amplitudes of coefficients of two mechanisms of distributed vortex receptivity: G_v^d (*a*) and G_{vs}^d (*b*) obtained in all studied regimes.

Typical results of direct comparison of quantitative receptivity characteristics obtained in the present experiments and in the Blasius boundary layer [2] are shown in Fig. 3 for regimes *R1f1* (*a,b*) and *R3f2* (*c,d*). It is seen that the amplitude of G_{vs}^d depend quite significantly on the streamwise pressure gradient which leads to a noticeable decrease of the receptivity coefficient amplitudes (Figs. 3*b,d*). On the contrary, the amplitudes of the receptivity coefficients on smooth surface $\overline{G}_v^d$ seem to be independent basically of the pressure gradient (Figs. 3*a,c*).

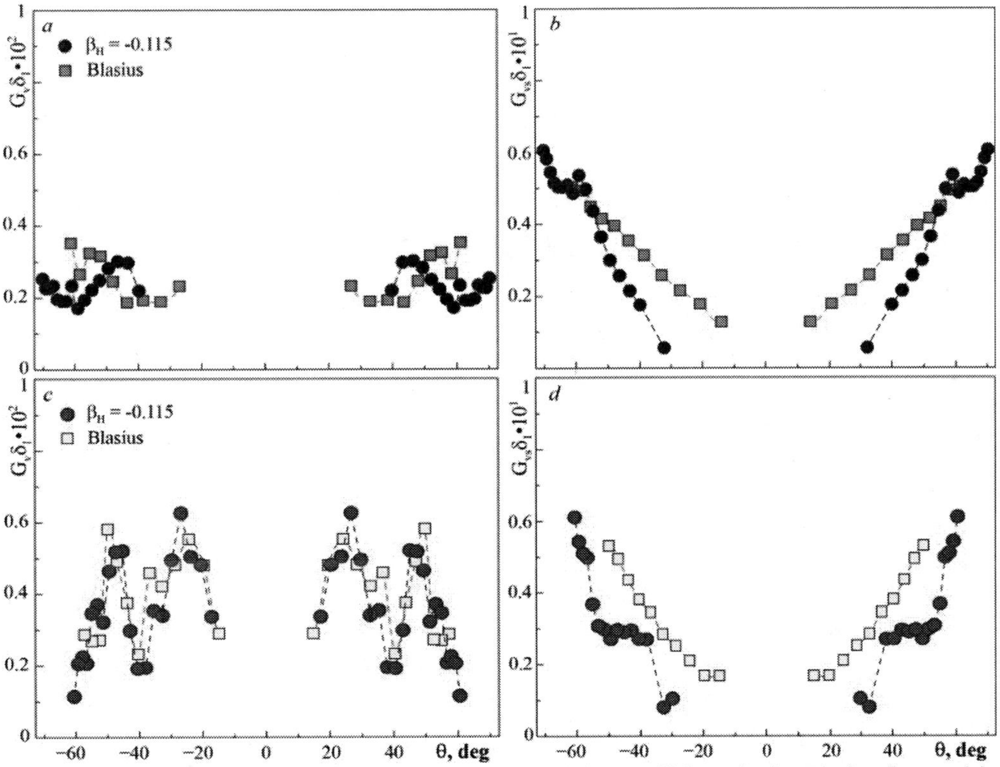

Fig. 3. Comparison of amplitudes of two types of receptivity coefficients in the Blasius flow and in the boundary layer with adverse pressure gradient (β_H = -0.115) for two regimes: *R1f1 (a,b)* and *R3f2 (c,d)*

This work is supported by the Russian Foundation for Basic Research, grant No 15-08-07818.

References

1. V.I. Borodulin, A.V. Ivanov, Y.S. Kachanov, and V.Y. Komarova, Thermophysics and Aeromechanics **13** (2), 199-224 (2006).
2. V.I. Borodulin, A.V. Ivanov, Y.S. Kachanov, and A.A. Fedenkova, *Proc. XIII Int. Conf. Methods of Aerophysical Research: Pt III*, Publ. House "Parallel", 45–50 (2007)

EPJ Web of Conferences **159**, 00006 (2017) DOI: 10.1051/epjconf/201715900006
AVTFG2016

The effect of stable stratification on vortex stretching in open channel flow

Konstantin Borynyak [1,*], *and Michael* Hrebtov[1]

[1]Institute of Thermophysics of SB RAS, 630090, Ac. Lavrentiev ave. 1, Novosibirsk, Russia

Abstract. We present the results of direct numerical simulation of stably stratified open channel flow. The main goal of the paper is to check the effect of thermal stratification on the dynamics of vortex interactions. By comparing the vortex stretching in neutral and stratified cases we show that imposing the stratification leads to an increase of the ratio between positive and negative vortex stretching magnitudes in the flow. That indicates the change in spectral transfer of turbulent energy.

1 Introduction

Stably stratified channel flow attracts increasing attention in scientific community over past decade ([1-3]). Due to the increase of available computational power more and more flow configuration are becoming ready for a direct numerical simulation. Stably stratified channel flow is a numerically low-cost model for a night atmospheric boundary layer, where the relatively hot air flows over the colder ground. Depending on the ratio between the shear and negative buoyancy the flow may become fully turbulent, laminar, or intermittent, with the dynamic alternation of turbulent and laminar spots.

The turbulent collapse in low Reynolds number channel flows, where the negative buoyancy completely suppresses the turbulence, was studied by several authors [3,4], showing that there exist a critical Richardson number preceding the collapse. With a slightly less than critical Richardson number the turbulent and laminar patches become very unstable and dynamic. In such conditions we might expect the largest influence of the stratification on the vortex interaction process. The internal gravity waves accompanying the transition from turbulent to laminar regimes were detected in the central part of the channel [2]. Their activity was shown to remain even after the transition to fully laminar regime. The comparison between the channel flow and open channel flow, where the upper boundary has free-slip condition [5] shows that the internal waves activity although sufficiently weaker is still present in open channel flows at the upper boundaries.

In this short paper we consider the effect of stable stratification on the vorticity dynamics, namely on the vortex stretching. The vortex stretching is considered the main source for the growth of 3D instabilities leading to the formation of forward turbulence cascade. The prevalence of positive stretching promotes the transfer of turbulent energy to the small scales. So the change of rate of occurrence of positive and negative vortex stretching events their relative strength should affect the spectral energy transfer process.

[*] Corresponding author: borisnsk07@gmail.com

DOI: 10.1051/epjconf/201715900006

2 Computational details

We simulate the open channel flow with a relatively low Reynolds number ($\mathrm{Re}_\tau = u_\tau h / \nu = 180$, where u_τ is a friction velocity, ν - kinematic viscosity and h - the open channel height). We studied two configurations: with a neutral stratification ($Ri_\tau = \alpha \Delta T g h / u_\tau^2 = 0$, where α is thermal expansion coefficient, g is gravitational acceleration and ΔT is temperature difference between top and bottom boundaries) and stable stratification ($Ri_\tau = 240$). The Prandtl number value was equal to 0.71.

The computational domain size was $8\pi \times 4\pi \times 1$ with $512 \times 50 \times 512$ grid points respectively. In longitudinal and transversal directions the grid was uniform, while in vertical direction the grid was clustered toward the wall with the grid step range: $\Delta y^+ = 0.1 - 5.5$.

We used a DNS solver based on an open-source CFD package called OpenFOAM. OpenFOAM has been verified for a number of scientific and engineering applications [6,7]. The governing equations of incompressible fluid flow with Boussinesq approximation for buoyancy force are sampled using the finite-volume method with second order accuracy in space and time.

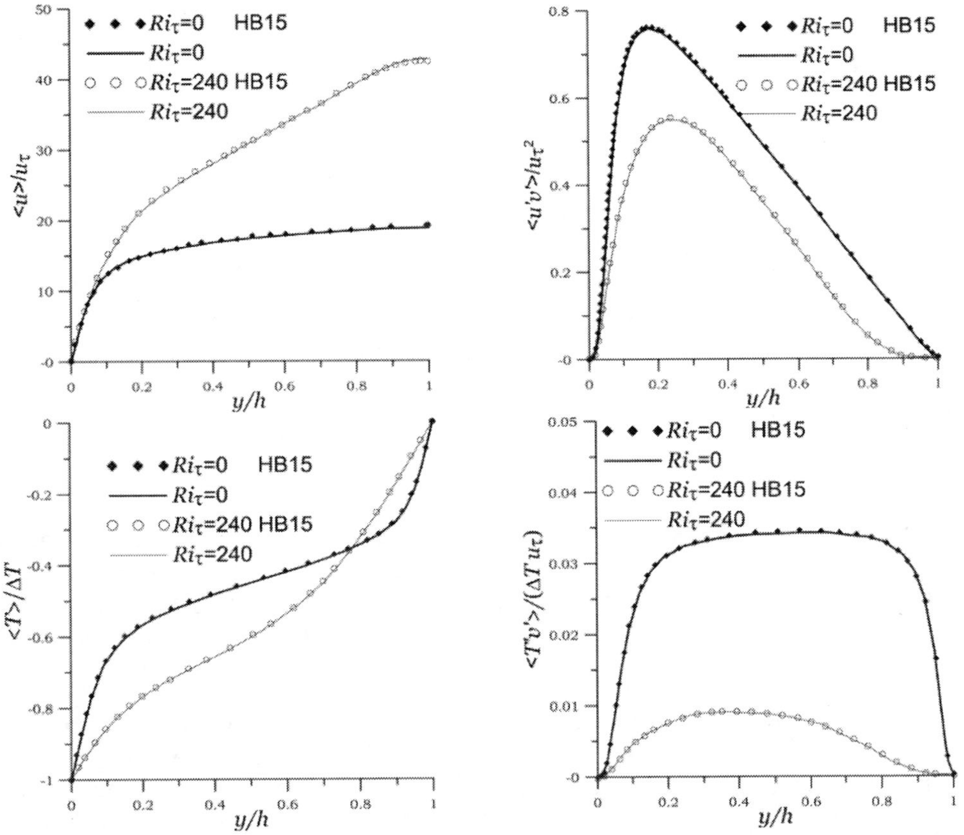

Fig. 1. Compaison of mean vertical profiles obtained in current simulation (lines) and taken from [3] (dots) for cases with stable and neutral stratification. Left: Mean longitudinal velocity and temperature. Right: Reynolds stress and turbulent heat flux.

DOI: 10.1051/epjconf/201715900006

Periodic boundary conditions were prescribed at the streamwise and spanwise boundaries. A no-slip boundary condition was set up for the bottom wall and a free-slip condition was set for a top boundary. Fixed temperature values (of 1 and 0) were used at top and bottom boundaries respectively.

3 Results and discussion

The selected Richardson number value (240) is showing a very strong intermittency with large triangle-shaped turbulent spots. The spots are advected downstream with the mean velocity and slowly changing their shape. This should make the effect of stable stratification on vortex dynamics as large as possible for the selected Reynolds number value (180). In the case of neutral stratification turbulent vortices occupy the entire volume of the channel.

3.1 The mean fields

Figure 1 shows the mean velocity and temperature fields together with Reynolds stress and turbulent heat flux profiles for both simulated cases. The effect of stratification on the mean fields is very strong leading to increase of the bulk velocity, decrease of Reynolds stress and turbulent heat flux. The mean temperature profile also becomes closer to the laminar one.

The flow configuration was similar to the one reported in [3], so the mean fields obtained in the simulation were compared with the ones from [3]. The comparison shows good agreement between the two simulations in both mean profiles and stresses.

3.2 The effect of stratification on vortex stretching

The vortex stretching term ($S_\omega = \omega_i S_{ij} \omega_j$, where S_{ij} is a strain rate tensor and ω is vorticity) in the enstrophy equation shows the production of enstrophy by vortex interactions. If S_ω is positive the vortex tube is stretched longitudinally and its cross-section area is diminished so that the volume of the tube remains unchanged. That in general leads to the spectral transfer of the energy toward the small scales. But it is possible for the S_ω to be negative, meaning that the vortex tube is shortened and widened, thus transferring the energy towards large scales of the flow and decreasing the local dissipation rate. It is worth noting that in two-dimensional flows S_ω is exactly zero.

To distinguish between these two processes the separate contributions of positive and negative vortex stretching into the net one were computed. For each horizontal plane the positive and negative parts of S_ω were integrated separately and then divided over the area of the computational part of plane. Then some averaging in time was performed to acquire the sufficient statistics.

Figure 2a,b shows the result of this operation. It can be seen that the negative contribution is sufficiently smaller than the positive one. Qualitatively the behaviour of positive and negative stretching profiles is the same, showing a maximum close to the laminar layer and rapidly diminishing toward the wall. In both cases this behaviour is similar, but in stratified case the magnitude of total vortex stretching maximum is about 1.5 times larger than in neutral one, while the mean enstrophy ($\Omega = <\omega^2 - <\omega^2>>$, the angular brackets mean the averaging operation) has the opposite behaviour ($\Omega_{neut}/\Omega_{sat}$=1.64).

The main question of this study is to compare the ratio between the positive and negative parts of S for both cases.

DOI: 10.1051/epjconf/201715900006

Figure 2c shows the S_{pos}/S_{neg} profiles. It can be seen that this ratio grows with distance from the wall and getting to its peak at y/h=0.3 in both cases. This point is well in the turbulent region of the flow and much further from the wall than the peak in vortex stretching. As for the values in Fig2c it can be seen that the stratified flow has about 0.75 times lower ratio of positive to negative stretching than the neutral one, which is a clear indication of the influence of stratification on the vortex dynamics.

However as it was mentioned earlier the mean enstrophy (Ω) is 1.64 times lower in a stratified case than in neutral one. Figure 2d. shows the profiles of vorticity variance normalized by the mean enstrophy over the whole volume. It is seen that in stratified case the profile has less steep decay with distance from the wall at y/h<0.4 and more steep decay than the neutral one further from the wall. It is interesting to note that this change in the behaviour occurs at the same distance from the wall where the decay of S_{pos}/S_{neg} starts.

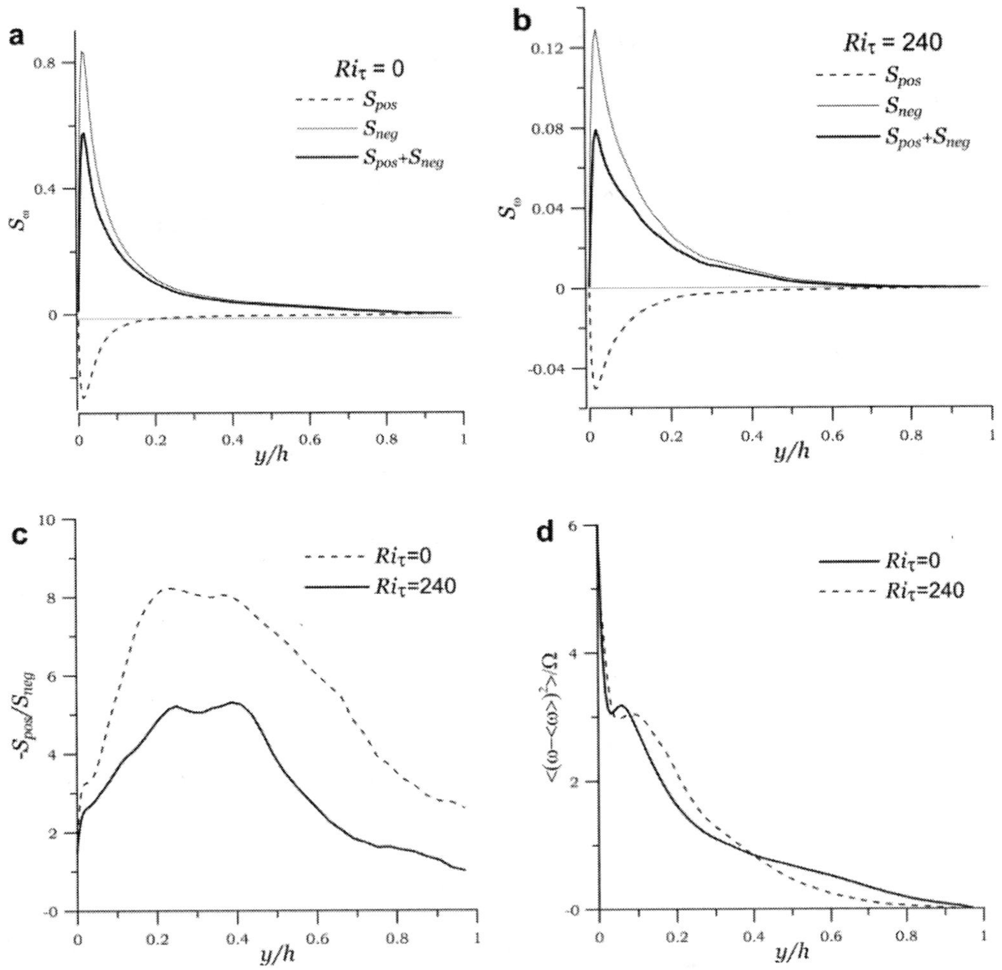

Fig. 2. The vertical profiles of: plane averaged positive, negative and total vortex stretching (a,b); the ratio of mean positive to mean negative vortex stretching (c); vorticity variance over horizontal plane normalized by total (volumetric) vorticity variance.

4. Conclusion

The present study shows that the stable stratification has significant effect on vortex dynamics, decreasing the ratio of magnitudes of mean positive and negative vortex stretching. That could be the indication of the change in cascade of turbulent energy transfer. Further Investigation may be in determining which mechanism is responsible for this increase of negative vortex stretching, excluding the effect of the interaction of vortices with the mean shear, and testing this with different *Re*/*Ri* pairs.

This work was supported by Russian Science Foundation (grant №16-19-00119).

References

1. K. Yeo, B. Kim and C. Lee, J. Turbul. ,**10**, 17 (2009)
2. M. García-Villalba and J.C. del Álamo, Phys. Fluids, **23** (2011)
3. P. He and S. Basu, Nonlin. Processes Geophys., **22** (2015)
4. O. Flores, J.J. Riley, Boundary-Layer Meteorol. **139** (2011)
5. E. Deusebio, P. Schlatter, G. Brethouwer, and E. Lindborg, J. Phys.Conf. Ser., **318** (2011)
6. F. Flores, R. Garreaud, and R. Muñoz, Comput. Fluids, **82** (2013)
7. D.A. Hoang, V. van Steijn, L. M. Portela, M. T. Kreutzer, and C. R. Kleijn, Comput. Fluids, **86** (2013)

EPJ Web of Conferences **159**, 00007 (2017) DOI: 10.1051/epjconf/201715900007
AVTFG2016

Investigation of air gasification of micronized coal, mechanically activated using the plasma control of the process

Evgenii Butakov[1*], *Anatoly* Burdukov[1], *Mikhail* Chernetskiy[1,2], and *Victor* Kuznetsov[2]

[1]Kutateladze Institute of Thermophysics, SB RAS, Novosibirsk, Russia
[2]Siberian Federal University, Krasnoyarsk, Russia

Abstract. Combination of the processes of coal combustion and gasification into a single technology of mechano-chemical and plasma-chemical activation is of a considerable scientific and technological interest. Enhancement of coal reactivity at their grinding with mechanical activation is associated with an increase in the reaction rate of carbon material, and at plasma-chemical effect, the main is an increase in reactivity of the oxidizing agent caused by the high plasma temperatures of atomic oxygen. The process of gasification was studied on the 1-MW setup with tangential scroll supply of pulverized coal-air mixture and cylindrical reaction chamber. Coal ground by the standard boiler mill is fed to the disintegrator, then, it is sent to the scroll inlet of the burner-reactor with the transport air. Pulverized coal is ignited by the plasmatron of 10-kW power. In experiments on air gasification of micronized coal, carried out at the temperature in the reaction chamber of 1000-1200°C and air excess α = 0.3-1, the data on CO concentration of 11% and H_2 concentration of up to 6% were obtained. Air and air-steam gasification of mechanically-activated micronized coals with plasma control was calculated using SigmaFlow software package.

1 Introduction

Coal is the energy fuel of the main resource both in our country and worldwide, so its effective use is certainly relevant from both the economic and ecological points of view. Enhancement of energy efficiency of its use, i.e., reduction of its consumption per production of an energy unit, is of a particular importance.

Combination of the processes of coal combustion and gasification into a single technology of mechano-chemical and plasma-chemical activation is of a considerable scientific and technological interest. Enhancement of coal reactivity at their grinding with mechanical activation is associated [1-3] with an increase in the reaction rate of carbon material, and at plasma-chemical effect [4], the main is an increase in reactivity of the oxidizing agent caused by the high plasma temperatures of atomic oxygen.

* Corresponding author: e_butakov@mail.ru

Due to an increase in the chemical activity of coals at their grinding with mechanical activation [5], determined at IT SB RAS, it is possible to develop the efficient burners on pulverized coals, whose parameters are close to the gas-oil installations. Thus, it becomes possible to develop the systems of pulverized coal gasification in the flow of mechanically-activated micronized coals with gasification parameters, which are considerably higher than the currently used ones. When developing the coal gasification system on micronized coal, the use of plasma-energy installations seems urgent to run the installations and regulate the thermal process. Since the problem of efficient use of coal as the energy fuel for the traditional thermal power technologies and gasification processes for the combined-cycle plants and synthesis gas production for gasification of settlements and industries are of the great interest; the relevance of the problem and research is unquestionable.

2 Experimental apparatus and procedure

To study the single-stage gasification process, the second stage of two-stage reactor with tangential scroll supply of coal-air suspension and cylindrical reaction chamber was used. Coal of DR grade with maximal amount of volatiles from Kuznetsk deposit (one of the main coal deposits of Siberia) was used in experiments. The technical composition of coal is as follows (mass %, dry): moisture W_{rl} - 19.0, ash A_r - 14.3, volatile content - 43.5, sulfur S_{dl} - 0.30, high heat value Q_{daf} - 7327 kcal/kg. Ground coal is fed to the disintegrator; then, it is fed by an ejector with transport air to the scroll burner. Pulverized coal is ignited by a low-power plasmatron (N~10 kW). Coal consumption is 17 kg/h, it is controlled by the voltage sensors of the feeders on the basis of preliminary calibration. Gas composition along the reaction chamber is measured by the optical-absorption gas analyzer at three points; temperature is measured in the same cross-sections of the chamber by the platinum-platinum-rhodium thermocouples.

The thermal setup of 1-MW power for coal combustion and gasification is shown in Fig. 1; location of temperature sensors and gas analyzers along the setup is presented in Fig. 2. Coal, ground in the standard boiler mill, is supplied by the feeder to the disintegrator; then, it is fed with transport air to the scroll inlet of the reactor-burner.

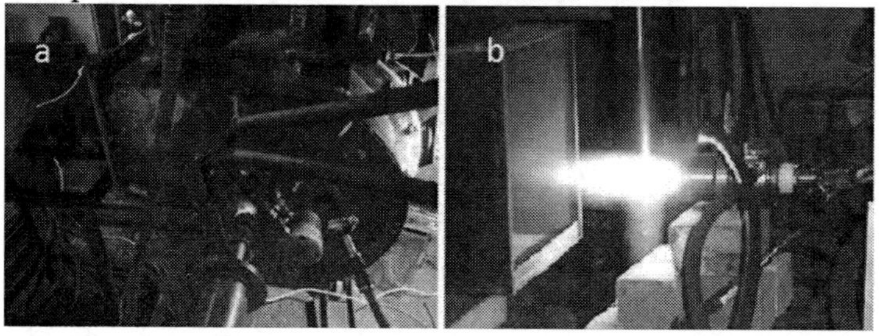

Fig. 1. a) Thermal 1-MW setup with plasma control of the process; b) Plasmatron during operation

Numerical simulation of the turbulent flow of an incompressible liquid in the volume of the burner used Reynolds equations with consideration of interfacial interaction. Reynolds equations were closed by means of the two-parameter model of turbulence: the standard k-ε model. To describe the processes of particle motion, the Lagrange method is used. The particle motion is described by equations of material point dynamics with consideration of resistance and gravity. Flow turbulence at particle motion is taken into account by introduction of random fluctuations of gas velocity into equation of motion for particles. The solution of radiative transfer equation is based on P1 approximation of the method of spherical harmonics for the gray two-phase two-temperature medium. The process of coal particle combustion is considered as the following successive stages: evaporation of moisture from the fuel, escape and combustion of volatile components, and combustion of coke residue [6]. Mechanical activation of coals is accounted by adding the multistage character of chemical reactions with partial overlapping of stages to the mathematical model. At that, the kinetic constants are taken from TGA data for mechanically activated coals [7]. Plasma activation of coal dust and formation of an additional plasma heat source are considered as an internal heat source with heat release characteristic of the plasmatron power. This source is an energy contribution of the electric arc. The model is supplemented with formation of the source terms for the gas components during the passage of carbon residue through the high-temperature region of a plasma jet. In addition to gasification reactions, chemical reactions of carbon, characteristic of plasma activation of coal, are taken into account. This model is implemented in the package CFD SigmaFlow applications [8].

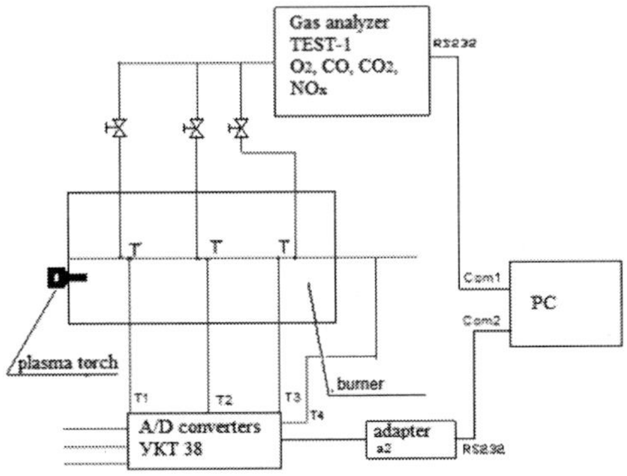

Fig. 2. Layout of temperature sensors and receiving probes of gas composition control.

3 Results and discussion

Experiments were carried out with the constant consumption of fuel: 17 kg/h at air excesses of 0.3; 0.4; 0.6; 1.0, and 35.7; 47.6; 71.4; 119 m^3/h and plasmatron power of 10 kW.

During the entire experiment, the maximal temperature in the center of reaction chamber was 1250 °C; at that, the temperature after the tangential scroll inlet did not exceed 750 °C. The temperature at the end of the reaction chamber was 1000 °C. The temperature does not depend on variations of the air flow rates, i.e., after pulverized coal ignition and beginning of stable combustion of the coal-dust flame, the time of pulverized coal particle stay in the chamber decreases with a change in the flow velocity, and this compensates the heat release growth with an increase in α in the chamber volume and does not lead to a significant increase in the flow temperature. This effect influences a change in CO concentration due to the reaction of CO reducing because the latter requires more time in comparison with the reaction of carbon oxidation, which occurs with a high rate.

It is necessary to note that oxygen concentration in the chamber at all α is close to zero, i.e., oxygen is consumed completely at all stages of combustion. Depending on the coefficient of air excess, concentration of carbon oxide varied within 12 -15%, and concentration of hydrogen varied within 7.5 - 10%, and CO concentration decreases with an increase in α, and hydrogen concentration increases; at that, with an increase in α, CO_2 concentration decreases from 12 to 10%. These effects relate to temperature changes and time of pulverized coal stay in the reaction chamber.

DOI: 10.1051/epjconf/201715900007

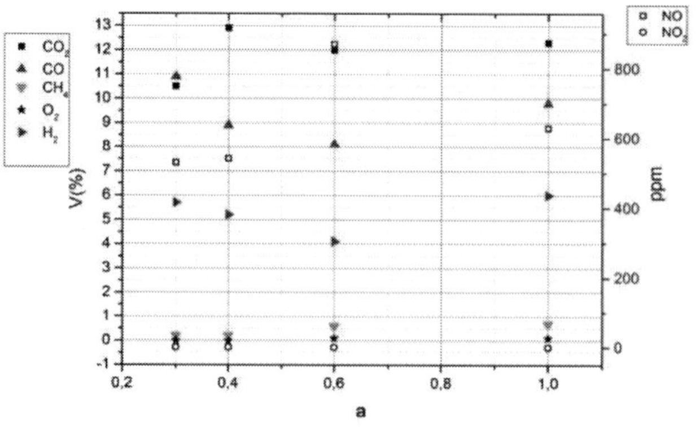

Fig. 3. Gas composition at air gasification of mechanically-activated micronized coals with plasma control of the process (plasmatron power – 10 kW; coefficient of air excess - 1, 0.6, 0.4, 0.3;)

Using calculations have been carried out Air and air-steam gasification of mechanically activated micronized coal with plasma control of the process was calculated using the SigmaFlow software package. The complex mathematical model was adapted by the data of experimental studies based on gasification of coal fuel, ground in the disintegrator at the excess air coefficient of 0.6. It can be seen (Table 1) that there is a satisfactory agreement with experimental results.

Table 1. Gas composition at the outlet of the second stage chamber at air excess coefficients α=0.6

Gas composition (vol. %)	CH_4	CO	H_2	H_2O	CO_2	N_2
Experimental data	0.5	8	4	-	12	-
Volumetric data	0.6	6.3	3.1	9	15	66

Results of numerical studies on the effect of plasmatron presence on coal dust gasification are shown in Table 2. The plasmatron was taken into account out by addition of a plasma heat source in the form of an internal heat source with heat release characteristic of plasmatron power of 10 kW. According to calculation results, plasma activation of mechanically activated coal fuel improves gas composition at the outlet. CO concentration increases from 31.7 to 33%, and hydrogen concentration rises from 8.9 to 10.5%.

Table 2. Gas composition at the outlet of the second stage chamber at excess air coefficients α=0.3 with plasma heat source of 10-kW power

Gas composition (vol. %)	CH_4	CO	H_2	H_2O	CO_2	N_2
Air gasification	4.9	31.7	8.9	1.3	1.7	51.5
Air gasification with plasmatron	4.4	33	10.5	0.8	1.5	49.8

4 Conclusion

In experiments on air gasification of micronized coal, carried out at 750-1200°C and air excess α=0.3-1 in the reaction chamber, first data on CO concentration of up to 11% and H2 concentration of up to 6% were obtained.

The complex mathematical model was adapted by the experimental results on air and air-steam gasification of micronized coals, mechanically activated at plasma control of the process. Satisfactory agreement of calculation and experimental data on gasification of coal fuel, ground in the disintegrator, is shown.

The run of setup using the plasmatron was tested for the first time as well as introduction of a plasma jet into the reaction chamber for controlling the thermal conditions of gasification process. Due to computational studies, it was shown that plasma activation of mechanically activated coal fuel improves composition of generator gas at the outlet.

The work was financially supported by the Russian Ministry of Education and Science, according to Subsidiary Agreement No. 14.613.21.0005 (unique project identifier RFMEFI61314X0005).

References

1. M.Yu. Chernetskiy, A.P. Burdukov, E.B. Butakov, I.S. Anufriev, P.A. Strizhak, Combustion, Explosion, and Shock Waves, **52**, 3, 326–328 (2016)
2. A.P. Burdukov, E.B. Butakov, V.I. Popov, M.Y. Chernetskiy, N.S. Chernetskaya, Thermal Science, **20**, 23-33 (2016)
3. A.P. Burdukov, V.I. Popov, G.V. Chernova, M.Yu. Chernetskiy, A.A. Dekterev, N.S. Chernetskaya, V.M. Markova, V.N. Churashev, T.S. Yusupov, Thermal Engineering, **60**, 889-894 (2013)
4. V.E. Messerle, A.B. Ustimenko, Y.E. Karpenko, M.Y. Chernetskiy, A.A. Dekterev, S.A. Filimonov, Thermal Engineering, **62**, 442-451 (2015)
5. A.P. Burdukov, V.I. Popov, M.Yu. Chernetskiy, A.A. Dekterev, K. Hanjalić, Applied Thermal Engineering, **74**, 174-181 (2014)
6. M. Yu. Chernetskiy, A. A. Dekterev, Combustion, Explosion, and Shock Waves, **47**, 3, 280–288 (2011)
7. M.Yu. Chernetskiy, A.A. Dekterev, A.P. Burdukov, K. Hanjalić, Fuel, **135**, 443-458 (2014)
8. M.Yu. Chernetskii, A.N. Alekhnovich, A.A. Dekterev, Thermal Engineering, **59**, 610-618. (2012)

EPJ Web of Conferences **159**, 00008 (2017)
AVTFG2016
DOI: 10.1051/epjconf/201715900008

The heat transfer crisis in shear driven liquid film of FC-72 by gas flow in a minichannel

Vyacheslav Cheverda[1,2*]

[1]Institute of Thermophysics, SB RAS, Lavrentiev Prospect, 630090, Novosibirsk, Russia
[2]Novosibirsk State University, Pirogova str. 2, 630090Novosibirsk, Russia

Abstract. The crisis in a FC-72 liquid film moving under the action of a gas flow through a smooth heater was experimentally investigated. It is shown that with gas and liquid flow rates increases a critical heat flux is growth.

1 Introduction

The rapid development of semiconductor technologies, including the reduction in the size of the transistor, that is, increasing their density, leads to higher heat fluxes and dissipated power from the chip surface. The forecast for high-performance chips suggests that the maximum power dissipation will exceed 500 watts [1, 2]. Moreover, in recent years, the performance requirements have increased, which leads to greater heat flux from the chip and the occurrence of "hot spots", the heat flux in which often exceeds 1 kW / cm^2, which is danger for microprocessor. The two-phase flow is a promising solution of this problem [3-5].

The authors' studies in [6-9] were focused on underheated thin fluid films flowing down through a locally heated plate. Forced flow of dielectric liquids, with intense evaporation in the minichannel, is a promising solution for cooling modern semiconductor devices for terrestrial and space applications [10]. Fluid FC-72 is designed to cool the electronics and is volatile with low surface tension [11]. The destruction of liquid films on a heated wall is of great technical importance, since this determines the allowable heat flux in a number of engineering applications. A comparison between the flowing fluid film and the liquid film moving under the action of the gas flow is shown in [12,13]. Surface temperature, heat flux for formation of the first dry spot and critical heat flux are shown and analysed in the paper [12]. The critical heat flux is measured and shown, the liquid film moving under the action of the gas flow is up to five times more efficient than the flowing fluid film. It was found in the number of papers that the thermocapillary foresees play a crucial role in shear driven cooling systems of microelectronic devices [14 - 16].

2 Experimental setup

* Corresponding author: slava.cheverda@gmail.com

The experiments were carried out using the experimental setup shown in (Fig.1). The setup includes: a system for supplying working fluids, a measuring system (sensors) and an optical (speed schlieren-system) system, a working fluid supply system, a data collection system and a remote-control system. The two temperature stabilization systems are based on the use of thermoelectric modules (Peltier elements) and are used to stabilize the temperature of the liquid and gas at the entrance to the work area. All thermal modules are cooled by a water circuit. The temperatures of liquid and gas at the inlet to the test cell are maintained between 10 and 30 ° C. Dielectric fluid FC-72 and dry nitrogen gas are used as working fluids. The operating pressure and gas flow rate are automatically controlled by the Bronkhost flow and pressure regulators. Degassed liquid is fed into the working area with a syringe pump from a flexible liquid container. From the working area, a mixture of steam, gas and liquid is pumped into the atmosphere. All parameters are remotely controlled using a control and data acquisition system developed on the LabView platform.

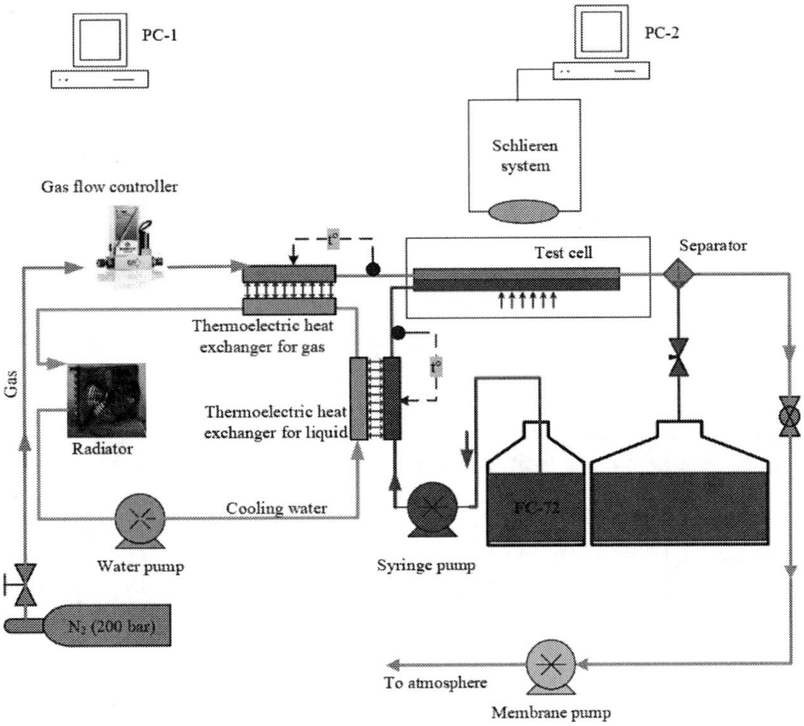

Fig. 1. Scheme of the experimental setup.

The experimental test cell (see Fig. 2) consists of: 1) a textolite plate, 2) stainless steel substrates with a copper pressed rod, 4) a liquid nozzle, 5) a textolite frame, and 6) a textolite cover with an optical window. The flow of the film is realized in the minichannel, using a liquid nozzle. The main element of the working area is a stainless-steel plate with a 10x10mm copper rod. Several calibrated K-type thermocouples are used to measure temperature. A stainless-steel plate is attached to a textolite plate together with a textolite frame. The frame with the plate is covered with a transparent window, forming a rectangular minichannel 67 x 30 x 1.4 mm (length, width and depth, respectively). Liquid film is formed with the help of a liquid nozzle and is carried along the channel by a gas stream. The temperatures of the substrate, as well as gas and liquid at the inlet and outlet of the working section, are measured with accuracy 0.4 ° C. The power of the heater is regulated and controlled by a power source.

DOI: 10.1051/epjconf/201715900008

The optical system makes it possible to visualize with great frequency and to measure the deformation angles of the gas-liquid interface. This system allows us to visualize at a frequency of up to 500 Hz for a full resolution of 1280x1024 pixels and up to 60 kHz for a reduced resolution of 1280 x 4 pixels.

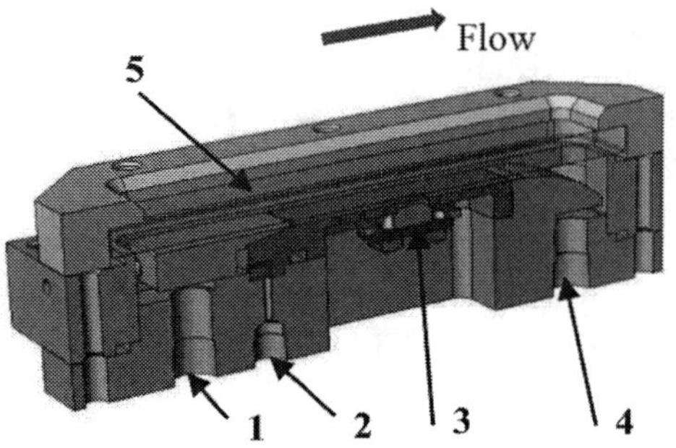

a) Scheme of the test cell

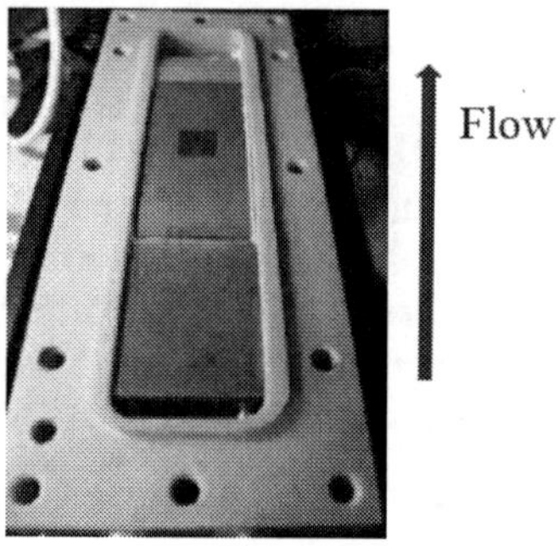

b) Photo of the test cell

Fig. 2. Scheme (a) and photo (b) of the test cell, 1 – gas inlet, 2 – liquid inlet, 3 – copper rod, 4 – two-phase outlet, 5 – optical window.

3 Experimental results

The development of crisis phenomena during the flow of a FC-72 film liquid under the action of a nitrogen gas flow is studding in details. When the heater is turned on all temperatures increase and deformations at the gas-liquid interface increase. With a further increase in the heat flux, formation of the first, unstable dry spots can be observed.

A dry spot forms near the right lower edge of the heating element due to the high temperature gradient at the gas-liquid interface. However, the heat flux is not high enough, and this first dry spot is moving from heater area. With increasing heat flux, the first stable dry spot appears. With a further increase in heat flux, the heater is practically not wetted by the liquid and the temperature of the heater slightly increases. At some point, the liquid begins to move around the heater and does not wet at all. All the temperatures of the heater and the substrate begin to increase with a relatively large gradient. In this case critical heat flux (Q_{CHF}) is measured. The results of measurements of the critical heat fluxes in a FC-72 liquid film moving under the action of a nitrogen gas flow are shown in Fig. 3. It can be seen that the critical heat flux increases with increasing gas and liquid Reynolds numbers (Re_G, L).

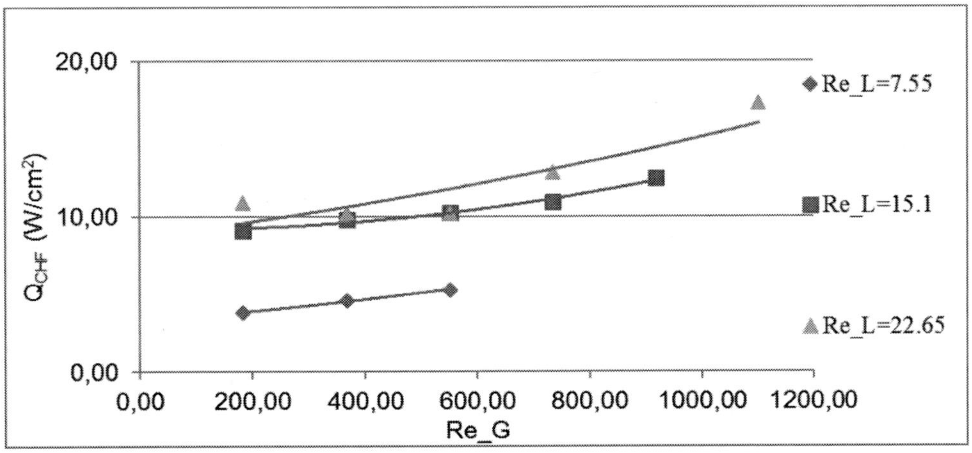

Fig. 3. The critical heat flux for shear driven FC-72 liquid film moving in a minichannel with smooth heater for different liquid and gas Reynolds numbers.

Conclusion

The work shows an increase of the critical heat flux for the FC-72 liquid film flow in a minichannel by a gas flow when using smooth heater when gas and liquid Reynolds number are increases. It is supposed that the dray patch formation and the crisis phenomenon are happening because of the strong thermocapillary effect. Thermocapillary foresees move liquid from the heater. The crisis phenomenon take place at temperatures of the gas-liquid interphase much smaller than the saturation temperature.

This study was performed in the framework of the Russian Ministry of Education (RFMEFI60414X0053).

References

1. Wang P., Bar-Cohen A., Yang B., Solbrekken G. L., Zhang Y. and Shakouri A., Proceedings of InterPACK '05, 73244 (2005).
2. Mahajan R., Chiu C., Chrysler G., Proc. IEEE, **94**, 1476–1486 (2006).
3. Chinnov, E. A.; Ron'shin, F. V.; Kabov, O. A., Technical physics letters. **41**, 817-819 (2015)
4. Chinnov, E. A.; Ron'shin, F. V.; Kabov, O. A., Thermophysics and aeromechanics. **22**, 621-629 (2015)
5. Chinnov, E. A.; Ron'shin, F. V.; Kabov, O. A., Thermophysics and aeromechanics. **22**, 265-284 (2015)
6. Kabov O.A., Thermophys. Aeromech. **7**, 513–520 (2000)
7. Kabov O.A., Scheid B., Sharina I.A., Legros J.-C., Int. J. Thermal Sciences, **41**, 664-672 (2002)
8. Chinnov E. A., Shatskii E. N, High temperature. **52,** 461-464 (2014)
9. Chinnov E. A., Shatskiy E. N., Technical physics letter. **40,** 7-9 (2014)
10. Bar-Cohen A., Rahim E., Heat Transfer Eng. **30**, 601, (2009).
11. Ing P., Sperry R., Philstorm R., Claybacker P., Webster J. and Cress S., Proceedings of the 1993 IEEE ECTC Conference, 219-237 (1993).
12. Kabov O.A., Lyulin Yu.V, Marchuk I.V., Zaitsev D.V., Locally heated shear-driven liquid films in microchannels and minichannels, Int. J. Heat Fluid Flow, **28**, 103-112 (2007)

13. Kabov O.A., Zaitsev D.V., Cheverda V.V., Bar-Cohen A., Experimental Thermal and Fluid Science, **35**, 825 (2011)
14. Hirokawa T., Murozono M., Kabov O. and Ohta H., Frontiers in Heat and Mass Transfer, **3**, 1-8 (2014).
15. Hirokawa T., Ohta H. and Kabov O. A., Interfacial Phenomena and Heat Transfer, **3**, 303-317 (2015).
16. Kabov O. A. and Kabova Yu. O., Thermophysics and Aeromechanics, **22**, 539-542 (2015).

The influence of the physical properties of the liquid on the formation of thermo-capillary structures

Evgeny Chinnov, *Andrey* Semenov

Kutateladze Institute of Thermophysics SB RAS, 630090, 1 Lavrentyev Ave., Novosibirsk, Russia
Novosibirsk State University, 630090, 2 Pirogova Str., Novosibirsk, Russia

Abstract. An experimental study of the formation of thermocapillary structures at low Reynolds numbers and temperature gradients at the surface of the film in a wide range of the viscosity of the liquid was carried out. The data on the distance between the rivulets are generalized. It is shown that the transverse dimension of structures is determined by the capillary constant and does not depend on either the viscosity of the liquid or the flow rate.

1 Introduction

The flow of a thin layer of liquid (film) over a solid surface belongs to the class of flows with a free surface. Various instability mechanisms play an important role in thees flows, leading to the development and interaction of waves of different scales. Film flows are used in various technological processes and apparatuses. Studies of heat transfer from flat heaters to a flowing liquid film began in the early 90s. In nonisothermal liquid films, there are two rivulet formation regimes [1]. In regime A, high temperature gradients (up to 10-15 K / mm) were observed on the surface of the film in the upper part of the heater. On the surface of the heater, a boundary condition close to T = const was realized. When the threshold heat flux density was reached, significant deformations appeared on the surface of the film in the upper part of the heater. The stream was divided into vertical rivulets with a certain wavelength λ. At low Reynolds numbers, high thermocapillary stresses directed against the flow led to a thickening of the film in the form of a horizontal roller [2]. Experimentally, structures of type A were studied for Re <25.

In regime B, the rivulet flow was formed gradually with increasing heat flow and distance from the upper edge of the heater [1]. The boundary condition q = const was realized on the surface of the heater. The inhomogeneities of the thickness of the film across the flow led to inhomogeneities in the temperature at the surface. The transverse temperature gradient (up to 1 K / mm) resulted in an increase in the deformation of the film surface. An experimental study of the evolution of hydrodynamic waves into thermocapillary-wave structures while heating of a vertically falling film was carried out in [3] for Re> 20. The fact of the appearance of temperature inhomogeneities at the front of a three-dimensional hydrodynamic wave is shown. Because of the action of thermocapillary forces, these inhomogeneities led to deformation of the film and the formation of rivulets.

To the present moment, the formation of structures of type B has been studied mainly on two liquids: water and FC-72, with significantly different surface tension coefficients. It is shown that the surface tension is significantly affected by the transverse size of the structures (the distance between the centers of the rivulets).

One of the latest theoretical papers [4] confirms the existence of two types of structures on the surface of a nonisothermal liquid film. It is shown that in the thermocapillary regime the wavy flow decays, and in the thermocapillary-wave regime, the waves propagate along the rivulets. However, the calculations are limited to the region Re <5-8 and do not describe the hydrodynamic instability regime of the liquid wave film. In [5], the thermocapillary rupture of a liquid film flowing under the influence of gravity along a vertical plate with a heater 150 × 150 mm is studied in a wide range of fluid properties. It has been established that the viscosity of a liquid has a significant effect on the threshold heat flux at which a film breaks. The effect of the viscosity of a liquid on the characteristics of thermocapillary structures remains undeveloped, especially in the context of small values of the Reynolds number of the film.

The task of this work is an experimental study of the formation of thermocapillary structures with temperature gradients on the surface of the film of less than 1 K / mm for small Reynolds numbers over a wide range of viscosity changes.

2 Experimental setup

The stand was a closed circulation loop, including a reservoir with a pump, a working section, a filter, rotameters, pipelines and shut-off valves. The working area consisted of a carrier plate, on which a film former, a heat stabilizer and a heater were installed. Using a pump the working fluid was supplied to the film former, which included a storage chamber, a dispenser and a nozzle with a calibrated flat slot. The liquid flowed down the plate and under the influence of gravity, returned to the reservoir with the electric pump. The heating element was an electric heater 150 mm wide and 150 mm long. On the surface of the heater, a condition close to q = const was realized. The following working fluids were used in the experiment: water, 60% glycerol in water, ethylene glycol. The properties of the liquids are presented in Table 1, where Ka is the Kapitza number ($Ka = \frac{\sigma^3}{g\nu^4\rho^3}$), Ma is the Marangoni number ($Ma = \frac{\gamma\Delta T}{\rho\nu^{4/3}g^{1/3}}$), where γ is the derivative of the surface tension coefficient on the temperature, the index *av* shows that the temperature difference is between the average temperature on the film surface and the initial temperature of the film, the index *l* shows that the temperature difference is calculated from the maximum temperature drop across the film flow.

Table 1. Working liquids and their properties.

Working liquid	T_0, ^{0}C	ρ, kg/m^3	ν*10^6, m^2/s	σ*10^3, N/m	$d\sigma/dT$*10^3, N/(m*K)	Ka	Ma_{av}	Ma_l
Water	24	997	0.912	72.16	0.191	5.6*10^{19}	59	11
60% glycerol	24	1151	7.665	67.8	0.131	6*10^{15}	3	2.4
ethylene glycol	24	1112	15.17	45.74	0.09	1.3*10^{14}	-	-

3 Results and conclusions

Fig. 1 shows the temperature distribution on the film surface of a 60% solution of glycerin in water (Reynolds number is 1, the heat flux density is 1.1 W/cm^2), where X is the coordinate counted from the upper edge of the heater, Z is the coordinate counted from the left edge of the heater. It can be seen that the temperature distribution near the upper edge of the heater is almost uniform. In the conditions of a uniform flow, thermocapillary structures are formed on the surface of the film in the form of rivulets with a thin film between them. Rivulets appear on the surface of the film gradually with increasing mean free path along the heater. The beginning of the formation of the rivulets is approximately 20 mm from the upper edge of the heater. The longitudinal and transverse temperature gradients on the surface do not exceed 0.5 K/mm, the temperature difference in the rivulet and the interrivulet region at X = 130 mm is of the order of 4 ° C, and the distance between the rivulets is 14 mm. In these experiments, the structures were also fixed for different Reynolds numbers from 0.1 to 10.

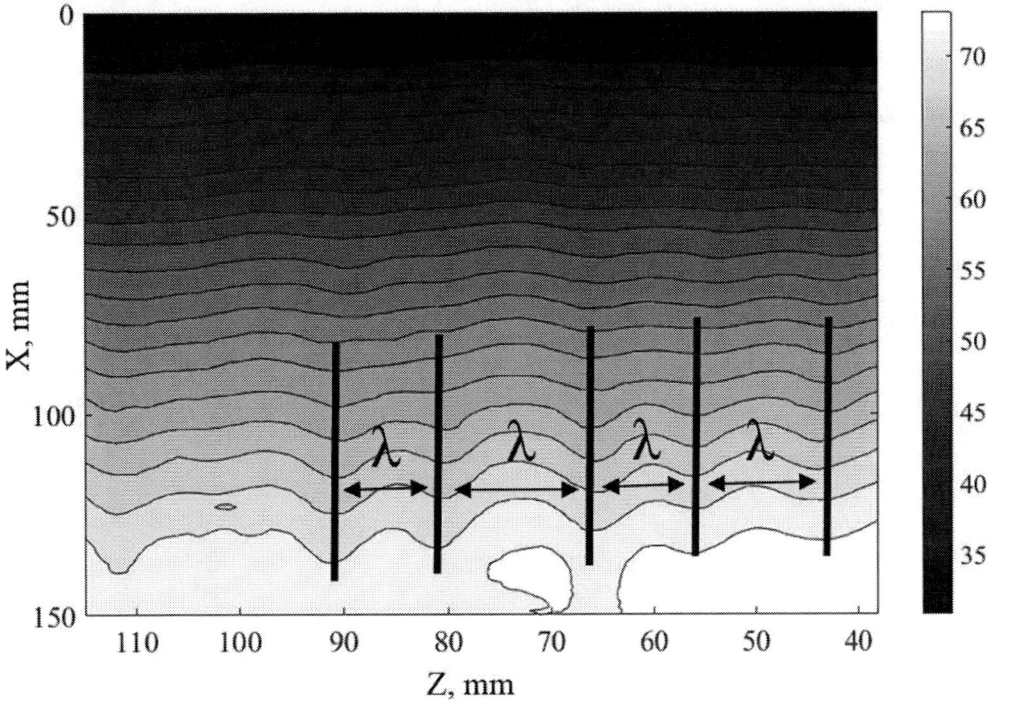

Fig.1 The temperature distribution on the film surface of a 60% solution of glycerin in water (Re = 1, q = 1.1 W/cm^2)

For the analysis of data, various dimensionless criteria and parameters were used. Distances were measured for liquids with different densities, surface tension coefficients, and over a wide range of viscosity and fluid flow rates. The distance between the rivulets is dimensioned by dividing by the liquid capillary constant ($l_\sigma = (\sigma/\rho g)^{0.5}$, where σ is the surface tension coefficients, g is the acceleration due to gravity). All data on the dimensions of structures of type B fall in the range from 4 to 6 and does not depend on either the viscosity of the liquid or the Reynolds number. The use of other parameters led to worse results. For example, using the parameter $l_\delta = (\sigma\mu^2/(\rho^3 g^2))^{1/5}$, which was used in [7], increased the spread of data by an order of magnitude.

The theoretical dependence obtained in [6] indicates an increase in the wavelength of the instability (distance between the rivulets) with a decrease in Ma, but the tendency of its increase with increasing Re remains unchanged, which contradicts the experimental data. For Re> 5-8, calculations based on the theoretical model [6] do not lead to the formation of rivulet flows, while the experimental data obtained indicate a transition from a three-dimensional wave flow of an isothermal film to a thermocapillary-wave regime B under a wide variation of the parameters. As follows from Table 1, the range of variation of the Marangoni number was almost 2 orders of magnitude.

It can be concluded that structures of type B are realized at relatively low Marangoni numbers under conditions of small values of the temperature gradients at the surface of the film over a wide range of the Reynolds number of the film and the physical properties of the liquid (Kapitsa number). The transverse size of the structures (the distance between the rivulets) is determined by the capillary constant and does not depend on either the viscosity of the liquid or the flow rate (Marangoni, Kapitsa and Reynolds numbers of the film).

The work was supported by the RF Ministry of Education and Science (Agreement No. 14.604.21.0053, project ID RFMEFI60414X0053).

References

1. Chinnov E.A., Kabov O.A. Journal of Applied Mechanics and Technical Physics **44** (2003)
2. Kabov O.A. Heat Transfer Research **27** (1996)
3. Kalliadasis S., Kiyashko A., Demekhin E.A. J. Fluid Mechanics **475** (2003)
4. Chinnov E.A., Kharlamov S.M., Saprikina A.V., Zhukovskaya O.V. Thermophysics and Aeromechanics **14** (2007)
5. Chinnov E.A. Technical Physics Letters **34** (2008)
6. Scheid B., Kalliadasis S., Ruyer-Quil C., Colinet P. Physical Review E **78** (2008)
7. Zaitsev D.V., Semenov A.A., Kabov O.A. Thermophysics and Aeromechanics **23** (2016)

EPJ Web of Conferences **159**, 00010 (2017) DOI: 10.1051/epjconf/201715900010
AVTFG2016

Mathematical modelling and optimization of biomass-plastic fixed-bed downdraft co-gasification process

Igor Donskoy[1,*]

[1]Kutateladze Institute of Thermophysics of SB RAS, 630090, Novosibirsk, Russia Lavrentieva st., 1

Abstract. Co-gasification of woody biomass and polyethylene is studied using mathematical modeling. The gasification process is downdraft fixed-bed. Comparison of modeling results with some experimental data is made. Influence of biomass/plastic ratio and air equivalence ratio on gasification efficiency is investigated.

1 Introduction

Low-grade fuels, such as forest and agricultural or municipal wastes, often cannot be converted into energy in efficient way. This is concerned with high content of non-combustible matter and possible hazardous contaminants formation (PAH, chlorine compounds, etc.). In this way, co-conversion of low-grade fuels with traditional fuels is under interest [1, 2].

Plastic utilization is an urgent problem because polymeric matter forms the main part of municipal wastes (both in mass fraction and in heat value). But thermal behaviour of plastic often leads to conversion process destabilization due to particles agglomeration [3, 4].

Co-gasification of woody biomass and plastic in downdraft fixed-bed reactor is studied at the present paper using mathematical modeling. Attainability of computed regimes is discussed.

2 Input data for modelling

Mathematical model of the gasification process was developed and used earlier in works [5, 6]. Geometric parameters were taken from [7]. Fuel consumption is 30 kg/h. Active bed height is 2 m, reactor diameter is 0.44 m. Air flow is varied from 30 to 60 Nm3/h. Plastic mass fraction in fuel mixture changes from 0% to 50%. Mean particle size is 5 cm for biomass and 5 mm for plastic. Properties of wood and plastic are listed at table 1. During pyrolysis tar formation occurs. Simplifications are made as following. Tar has uniform composition ($CH_{0.6}$), and tar fraction in pyrolysis products is constant for each fuel. Ethylene in pyrolysis gas is formed only from plastic and its fraction in pyrolysis products also is constant (30% on organic matter basis). Two compositions of wood were used: the

* Corresponding author: donskoy.chem@mail.ru

first was taken from paper [7] to compare modeling results with experiment (wood-1); the second is averaged wood composition based on our previous researches [8] (wood-2).

Some heterogeneous reactions are taken into account in computations. Kinetic parameters of these reactions are listed at table 2 [8, 9]. Decomposition of tar has the same kinetic parameters for both fuels.

Key process characteristics are gas composition, temperature and cold gas efficiency (CGE).

Table 1. Composition of wood and plastic.

Fuel property	Wood-1	Wood-2	Plastic
A^d, %	29	20.0	0.1
W^r, %	0.45	0.45	0.0
V^{daf}, %	85.9	84.65	99.9
C^{daf}, %	35.3	41.04	85.0
H^{daf}, %	7.60	7.01	14.0
O^{daf}, %	57.2	51.93	0.0
Tar fraction in pyrolysis products	0.2	0.2	0.9

Table 2. Kinetic parameters for heterogeneous reactions of biomass and plastic.

Reaction	Biomass		Plastic	
	k_0	E_a, kJ/mol	k_0	E_a, kJ/mol
Pyrolysis	$5.4\cdot10^3$, 1/s	96	$1.7\cdot10^{14}$, 1/s	220
$C + O_2$	$2.4\cdot10^8$, m/s	142	-	-
$C + CO_2$	$1.32\cdot10^9$, m/s	259	-	-
$C + H_2O$	$9.3\cdot10^7$, m/s	175	-	-
Tar decomposition	$5\cdot10^8$, 1/s	150	$5\cdot10^8$, 1/s	150

3 Comparison with reference data

Computations were made using mathematical model and set of kinetic parameters listed above. Input data for stationary regimes from [7] was used. Wood composition used is wood-1 at table 1. Comparison results are shown at table 3. One can see that the model could predict accurately a few number of process characteristics such as output syngas volume rate and concentration of several species (for example, CO_2). Other characteristics could be calculated with qualitative agreement only. Gasification regimes of biomass without plastic are reproduced better. The difference between measured and calculated values of temperature is most likely concerned with underestimation of heat losses.

To make modeling results more close to experiment it is necessary to improve pyrolysis stage modeling block. This process is considered as one-step overall-reaction. The most of modeling error is a result of incorrect estimate of hydrocarbons concentration.

4 Results of optimization study

Computations were also made to estimate optimal co-gasification conditions. Wood-2 was used as a wood composition here. Variable parameters are air equivalence ratio and plastic mass fraction in fuel mixture. The results are presented at fig. 1.

Table 3. Comparison of modelling results and experimental data (Garcia-Bacaioca et al., 2008).

No.	P1-1	P1-2	P1-3	P2-1	P2-2	P2-3	P3-1	P3-2
G_{air}, Nm³/h	43	50	60	38.7	51.9	59.3	40	52.5
G_{fuel}, kg/h	34.4	40.4	46	29.6	35.3	37.9	33.4	30.9
Plastic fraction, %	0	0	0	9.6	9.8	9.5	15.2	17.4
t_{air}, °C	75	70	113	85	116	139	109	126
G_{gas}, Nm³/h	56.8	66.9	82.4	59.5	80.3	94.3	61.9	81.5
	63.6	74.6	85.9	62.4	76.7	83.3	72.9	75.6
N_2, %	59.3	57.4	56.0	50.9	50.5	49.2	50.4	50.8
	52.8	52.4	54.6	48.5	52.9	55.6	42.8	54.3
O_2, %	0.5	0.7	0.4	0.3	0.5	0.4	0.1	0.6
	0.0	0.0	0.0	0.0	0.0	0.0	0.0	0.0
CO_2, %	17.3	14.7	17.2	12.0	10.6	10.8	9.6	7.3
	16.4	16.6	16.2	14.2	13.8	13.6	14.9	12.2
CO, %	10.1	10.2	9.4	19.2	20.1	20.2	19.6	22.5
	11.9	11.8	11.9	13.3	13.0	12.8	12.4	13.5
H_2, %	11.2	15.1	14.8	13.9	13.7	15.1	14.4	14.6
	18.3	18.7	16.7	22.3	18.6	16.3	26.8	17.7
CH_4, %	0.8	0.9	1.2	2.0	2.5	2.4	3.1	3.0
	0.0	0.0	0.0	0.0	0.0	0.0	0.6	0.0
C_2, %	0.7	0.9	1.1	0.8	1.2	0.9	1.7	1.6
	0.0	0.0	0.0	1.1	1.1	1.0	1.6	1.7
Q_{gas}, MJ/Nm³	3.2	3.7	3.8	5.6	6.1	6.1	6.7	6.9
	3.5	3.5	3.3	4.7	4.3	4.0	5.6	4.6
CGE, %	44.5	48.5	52.1	66.3	81.9	90.2	65.9	93.6
	50.8	51.2	48.9	63.6	59.2	56.1	70.8	62.3
t_{out}, °C	787	687	687	620	806	957	779	811
	888	875	939	797	888	962	641	912

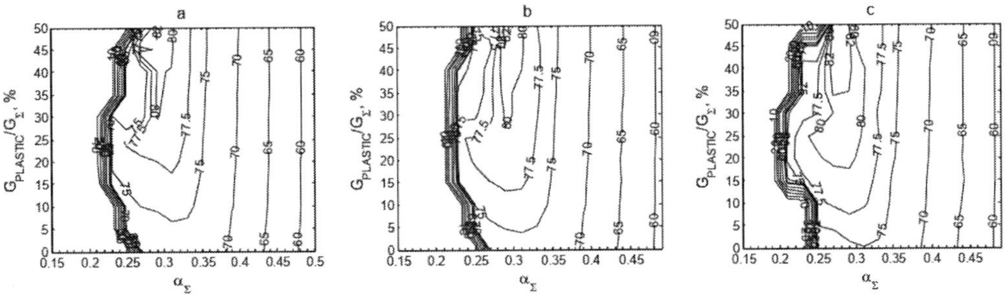

Fig. 1. Cold gas efficiency (%) dependence on summary air equivalence ratio (α_Σ) and mass fraction of plastic ($G_{PLASTIC}/G_\Sigma$): a – initial air temperature is 30°C; b – 100°C; c – 200°C.

One can see that cold gas efficiency has well-known dependence on air equivalence ratio [10]. It should be noted that line of maximum CGE for mixtures of fuels with high-volatiles content (>80%) is very close to a boundary of thermally stable gasification conditions. The more is initial air temperature the wider are boundaries of efficient gasification regimes. Plastic addition increases CGE too. But some regimes at fig. 1 cannot be attained, because the more plastic fraction the more feasibility of the bed agglomeration. So the co-gasification efficiency limit would be estimated more precisely using correlations

 DOI: 10.1051/epjconf/201715900010

of the critical plastic fraction (e.g. fraction that provide bed structure stability) with process conditions. This task is a possible topic of future works.

This work was carried out at the Kutateladze Institute of Thermophysics SB RAS and financially supported by the Russian Science Foundation (project number 15-19-10025).

References

1. U. Arena, Waste Manag., **32**, 625-639 (2012).
2. D.R. McIlveen-Wright, F. Rinto, L. Armesto et al., Fuel Proc. Tech. **87**, 793-801 (2006).
3. M.L. Mastellone, U. Arena, Polymer Degradation and Stability **85**, 1051-1058 (2004).
4. S.M. Lera, Tesis Doctoral (PhD) (Univ. de Zaragoza, 2015)
5. I.G. Donskoi, A.V. Keiko, A.N. Kozlov et al., Thermal Eng. **60**, 904-909 (2013).
6. I.G. Donskoy, MATEC Web of Conf. **72**, 01026 (2016).
7. P. Garcia-Bacaioca, J.F. Mastral, J. Ceamanos et al., Biores. Tech. **99**, 5485-5491 (2008).
8. A. Kozlov, D. Svishchev, I. Donskoy, A.V. Keiko, JTAC **109**, 1311-1317 (2012).
9. H. Bockhorn, A. Hornung, U. Hornung, D. Schawaller, JTAC **48**, 93-109 (1999).
10. M.J. Prins, Proefschrift (Tech. Univ. Eindhoven, 2005).

Experimental determination of change in the liquid layer thickness on the heated surface by confocal sensor

Dmitry Feoktistov[1,*], *Sergey* Misyura[1] , *Anastasia* Islamova[2] , and *Kseniya* Batishcheva[2]

[1]Kutateladze Institute of Thermophysics, Siberian Branch, Russian Academy of Sciences, Novosibirsk, Russia
[2]National Research Tomsk Polytechnic University, Tomsk, Russia

Abstract. Change of the distilled water layer thickness in time under the evaporation from stainless steel surface was studied experimentally. To register a change in the liquid layer thickness an experimental technique with the use of high-precision confocal method was developed. Change of the thickness of thin layers (less than 3 mm) in time is found to have a linear nature. At decreasing the thickness up to 0.5 mm, linear nature changes to the polynomial. It can be explained by the change in surface tension.

1 Introduction

It is necessary to provide heat removal from individual thermal loaded components and integrated modules [1-3] for reliable and accident-free operation of modern electronic equipment. The use of traditional cooling ways [4-8] based on the application of high heat flows of different heat transfer agents circulating through closed circuits are not always implemented technically and high priced. So to remove high heat flows around 1000 W/cm2 [9-11], it is necessary to use enhanced heat exchange surfaces exceeding the area of heat source. In this case, one of the most effective ways of removing heat from the heat-loaded elements is the use of cooling systems based on evaporation of thin liquid layers [12, 13]. Compared to traditional systems, such systems use high energy potential of the phase transition [8, 9, 12, 14]. Due to this effect, high intensity of heat removal during evaporation of thin liquid layers provides keeping of the required temperature mode of the element. There are many studies in this scientific field [15-19]. Scientific and technological progress contributes to the emergence of inexpensive high-precision equipment, such as confocal sensors. Their usage for study of thin liquid layer evaporation allow to clarify previously known physical mechanisms, and probably, to establish new ones.

The purpose of this work is to determine experimentally evaporation rates of thin liquid layers by using confocal sensor.

2 Experimental apparatus

* Corresponding author: fdv@tpu.ru

Experimental studies have been conducted on the experimental setup (Fig.1). The main elements are registration system of the layer thickness (by using confocal sensor confocalDT 2451), temperature registration system (by using controller National Instruments and mini thermocouples Omega with a thermojunction of 80 μm), and test cell (Fig. 2).

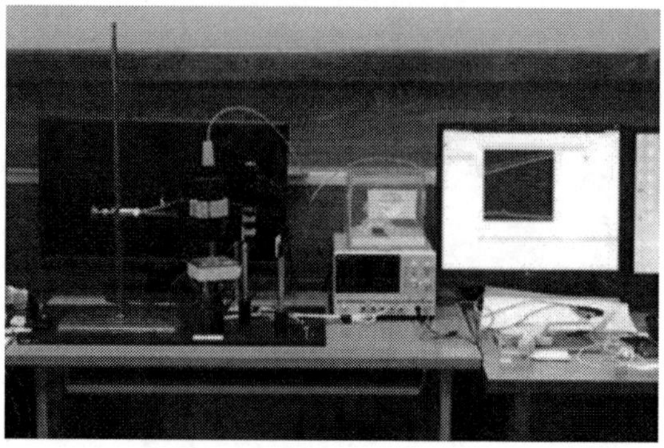

Fig. 1. Photographic image of the experimental setup.

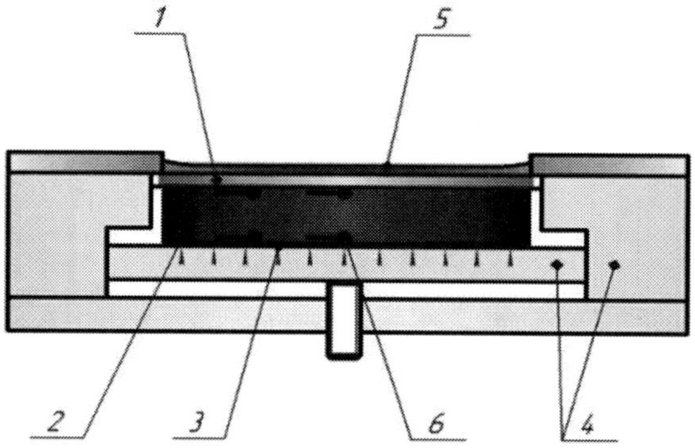

Fig. 2. The test cell: 1 – substrate; 2 – copper plate; 3 – piezoelectric element; 4 – case of the test cell made of thermal insulator; 5- liquid layer; 6 – thermocouples.

A more detailed description of the test cell is given in [19].

Experiments were conducted at constant heat flow; the initial temperature corresponded to 80, 90 and 100°C. As a working liquid we chose the most common coolant, it is distilled water. Liquid from 1 mm to 3 mm thick was dosed on the substrate uniformly.

The layer thickness was recorded from the moment of placing the liquid on the substrate made of stainless steel to break of layer due to the phase deformation.

3 Results and discussion

According to results of conducted experiments the dependences of layer thickness from time were obtained (Fig.3).

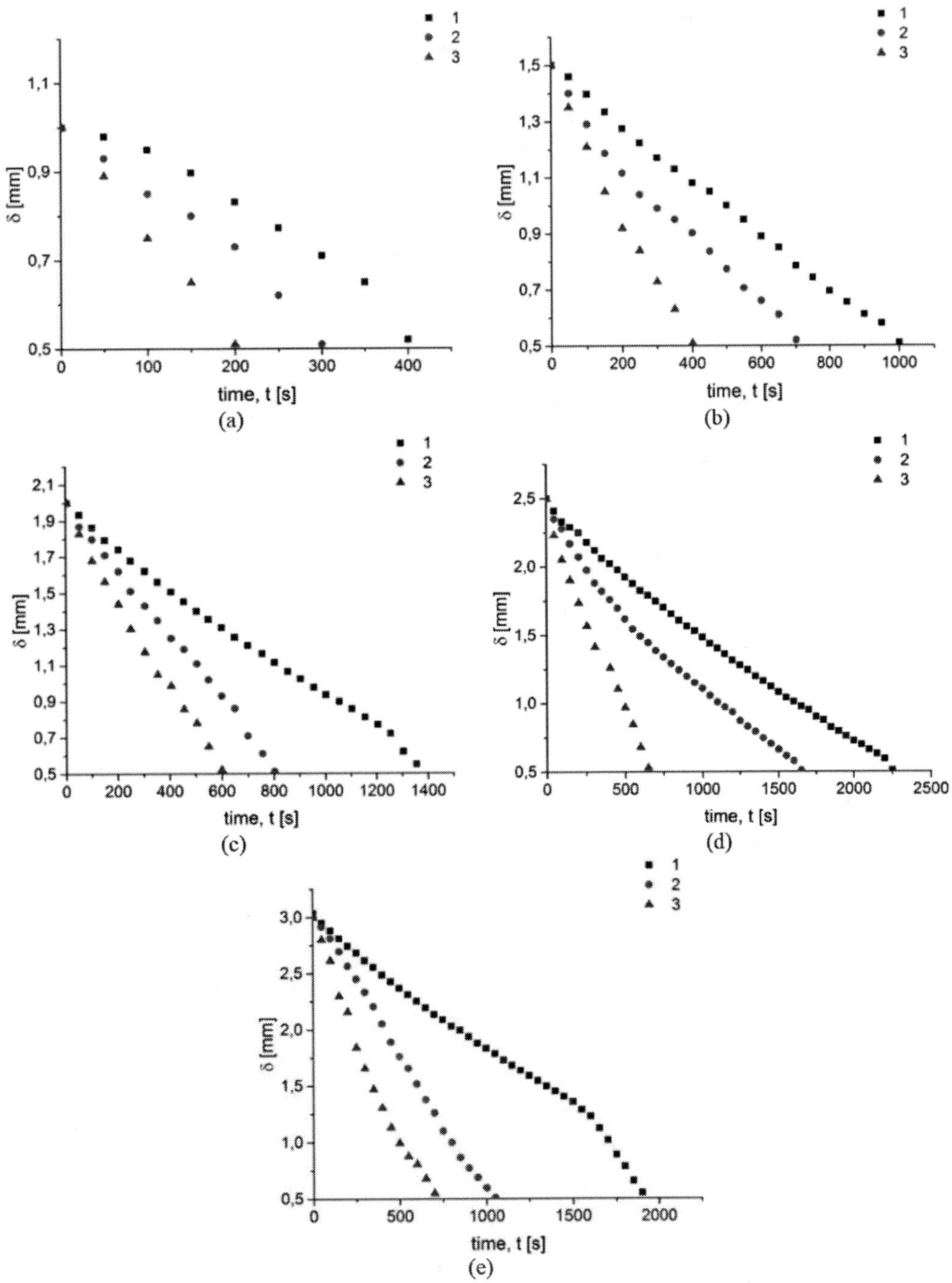

Fig. 3. Time dependences of distilled water layer thickness. The thickness of layer, mm: a) 1; b) 1.5; c) 2; d) 2.5; e) 3. Substrate temperature, °C: 1) 80; 2) 90; 3)100.

According to change of distilled water layer thickness it is found that the dependence has a linear nature. However, when the thickness decreases up to 0.5 mm, linear nature changes to the polynomial. Probably, it can be explained by the following way. During decrease in the liquid thickness up to 0.5 mm there is a deformation of the interface. The flat shape of the surface transforms to wavy shape. It happens due to the surface tension.

DOI: 10.1051/epjconf/201715900011

Interface energy on the solid/liquid boundary tends to compress the layer, i.e. the surface energy decreases due to decrease in surface area. Spreading is prevented by cohesive forces acting inside the liquid layer.

This work was carried out at the Kutateladze Institute of Thermo-physics SB RAS and financially supported by the Russian Science Foundation (project number 15-19-10025).

References

1. S. Misyura, Int. J. Therm. Sci., **92**, 34 (2015)
2. V. Maksimov, T. Nagornova, I. Shestakov, EPJ Web of Conferences, **82** (2015)
3. V. Nakoryakov, S. Misyura, S. Elistratov, R. Dekhtyar, J. Eng. Thermophys., **23**, 257 (2014)
4. G. Kuznetsov, A. Zakharevich, N. Bel'kov, Chemical and Petroleum Engineering, **50**, 1 (2014)
5. A. Zakharevich, N. Bel'kov, EPJ Web of Conferences, **76** (2014)
6. V. Maksimov, T. Nagornova, O. Stepanova, MATEC Web of Conferences, **23** (2015)
7. V. Lebedev, V. Lemanov, S. Misyura, V. Terekhov, Fluid Dynamics, **28**, 624 (1993)
8. D. Zaitsev, D. Rodionov, O. Kabov, Tech. Phys. Lett., **35**, 680 (2009)
9. D. Zaitsev, D. Kirichenko, O. Kabov, Tech. Phys. Lett., **41**, 551 (2015)
10. D. Zaitsev, M. Aviles, H. Auracher, O. Kabov, Microgravity Sci. Tec., **19**, 71 (2007)
11. D. Glushkov, J. Legros, P. Strizhak, A. Zakharevich, Fuel, **175**, 105 (2016)
12. Y. Lyulin, D. Feoktistov, I. Afanas'ev, E. Chachilo, O. Kabov, G. Kuznetsov, Tech. Phys. Lett., **41**, 665 (2015)
13. E. Gatapova, R. Filipenko, Y. Lyulin, I. Graur, I. Marchuk, O.Kabov, Thermophys. Aeromech., **22**, 701 (2015)
14. D. Glushkov, G. Kuznetsov, P. Strizhak, R. Volkov, Therm. Sci., **20**, 131 (2016)
15. D. Glushkov, G. Kuznetsov, P. Strizhak, Math. Probl. Eng., **2014**, 920480 (2014)
16. D. Feoktistov, E. Orlova, A. Islamova, MATEC Web of Conferences, **23**, 01054 (2015)
17. D. Feoktistov, E. Orlova, A. Islamova, EPJ Web of Conferences, **110**, 01018 (2016)
18. A. Semenov, D. Feoktistov, D. Zaitsev, G. Kuznetsov, O. Kabov, Thermophys. Aeromech., **22**, 771 (2015)
19. K. Ponomarev, E. Orlova, D. Feoktistov, EPJ Web of Conferences, **110**, 01060 (2016).

EPJ Web of Conferences **159**, 00012 (2017) DOI: 10.1051/epjconf/201715900012
AVTFG2016

Selection of technology for the low calorific synthetic gas combustion in the gas turbine combustion chamber

Prokopy Filippov*, *Evgeny* Levin, and *Alexander* Ryzhkov

Federal State Autonomous Educational Institution of Higher Education «Ural Federal University named after the first President of Russia B.N. Yeltsin», 620002, 19 Mira street, Ekaterinburg, Russia

Abstract. The leading gas turbines manufacturers are developing the technologies of the environmental friendly combustion of industrial and synthetic gases of low calorific values. In this case they are faced with critical problems concerning combustion stability assurance and the necessity of the gas turbines significant modernization due to the differences between the low calorific and natural gases. The numerical simulation results of the low calorific value synthetic gas combustion in the combustion chamber by means of different technologies are considered in the paper.

1 Introduction

During development of the combined cycle power plant (CCPP) operated on the synthetic gas it is required to consider the significant difference between the fuel gas and the natural gas properties such as the calorific value, the Wobbe Index, the flame propagation rate as well as the presence of different pollutants. For steady combustion and CCPP cost-effective operation assurance the changes in the standard gas turbine structure and the CCPP scheme can be required [1]. The greater the differences between the fuel gas and the natural gas properties, the greater the modernization depth. In this case the leading manufacturers of the synthetic gas turbine often follow the path of the fuel gases unification by hydrodynamic, thermodynamic and kinetic characteristics by means of the gases dilution or enrichment, and during the synthetic gas combustion scheme development the thermodynamic performance is associated with the corresponding kinetic diagram providing the achievement of the technological and environmental indexes of the process.

However, in case of the air gasification, where the gas calorific value is approximately 3.5 MJ/m^3, the applied combustion technology is inoperative, and consequently the additional measures for the synthetic gas correction with the purpose of the calorific value increase should be taken or another combustion technique development is required [2]. The low calorific synthetic gas combustion without the enrichment procedures is considered below.

* Corresponding author: fps_proxi@mail.ru

On the basis of analysis of the most efficient combustion chambers of the gas turbines operated on the synthetic gases applied at the IGCC, the analog was found and the numerical variation studies of the gas turbine combustion chamber with the power of 300 MW were performed with consideration of the regulatory compliance in the field of the NO_X emissions.

2 Selection of the gas turbine combustion chamber design for the synthetic gas combustion

For selection of the optimal gas turbine combustion chamber design the numerical simulation of the operation of combustion chamber of three types (Siemens, GE и MHPS) was performed on the low calorific value fuel gases, obtained after correction of the oxygen gasification gases and blast-furnace gases. The gas fuel combustion model was selected among the following four models being most widely used in CFD software packages: the Eddy Dissipation (ED) model, the Finite Rate Chemistry model (FRC), the integrated model Eddy Dissipation/Finite Rate Chemistry (ED/FRC) and the Laminar Flamelet model (LF). Model selection was performed on the basis of comparison of the calculated data, obtained by means of the models, with the experimental data [3], which is widely used for verification of different models [4, 5]. The calculation results, obtained in the CFD software package with the use of the ED/FRC model, were the closest to the experimental data along the whole combustion chamber length. Therefore the ED/FRC model was used as the gas fuel combustion model.

For the combustion chamber operation simulation the MHPS type combustion chamber was used as the analog which provided the permissible temperature non-uniformities and low emissions of nitrogen oxides (19.02 ppm = 24.73 mg/nm^3) and CO (0.8 ppm = 1 mg/nm^3).

3 Determination of the low calorific value gases combustion technique

The NOx emissions during the synthetic gas combustion in the gas turbine combustion chamber depends on the temperature level and temperatures distribution, the local excess air factors, the fuel composition and the calorific value and the combustion products residence time in combustion area.

The impact of the fuel gas characteristics and the temperature regimes during combustion on the NOx emissions was studied in the present paper with the use of the numerical simulation methods with the purpose of determination of the optimal conditions for the low calorific synthetic gas combustion in the gas turbine.

The following temperature regimes of the synthetic gases combustion in the gas turbine can be distinguished:

1) the cold gas (100-180°C) combustion in the cycle air (400-500°C). Combustion of the reference gas turbine fuel (natural gas) by the present scheme within the range of α=1-3 provides the environmental friendly operation modes of the modern gas turbines of the temperature class 1300-1400°C. The synthetic gases with the unified calorific value and the high hydrogen concentration H_2>20% are combusted in the gas turbines of the temperature class 1100-1300°C in a similar way.

2) the hot gas (300-500°C) combustion in the cycle air (400-500°C). At the present time it is the basic method for the synthetic gases combustion in the gas turbine of the temperature class 1100-1400°C. This method is used for the oxygen and air blown integrated gasification combined cycles (IGCC) systems and also for industrial (blast-

furnace) gases combustion. The gas composition unification by the Wobbe Index, the calorific value and the hydrogen composition is performed in advance. It is the future technology for the IGCC with the hot gas treatment system.

3) the hot gas (300-500°C) combustion in the high-heat (900-1000°C) cycle air. In case of the moderately low heat of the fuel gas (up to 500°C) the low calorific gases combustion temperatures become entirely sufficient for the modern gas turbines of the temperature class 1300-1400°C. The majority of the gas turbines at the existing operating oxygen-blown IGCC with the cold synthetic gas cleanup process operate according to the similar technique, in this case the gas preheating to the temperature of ~300°C is performed in the water heaters. The heating to the higher temperatures (up to 420°C) is performed in case of operation on the enriched blast-furnace gas by means of the booster compressor operation. The implementation of the hot synthetic gas cleanup systems in the newly built IGCC will result in the shift of the present technology application field to the more low calorific gases.

The temperature regimes No 2 (the gas temperature of 300°C, 500°C and the air temperature of 400°C, 500°C) and No 3 (the gas temperature of 500°C the air temperature of 900°C) hereinafter referred to as 300/400, 500/500 and 500/900 were used during the calculation of the synthetic gas combustion in the gas turbine combustion chamber.

The exhaust gas flow rate at the combustion chamber outlet was kept constant in all regimes during the calculations. The average exhaust gas temperature is set and maintained at 1400°C ± 50°C constant. The temperature deviation impact was considered by means of the temperature correction. The exhaust gas temperature dependence of the NOx emissions, used for the temperature correction determination, is presented in Figure 1.

The characteristics of the synthetic gas used for determination of correction for the temperature deviation are as follows: CH_4=0.7%; CO=22.7%; CO_2=4.78%; H_2=22.7%; H_2O=5%; N_2=44%; H_2S=0.0015%; NH_3=0.1%; COS=0.0015%; HCN=0.003%; LHV=5.575 MJ/nm^3; H_2/CO=1. The correction curve is plotted for the A point conditions (Figure 2).

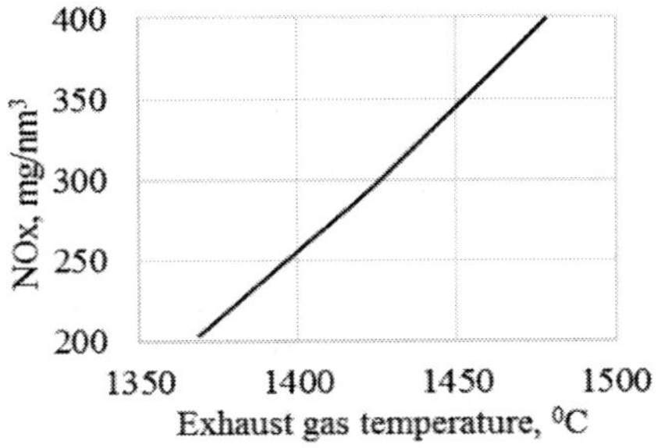

Fig. 1. The exhaust gas temperature dependence of the NOx emissions

During the linearization of the correction curve, presented in Figure 1, the NOx temperature sensitivity was 1.8 mg/(m^3·K).

The impact of the combustion regime of the synthetic gases of different LHV on the NOx emissions is presented in Figure 2.

Characteristics for the used synthetic gases (vol.%): CH_4=0÷0.7; CO=9.8÷31.4; CO_2=4.8÷15.9; H_2=7.7÷34; N_2=42.3÷58; H_2O=5; H_2S=0.0015; COS=0.0015; HCN=0.003; NH_3=0.1.

In the 300/400 regime the decrease in the calorific value from 5-6 MJ/m^3 to 4.2 MJ/m^3 results in the NOx emissions decrease from 140-185 mg/m^3 to 50-60 mg/m^3. In this case the emissions are at the standard values level even without the special measures implementation. It can be explained by the decrease in the theoretical value of the synthetic gas combustion temperature, and consequently the decrease in the local temperature maximums in the combustion chamber, which are the points of the maximum NOx emissions.

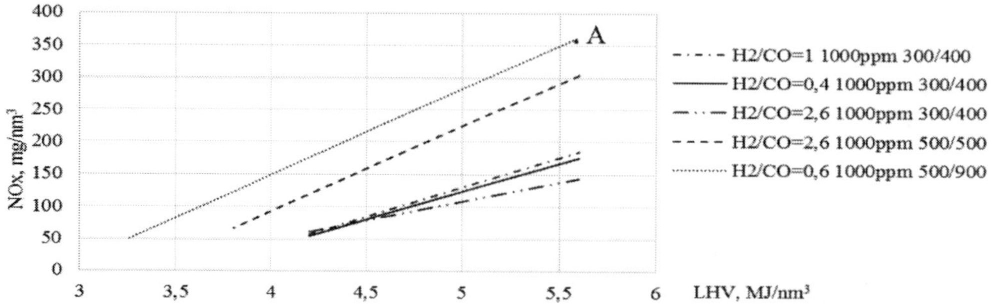

Fig. 2. The impact of the combustion regime of the synthetic gases of different heat capacities on the NOx emissions

In the 300/400 regime the decrease in the calorific value from 5-6 MJ/m^3 to 4.2 MJ/m^3 results in the NOx emissions decrease from 140-185 mg/m^3 to 50-60 mg/m^3. In this case the emissions are at the standard values level even without the special measures implementation. It can be explained by the decrease in the theoretical value of the synthetic gas combustion temperature, and consequently the decrease in the local temperature maximums in the combustion chamber, which are the points of the maximum NOx emissions.

Switch from the hot gas combustion in the cycle air (300/400) to the higher temperature regimes (500/500 и 500/900) results in the significant increase in the NOx emissions. However, in case of the low calorific value of the synthetic gas (3.3 MJ/m^3), even combustion in the highest temperature regime (500/900) permits to provide the low NOx emissions.

For the hot synthetic gas with the calorific value of 3.2 MJ/m^3 in case of the high temperature air heating the presence of ammonia with concentration of 1000 ppm has no sufficient impact on the NOx emissions, which permits to use the synthetic gas uncleaned from ammonia in the gas turbine and exclude the expensive unit of cathalytic cleaning from the nytrogen-bearing impurities from the gas cleanup system.

For the NOx emissions abatement in the regimes with NOx>50 mg/m^3, the measure widely used by the leading gas turbine manufacturers was taken [6, 7], namely the steam supply to the combustion chamber. Usually the steam injection is less than 5% of the air volume inhausted by the compressor [6]. In case of NOx<100 mg/m^3 (for dry process) the 5 % condition is met, but if NOx>100 mg/m^3 the steam injection exceeds 13% of the exhaust gases flow rate downstream of the combustion chamber. In the lower initial NOx range for dry process (<100 mg/m^3), the NOx emissions decreased slowly with the rise in steam injection. The use of the synthetic gas cleaned from ammonia causes slightly reduction of the NOx emissions level, as far as the fuel NOx emissions are reduced. The increase in the H_2/CO ratio in the fuel gas composition during the steam supply to the combustion chamber results in the increase in the NOx emissions.

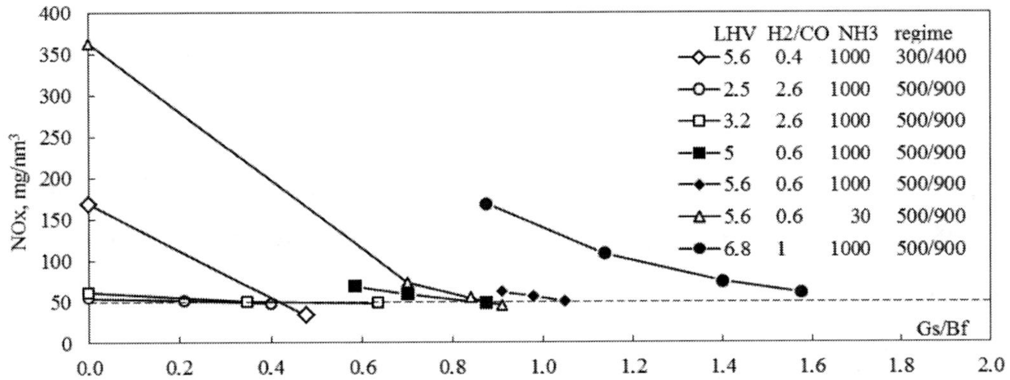

Fig. 3. Dependence of the combustion chamber steam quantity on the NOx emissions. Gs is the steam flow rate, t/h; Bf is the synthes gas flow rate, t/h.

General characteristics for the used synthetic gases: H_2O=5%; H_2S=0.0015%; COS=0.0015%; HCN=0.003%, the distinguishing characteristics are presented in Table 1.

Table 1. The synthetic gases characteristics

№	1	2	3	4	5	6	7
LHV, MJ/m³	5.6	2.5	3.2	4.97	5.58	6.78	5.58
CH_4, vol.%	0.7	0	0	0.7	0.7	0.7	0.7
CO, vol.%	31.4	18	8	27.8	27.8	27.8	27.8
CO_2, vol.%	4.8	20	15.9	4.8	4.8	4.8	4.8
H_2, vol.%	12.5	2.5	20.8	16.7	16.7	27.8	11.1
N_2, vol.%	45.5	54.5	50.1	50.4	44.8	33.7	44.9
NH_3, vol.%	0.1	0.1	0.1	0.1	0.1	0.1	0.003

Conclusions

The use of the high temperature air preheating and the hot gas treatment for the low calorific synthetic gas (with the calorific value of 3.3 MJ/m³) permits to increase the combustion temperature to the level providing the stable combustion and the NOx emissions value at the standard value. The present technique offers an opportunity for consideration of the synthetic gas application for the turbunes of the temperature class 1700°C.

This work was carried out at the Ural Federal University and financially supported by the Russian Science Foundation (project no. 14-19-00524).

References

1. T. Komori, H. Hara, H. Arimura, Y. Kitauchi, *Proc. of IGNC*, **TS-103**, 1-8 (2003)
2. T. Hasegawa, Energies, **3**, 335-449 (2010)
3. D. Garreton, O. Simonin, *ASCF Ercoftac CFD Workshop* (1995)
4. H. C. Magel, U. E. Schnell, K. R. G. Hein, *Proc. of 26 Symp. On Comb.*, 67-74 (1996)
5. C. V. da Silva, C. A. Segatto, A. V. de Paula, *Proc. of COBEM* (2013)
6. S. V. Tsanev, V. D. Burov, A. N. Remezov, *Gas turbine power plant and combined cycle power plant*, 3rd ed., (2009)
7. K. Tanaka, K. Nishida, W. Akizuki, MHI Tech. Review, **46**, 6-12 (2009)

EPJ Web of Conferences **159**, 00013 (2017) DOI: 10.1051/epjconf/201715900013
AVTFG2016

Measurements of the evaporation rate upon evaporation of thin layer at different heating modes

E.Ya. Gatapova[1,2,*], *E.G.* Korbanova[1,2]

[1]Kutateladze Institute of Thermophysics SB RAS, Novosibirsk, Russia
[2]Novosibirsk State University, Novosibirsk, Russia

Abstract. Technique for measurements of the evaporation rate of a heated liquid layer is presented. The local minimum is observed which is associated with the point of equilibrium of the liquid–gas interface. It is shown when no heat is applied to the heating element temperature in gas phase is larger than in liquid, and evaporation occurs with the rate of 0.014–0.018 μl/s. Then evaporation rate is decreasing with increasing the heater temperature until the equilibrium point is reached at the liquid–gas interface. Further increasing of the heater temperature leads to increasing of the evaporation rate.

1 Introduction

Investigation of evaporation process in a two–layer system under non–isothermal conditions is important for many applications, such as microelectronics cooling [1]. The advancement in high performance heat exchangers, vapor chambers and heat pipes [2] with microscale passages stimulates also development of a new fundamental knowledge of the processes with phase change [3,4].

If the liquid and the vapor–gas mixture are stationary it is usually considered that the movement of the vapor in the gas phase occurs mainly as a result of molecular diffusion. However, if the molecular mass of the evaporating liquid is less than that of the gas contacting with the liquid, then the moment comes when the buoyancy forces remove the gas layer from equilibrium state and the diffusion regime is replaced by a convective one. The resulting convection can significantly increase the evaporation rate [5]. Investigations of the temperature profile near liquid–gas interface and the conditions when the equilibrium state at the interface is reached for complex system are important from the practical and fundamental point of views. Evaporation occurs under non–equilibrium conditions and the continuity of the temperature at the interface can not be ensured. Evaporative flux depends on the thermal difference at the interface [4], and the precise measurements of these two quantities (temperature and evaporation rate) can give information on the boundary

* Corresponding author: gatapova@itp.nsc.ru

DOI: 10.1051/epjconf/201715900013

condition at the liquid–gas interface as well as data on the evaporation/condensation coefficients.

In this work, we present the results of the evaporation rate measurements of the locally heated liquid layer.

2 Experimental setup and measurements technique

Experiments are conducted at normal atmospheric condition. Ultrapure deionized and degased water (MilliQ) is used as a working fluid for experiments. The liquid is injected using a syringe pump (Cole Parmer) to cuvette with local heater, which is placed in microbalance system (Fig. 1). Calculation of the mass of evaporated liquid is carried out using BM-252 microbalance of accuracy class I (special), with the smallest weighing limit of 0.01 mg and control of the ambient gas humidity and temperature. Microbalance is connected to PC and measurements are controlled by special software. The experiments are performed at quasi–stationary heating mode. The temperature of the heater is controlled by three thermocouples. The pressure is measured by barometer. Calibrations of all thermocouples are carried out in the temperature range 5–100 °C. Two reference thermometers (ETS-100) are used. Data registration is carried out by means of temperature control and measurement system consisting of data collection (NI 9214) and the corresponding software.

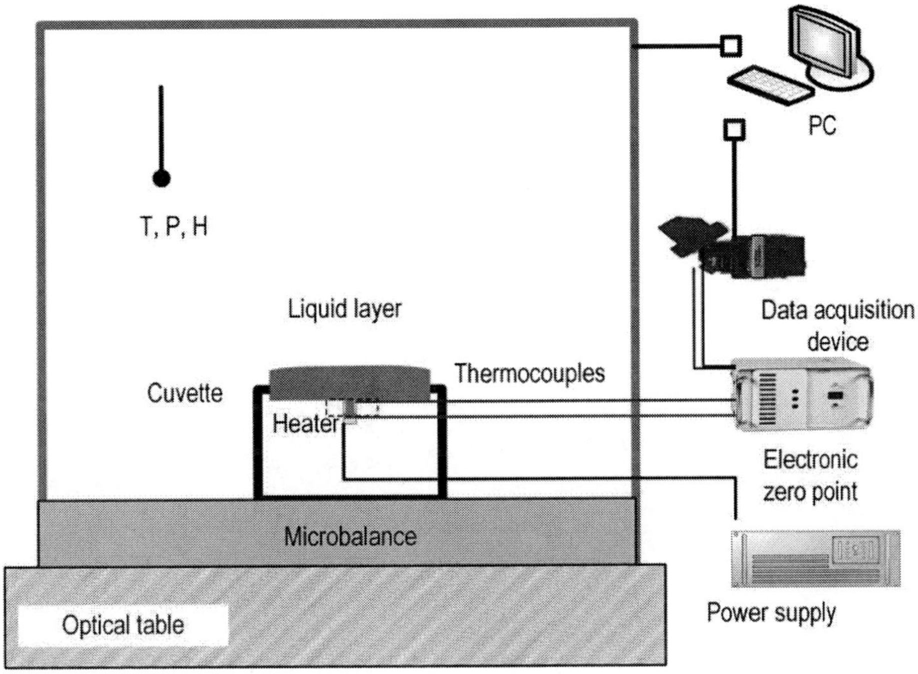

Fig. 1. Scheme of the experimental setup.

3 Results and discussion

Results of the evaporation rate measurements for water–air system are presented in Fig. 2. The local minimum is observed which is connected with the point of equilibrium of the liquid–gas interface. Let us discuss the data on the temperature profile across water–air system near interface obtained in Ref. [4]. When no heat is applied to the heating element the negative value of the temperature jump is measured, i.e. the temperature in the air near the liquid–gas interface is larger than in liquid. By increasing of the heat flux in heater some kind of equilibrium point can be observed, where temperature jump is absent. The measured temperature profiles have the classical continual shape. The fracture in the temperature profile is observed, which is due to different thermal conductivities of liquid and gas phases. By further increasing of heat flux in heater, the slope of the temperature profile becomes steeper, and thus the temperature gradient and Fourier heat fluxes in gas and liquid phases. The further increase of the heating leads to increase of the temperature jump and evaporation rate [4].

Basing on our previous temperature profile measurements [4], we can conclude that when no heat is applied to the heating element temperature in gas phase is larger than in liquid, and evaporation occurs with the rate of 0.014–0.018 μl/s. Then evaporation rate is decreasing with increasing heater temperature until the equilibrium without temperature jump is reached at the liquid–gas interface. Further increasing of the heater temperature leads to increasing of the evaporation rate.

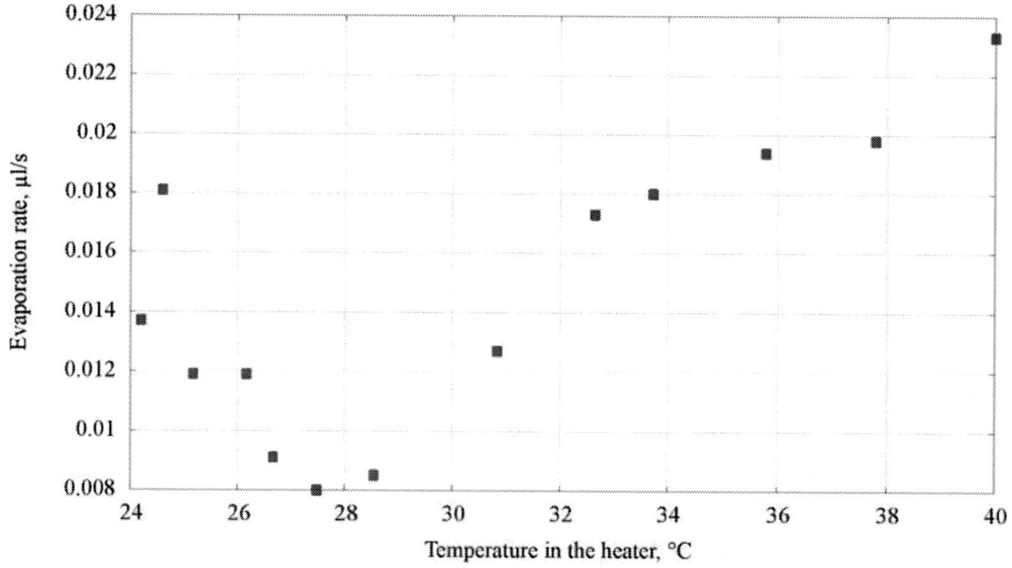

Fig. 2. Evaporation rate dependence on the average temperature of the heater.

In conclusion, results of the evaporation rate measurements of a heated liquid layer are presented. It is shown that there is local minimum of the evaporation rate at the point of equilibrium of the liquid–gas interface.

The authors gratefully acknowledge the support of this work by the Ministry of Education and Science of the Russian Federation (Agreement 14.616.21.0016, project identifier RFMEFI61614X0016).

References

1. M.H. Nasr, C.E. Green, P.A. Kottke, X. Zhang, T.E. Sarvey, Y.K. Joshi, M.S. Bakir, A.G. Fedorov, International Journal of Heat and Mass Transfer, **108**, 1702 (2017)
2. M.A. Chernysheva, S.I. Yushakova, Y.F. Maydanik, International Journal of Heat and Mass Transfer, **81**, 297 (2015)
3. E.Y. Gatapova, I.A. Graur, F. Sharipov, O.A. Kabov, International Journal of Heat and Mass Transfer, **83**, 235 (2015)
4. E.Y. Gatapova, I.A. Graur, O.A. Kabov, V.M. Aniskin, M.A. Filipenko, F. Sharipov, L. Tadrist, International Journal of Heat and Mass Transfer, **104**, 800 (2017)
5. L. Prandtl, O. Tietjens, Applied hydro– and aeromechanics, Dover publications, inc. (1934)

EPJ Web of Conferences **159**, 00014 (2017) DOI: 10.1051/epjconf/201715900014
AVTFG2016

Lift and Drag Forces for a Sphere on a Flat Wall in Non-Newtonian Shear Flow

Yaroslav Ignatenko[1,*], *Oleg* Bocharov[1] , and *Roland* May[2]

[1]Scientist, Novosibirsk Technology Center of Baker Hughes Incorporated, 630090, Kutateladze 4a, Novosibirsk, Russia
[2]Mechanical Engineer, Celle Technology Center of Baker Hughes Incorporated, Baker-Hughes-Strasse 1, Celle, Niedersachsen 29221, Germany

Abstract. Parametrical study of perfect steady rolling and steady state freely motion of spherical particle on the wall in shear flow of non-Newtonian fluid was carried out. In the case of perfect rolling the jumps of forces acting on a sphere in the moment of changing particle velocity from zero to positive were observed. Non-Newtonian properties of fluid lead to decreasing of drag force coefficient for low particle's velocity and to opposite effect for high velocity relative to Newtonian fluid. There is no such significant effect for a lift force coefficient. Steady state particle's velocity ($F_d(V_{pss})$=0) increases for non-Newtonian fluid for both considered types of motion. Steady state particle's velocity and lift force coefficient are lower for the case of freely particle motion. Steady state rotational velocity decreases for non-Newtonian fluid.

1 Introduction

One of multiphase flow regimes is stirring or lifting of particles over a wall or a bed of particles. The movement of spherical particles in a shear flow along a wall is a common phenomenon in a nature and technical design. For example, the movement of river drift and the transport of cuttings during horizontal drilling can be described by this simple model. In these situations, the determination of forces acting on the particle is an important and fundamental task.

One of the first investigations about flow around a sphere near a wall was made by Goldman et al. 1967 [1, 2]. An analytical solution was obtained for the drag force and torque acting on the sphere near the wall using the Stokes approximation. Two scenarios were considered: a resting sphere in shear flow and a moving sphere in a non-moving fluid. In the latter scenario, the drag force and torque increases logarithmically with decreasing wall distance. Sphere motion on the wall was investigated numerically by Lee et al. [3, 4] using Navier-Stokes equations. They considered perfect rolling and free motion with creeping for a Newtonian fluid.

* Corresponding author: yaroslav.ignatenko@gmail.com

Most articles presented in literature about rolling/sliding particles on a wall in shear flow are made for Newtonian rheology. Current study is concentrated on effect of non-Newtonian fluid rheology for the problem of single particle motion on a wall.

2 Problem statements

We consider a spherical particle with diameter d, rotating at an angular velocity ω about z-axis and translating parallel to a wall with velocity v_p in x-axis direction (see Fig. 1a). The far-field ambient flow is a linear shear flow directed in the x-axis $u_\infty=Gy$, where G is shear rate of undisturbed flow ($G=du_\infty/dy$). On the wall and sphere non-slip boundary conditions was set up. Steady state flow around the sphere can be described by Navier-Stokes equations

$$\nabla\cdot\mathbf{u}=0, \quad (1)$$

$$\mathbf{u}\cdot\nabla\mathbf{u}=-\frac{\nabla p}{\rho}+\nabla\cdot\boldsymbol{\tau}, \quad (2)$$

here p – pressure, u – flow velocity, τ - viscous stress tensor, that depends on shear rate and is defined by rheological model. Here viscous stress tensor is described by the Herschel-Bulkley rheology law:

$$\boldsymbol{\tau}=(k\dot{\gamma}^{n-1}+\tau_0/\dot{\gamma})\dot{\boldsymbol{\gamma}}, \quad (3)$$

where $\dot{\boldsymbol{\gamma}}$ - shear rate tensor, $\dot{\gamma}=\sqrt{2\dot{\boldsymbol{\gamma}}\dot{\boldsymbol{\gamma}}}$ - shear rate, k – consistency index, n – power index, τ_0 – yield stress.

Simulation was carried out with Reynolds number $\mathrm{Re}_s=\dfrac{\rho G^2 d^2}{2\left(kG^n+\tau_0\right)}$ of free stream up to 200 and positive particle's velocity $V_p=2v_p d/G<1.4$ and made dimensionless by dividing the velocity of free stream on the opposite to particle center, power index $0.4\le n\le 1.0$, Bingham number $Bn=\tau_0/kG^n<10$. The dimensionless rotational velocity of the particle is $\Omega_p=\omega/G$.

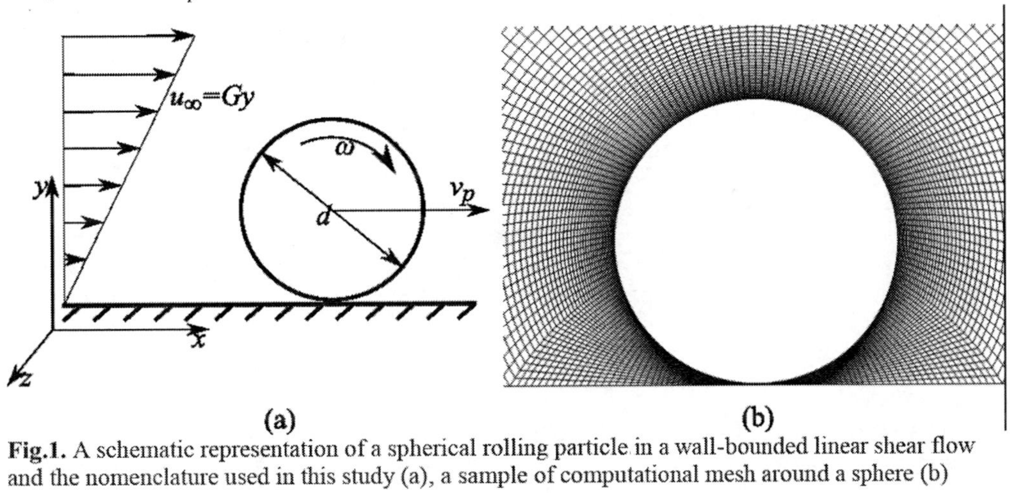

Fig.1. A schematic representation of a spherical rolling particle in a wall-bounded linear shear flow and the nomenclature used in this study (a), a sample of computational mesh around a sphere (b)

All simulations were made in the coordinate system translating with the particle using the open source CFD package OpenFOAM 3.0.1 [5] and the simpleFoam solver that is based on a SIMPLE-C algorithm [6]. The simulation volume was cuboid. Its surface was mapped isomorphically on the surface of a hemisphere, and then the surface of hemisphere was mapped isomorphically on surface of a particle (sphere). Using this technique a structural hexahedral mesh was created (Fig. 1b). The distance between a sphere and the boundaries of the simulation volume was equal to 25 sphere diameters. To avoid numerical issues, a small gap $d/200$ between the sphere and the wall was presumed [2, 3].

3 Results and Discussion

We considered two cases – perfect rolling and freely motion. Perfect rolling means that there is no creeping between particle and wall, angular velocity of sphere in this case is $\omega=2Vp/d$. Freely motion indicates terminal motion of the particle when drag force and moment of forces acting on a sphere are equal to zero.

3.1 Perfect rolling

The simulation investigated the forces acting on the sphere and their dependence upon the flow regime and fluid rheology properties while a perfect rolling of a sphere on the wall. The results of our simulation are in a good agreement with results presented in [3, 4] for Newtonian fluids. When the particle translational velocity V_p is low, the drag force coefficient C_d in non-Newtonian fluid is lower than in Newtonian one (Fig. 2a). The opposite behavior of the drag force coefficient C_d is observed while the translational velocity V_p increases. This result can be explained by increasing of viscosity with decreasing of relative speed of the sphere. Lift force coefficient behavior was non-monotonic versus rheology parameters (Fig. 2b). Lift C_L and drag C_d forces increase in the vicinity of zero velocity V_p. This can be explained by significant changes in the flow regime: the transition from rest to a moving status of the particle.

In the article [3] the authors showed the existence of steady state velocity V_{pss} when the drag force C_d is zero (free motion). The velocity V_{pss} increases with the Reynolds number. The results of our simulation show that in a non-Newtonian fluid the free motion velocity V_{pss} is more than in the Newtonian (Fig. 2a) and it can be greater than unity. In the case of large particle's velocity V_p, the perfect rolling lift force C_L became negative (Fig. 2b). Consequently, friction between the particle and the surface prevent particle take-off in some cases.

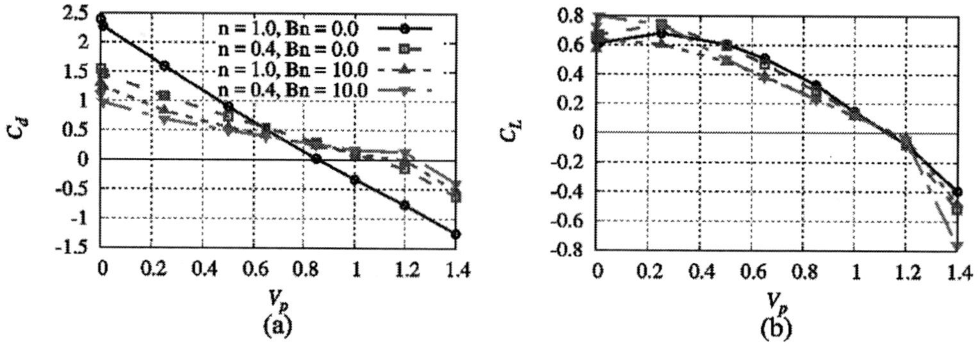

Fig.2. Forces dependence on translational velocity V_p for Re_s=50 and for different fluid properties (a) C_d ; (b) C_L

3.2 Freely motion

We examined also the terminal steady-state of a particle when it is allowed for freely move in wall-bounded linear shear flow. This terminal state is characterized by zero-drag and zero-torque conditions. There is no friction between the sphere and wall because the gap is equal to $d/200$. A simulation terminal angular velocity Ω_{pss} and translational velocity V_{pss} were found (see Fig. 3). Terminal translational velocity V_{pss} increased with the increase of the Bingham number Bn and the decrease of the power index n (Fig. 3a). In a friction-free case the terminal translational velocity V_{pss} can be greater than one, as in the case of perfect rolling.

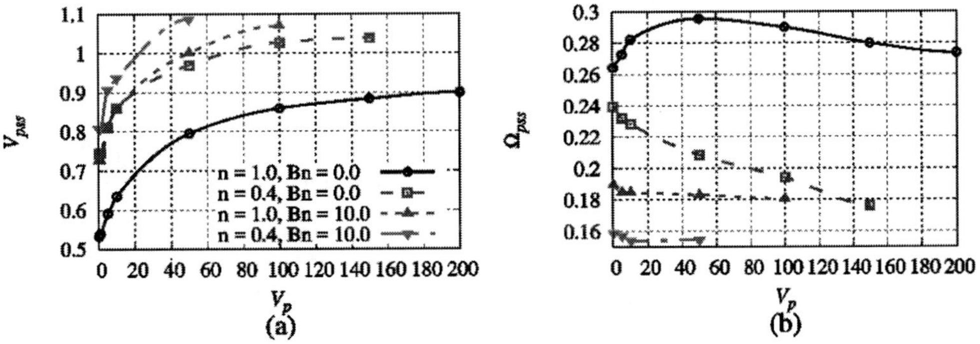

Fig. 3. Terminal translational velocity V_{pss} and rotational velocity Ω_{pss} dependence on Reynolds number Re_s for different fluid properties

The terminal angular velocity Ω_{pss} decreased with the decrease of the power index n and with the increase of the Bingham number Bn (Fig. 3b). In addition, the increase of the Bingham number Bn flattened the angular velocity Ω_{pss} dependency on Reynolds number Re_s.

Dependence lift force C_{Lss} on Reynolds number Re_s and rheology parameters were obtained (Fig. 4). This provides the possibility to formulate take-off conditions for a spherical particle. Because the terminal velocity V_{pss} was greater in the non-Newtonian fluid, the lift force C_{Lss} was lower (Fig. 3a, Fig. 4) because lift force decreases with translational velocity (Fig. 2b).

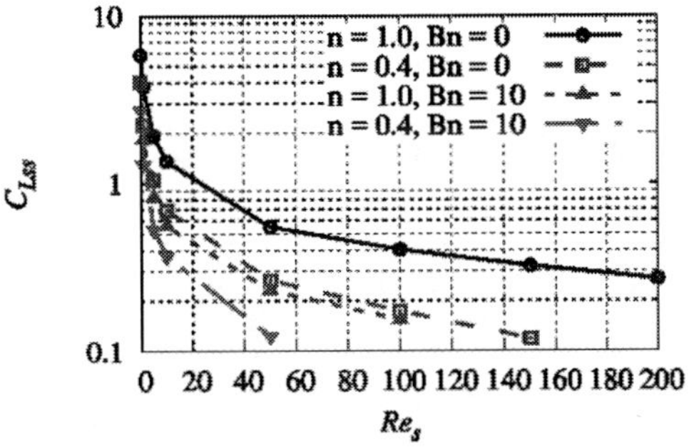

Fig. 4. Terminal lift force coefficient dependence C_{Lss} on Reynolds number Re_p for different fluid properties

Non-Newtonian rheology of fluid leads to an increase of terminal velocity, to a decrease of terminal angular velocity and lift force.

The convergence of the numerical algorithm becomes worse with the increasing Reynolds Re_s and Bingham Bn numbers, the particle's velocity V_p and the decreasing power index n. This is most likely due to the transient flow.

DOI: 10.1051/epjconf/201715900014

References

1. A. J. Goldman, R. G. Cox, H. Brennen, Chem. Eng. Technol. **22**, 637–651 (1967)
2. A. J. Goldman, R. G. Cox, H. Brennen, Chem. Eng. Technol. **22**, 653–660 (1967)
3. H. Lee, S. Balachandar, J. Fluid Mech. **657**, 89-125 (2010)
4. H. Lee, M. Y. Ha, S. Balachandar, Int. J. Multiphase Flow **37**, 108–124 (2011)
5. Official cite of OpenFOAM URL: openfoam.org/
6. J. H. Ferziger, M. Peric, *Computational Methods for Fluid Dynamics* (Springer, Berlin, 2002)

EPJ Web of Conferences **159**, 00015 (2017) DOI: 10.1051/epjconf/201715900015
AVTFG2016

The effect of surface temperature on dynamics of water droplet in minichannel with gas flow

Ekaterina Isachenko[1,2,*] and *Evgeniy* Orlik[1,2]

[1]Kutateladze Institute of Thermophysics SB RAS, 1 Lavrentyev Ave., 630090, Novosibirsk, Russia
[2]Department, Novosibirsk State University, 2 Pirogova Str., 630090, Novosibirsk, Russia

Abstract. The experiments have been carried out to study dynamics of liquid droplets, blown by the gas flow in a mini-channel. The mean velocity at which the droplet motion over the substrate starts was determined depending on the surface temperature at different droplet volumes. The shadow method was the main method of measurement. The advancing and receding contact angles were measured depending on the gas flow rate. The friction force was determined using the advancing and receding contact angles and droplet size. A motion of a droplet was also observed from the top. The local velocity and acceleration of droplet were calculated.

1 Introduction

The development of microelectronics is closely related to the problem of heat removal from the semiconductor chip. To date, the most affective cooling systems are the systems based on impact jets and sprays, two-phase flows in minichannels and fluid flows in micro-channels. These systems enable removal of heat fluxes of up to 500 W from the chip size 10 to 10 mm [1, 2, 3, 4]. However, already now the electronic industry is ready to produce components where the heat flux density can reach values of 1 kW/cm^2 and higher [6]. Even the most effective systems that use the two-phase flows are unable to cool such components, which is a technological barrier to the further development of microelectronic systems. One of the technical solutions to achieve significant intensification of heat transfer is a device, forming near-wall droplet flows in micro - and minichannels. The transition from a continuous film flow to the droplet flow with an increase in the length of the contact lines leads to the intensification of heat transfer at evaporation [7].

Above all the dynamics of a droplet in the gas flow is studied in the absence of evaporation. There are studies about dynamics of liquid droplet in turbulent flow in a sufficiently high channel comparing with height of droplet [8] and including vibration effect [9]. Numerical work [10] reports the dynamic and interaction of two droplets in a channel with gas flow, where the key parameter is the distance between the droplets. This interaction between the droplets has to be considered for designing cooling systems with forming near-wall droplet flows.

* Corresponding author: isach-e@yandex.ru

DOI: 10.1051/epjconf/201715900015

Significantly, changing viscosity and surface tension in liquid droplet on heated substrate may have a significant impact on its dynamics. But there is lack of data in the literature devoted to this aspect. In the scope of present work the dynamics of single droplet has been explored depending temperature of substrate and size of droplet in mini-channel with gas flow.

2 The experimental setup

The scheme of the experimental setup is shown in figure 1. It includes a channel with height 6 mm, removable substrates, the substrate temperature controlling and the air supply systems. Optical access for visualisation provided through the windows from side walls and the top cover. Ultrapure water obtained with the use of Milli-Q system is used as working fluid. The droplet is placed on surface into the channel with a syringe. The experiments have been carried out at the initial droplet volume of 60-150 μl. All experiments were realized on a substrate of stainless steel with polished surface. The Reynolds number of the gas flow was varied from 0 to 7000. The source gas is dry air from special air compressor. The temperature is maintained 25°C using a thermoregulation system. The substrate was heated by Peltier elements, installed in contact with the substrate. The substrate temperature during the experiment was regulated with a precision of 0.2°C using the PR-59 controller.

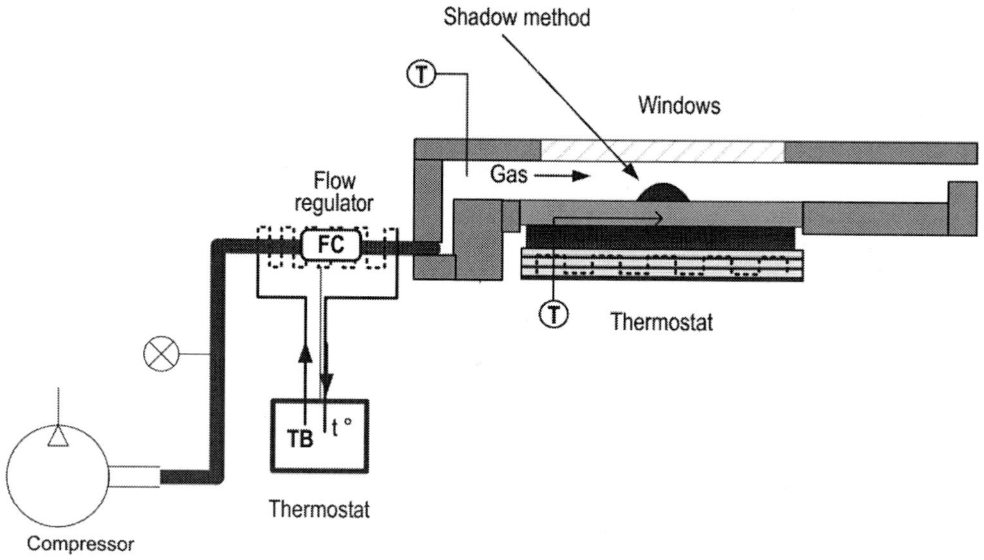

Fig. 1. Scheme of the experimental setup.

To register the contour of a sessile droplet the shadow method was used. Its principle is based on the fact that the physical object is illuminated by a parallel light beam, and its shadow is recorded by the camera, as shown in Fig. 2. Optical equipment allows obtaining images with a resolution of 6 μm/pixel (see Fig. 3). The obtained images were processed by different methods with the help of software (the Drop Shape Analysis by KRÜSS). Before the experiments, the hysteresis of the contact angle on the substrate was measured. The advancing angle of wetting was around 80 degrees, and the receding angle was around 50 degrees. Images obtained above droplet were used for calculation of the local velocity and acceleration (see. Fig. 6).

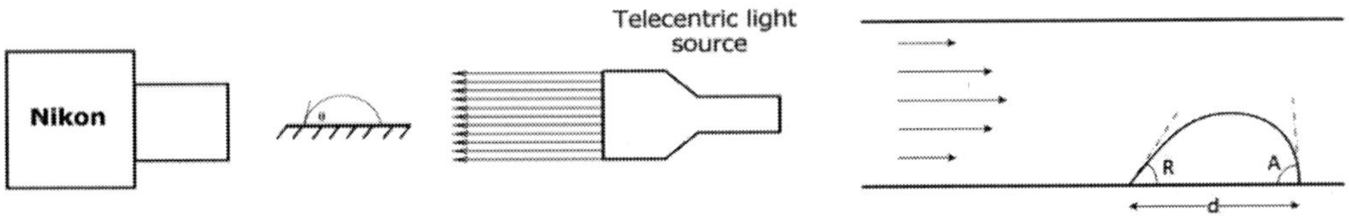

a) b)

Fig. 2. Shadow method –a), schematic drawing of the droplet in the channel –b).

3 Results

At an increased velocity of the gas flow, a droplet lost symmetry (see Fig. 3), and at some velocity values it started moving over the substrate (see Fig. 6). Flow velocity required for moving droplet decreasing with increasing of droplet volume that evidenced by the data which are presented in Figure 4. It should be noted that increasing temperature of the substrate surface, leads a decrease in the receding contact angle, in other words, the stack of droplet to the surface. One of the explanations that must be taken into consideration is that the viscosity and the surface tension of water significantly depend on temperature.

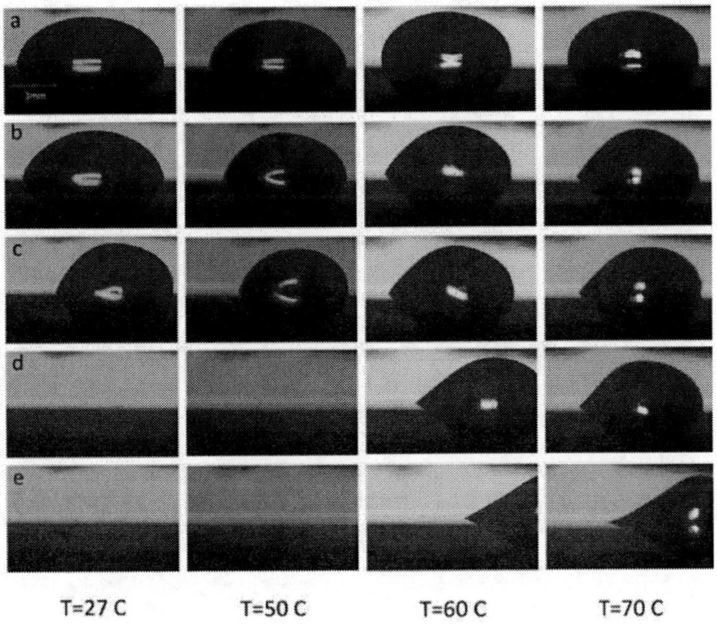

Fig. 3. Droplet dynamics. Mean velocity of air in m/s : a) 0,b) 4.2, c) 4.6, d) 5.5, e) 6.3.

 DOI: 10.1051/epjconf/201715900015

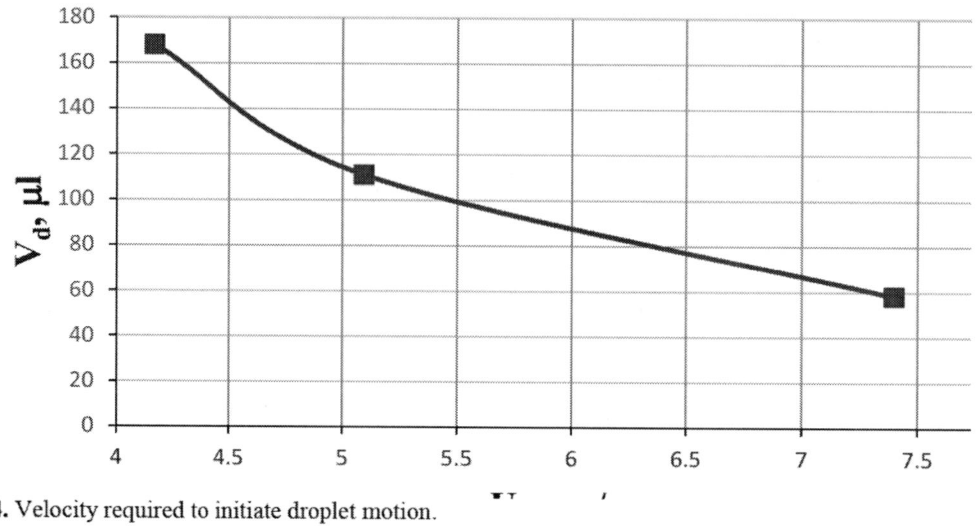

Fig. 4. Velocity required to initiate droplet motion.

Figure 5 shows the friction force affecting the contact line. It was determined by the formula 2σr(cos(A)-cos(R)), where σ is the surface tension, and the angles A and R, r is radius as are shown in Fig. 2 b). The larger symbols correspond to the values at which the droplet began its motion. In the work [8] the dynamics is studied for different droplets on the silane substrates. The values of velocity required to move droplet and correspondent friction forces in [8] were very similar to those obtained in our work. In our work it was found that with increasing temperature the droplet motion starts at somewhat larger gas flows, and the friction force affecting the droplet on the heated surface, calculated by the above formula, is a few times larger than in the case without heating.

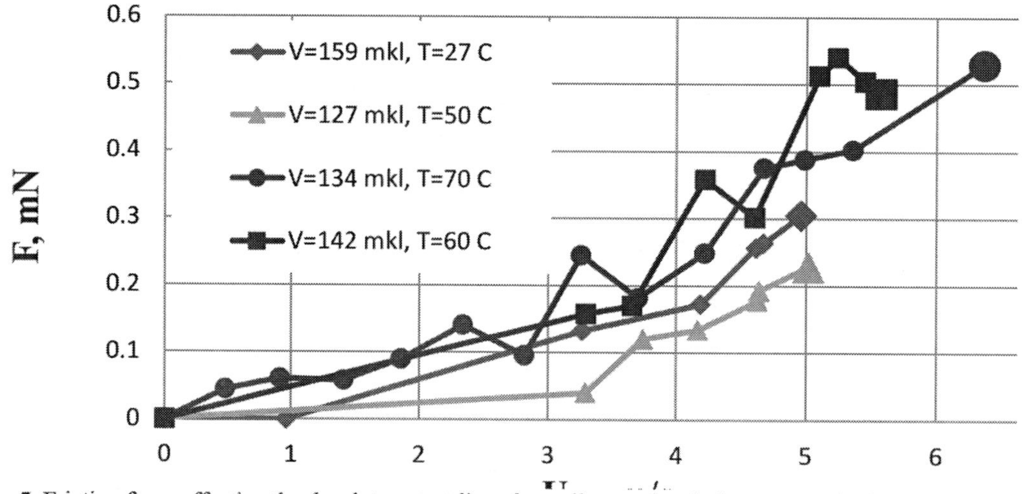

Fig. 5. Friction force affecting the droplet contact line, depending on the air flow rate and substrate temperature.

A local velocity rate of water droplet was determined using analysis of images, which were recorded from top. After droplet was starting movement, its velocity reaches about 1~2 mm per second for 50 mm path, as shown in Figure 7. Droplet can locally stick during motion, that reduces its velocity (that is visible from figure, 40 seconds, T = 27, Figure 6), but the shape of droplet stays sufficiently round

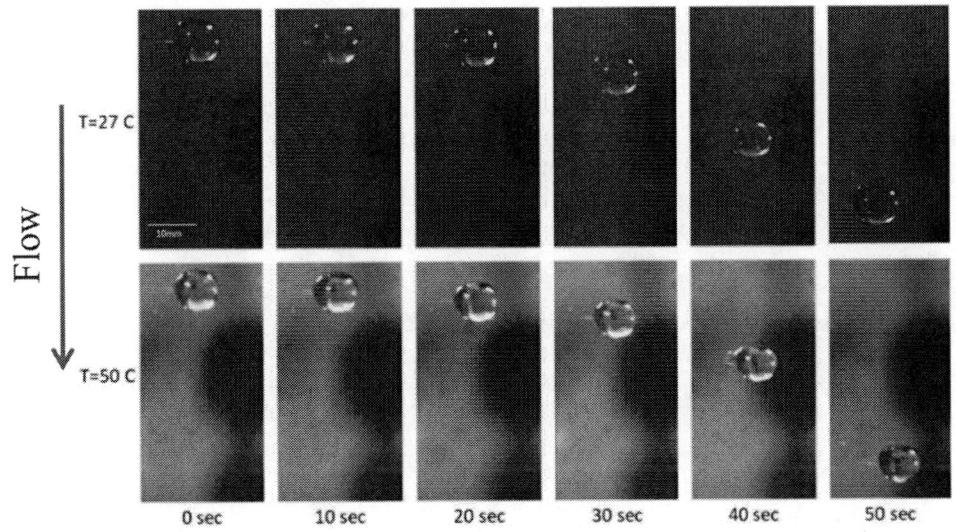

Fig. 6. Series of images of the movement of droplet on surface at temperature T = 27 C and T = 50 °C (top view). The arrow indicates the direction of the gas flow.

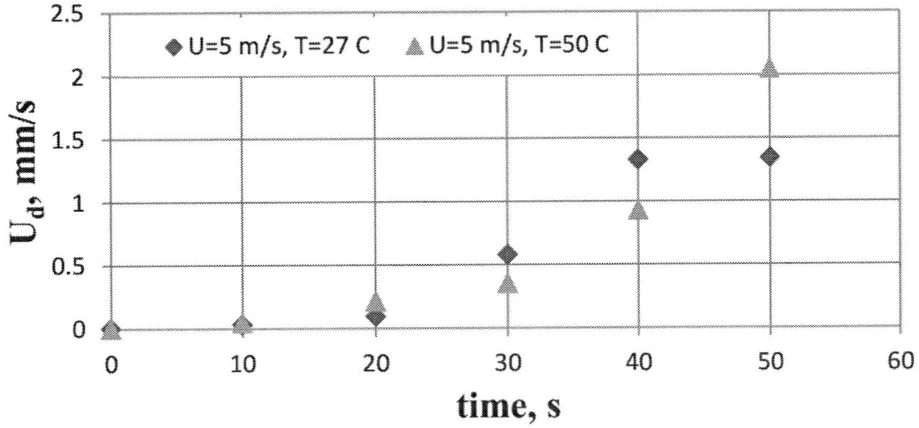

Fig. 7. =Velocity of droplet calculated from figure 6.

4 Conclusion

In this paper, we have carried out a detailed study of the dynamics of a single droplet in the 6 mm channel with gas flow. New data on the dynamics of water drops on a heated surface have been obtained. It has been found that with increasing temperature, the droplet starts to stick to the substrate surface.

The authors express their gratitude to the Russian Science Foundation for the support of this work (project No. 14-19-01755).

 DOI: 10.1051/epjconf/201715900015

References

1. A. Chinnov and O. Kabov, High Temperature, **42,** 2 (2004)
2. D. Zaitsev, O. Kabov, V. Cheverda and N. Bufetov, High temperature, **42,** 3 (2004)
3. T. Hirokawa, H. Ohta and O. Kabov, Interfacial Phenomena and Heat Transfer, **3,** 3 (2015)
4. D. Zaitsev, D. Rodionov and O. Kabov, Technical Physics Letters, **35,** 7 (2009)
5. K. Jungho, International Journal of Heat and Fluid Flow, **28,** 4 (2006)
6. A. Bar-Cohen, Intrachip/interchip enhanced cooling fundamentals DARPA, Microsystems Technlogy Office (2012)
7. V. Cheverda, I. Marchuc, A.Karachevsky, A. Orlik and O. Kabov, Thermophysics and Aeromechanics, **23,** 3 (2016)
8. J. Fan, M. Wilson and N. Kapur, Journal of Colloid and Interface Science, **356** (2011)
9. T. Maurer and U. Janoske, Computational Methods in Multiphase Flow, **8** (2015)
10. Jie Wu and Jun-Jie Huang, Journal of Applied Physics, **118,** 044902 (2015)

Shear driven FC-72 liquid film flow on the heater with microgrooves

Vyacheslav Cheverda[1,2*]

[1] Institute of Thermophysics, SB RAS, Lavrentiev Prospect, 630090, Novosibirsk, Russia
[2] Novosibirsk State University, Pirogova str. 2, 630090Novosibirsk, Russia

Abstract. The crisis in a FC-72 liquid film moving under the action of a gas flow through a smooth and micro-grooved heater was experimentally investigated. It is shown that the use of micro-grooves leads to an increase of the critical heat flux.

1 Introduction

When the microprocessor is cooled, the local heat flux is estimated at 1 kW/cm^2 [1]. A two-phase flow in a channel is promising solution of this problem. The flow regimes map inside the microchannel has been constructed in the papers []. The formation of dry spots in the horizontal layer of the isothermal liquid depends on the contact angle of the wetting and the thickness of the film has been studied in [2]. The heat transfer and rupture of a liquid film on horizontal, slightly inclined and vertical surfaces with local heating is investigated in [3, 4, 5]. It is shown that the destruction of the water film is carried out in two stages: after the film is ruptured, a film with a thickness much less than the initial film is observed on the heating element. Then, this thin film of the liquid breaks and the surface is completely dried.

Improvement of heat transfer is observed due to the creation of various structures on heated surfaces, in particular, due to improved wetting of the working surface [6]. The use of different structures (for example, micro-finning) on a heated surface is a good alternative to more complex factors of stabilization (gas flow, electric field, etc.) of the film flow on heating surfaces [7]. Heat transfer in a liquid film flowing along the surface with micro-grooves was experimentally investigated by the authors [8]. It is shown that the heat transfer crisis on the surface with microgrooves is tightened in comparison with the smooth surface. The deformation of a locally heated liquid film (25% ethyl alcohol in water) moving under the action of a gas stream under ground conditions and with different level of gravity was investigated numerically in [9]. The three-dimensional nonstationary problem was solved. The formation of thermocapillary deformations is shown.

2 Experimental setup

* Corresponding author: slava.cheverda@gmail.com

 DOI: 10.1051/epjconf/201715900016

An open contour is used for the experiments (Fig.1). During the experiment, liquid and gas nitrogen are supplied from the tanks. With the help of Bronkhorst regulators, gas flow and pressure are maintained in the test cell. With the help of a syringe pump, degassed liquid is supplied into the work area. To maintain a constant temperature of the liquid and gas at the entrance to the work area, heat-stabilizing systems based on Peltier elements are used. Using a membrane pump, a two-phase mixture is evacuated to the atmosphere. Separation and recovery systems using liquid nitrogen are used to separate the FC-72 liquid from the gas phase. Using the program developed in the LabView system, all parameters of the experiment are recorded and monitored.

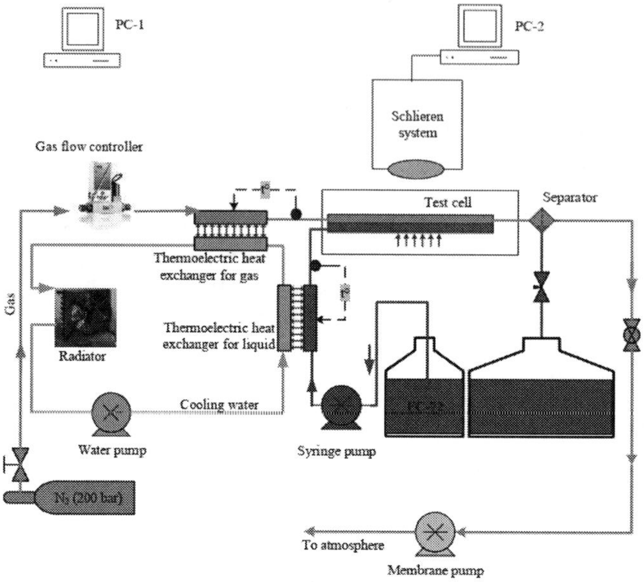

Fig. 1. Scheme of the experimental setup.

The test cell (Fig. 2a) consists of a base made of stainless steel and a textolite plate. The side of the substrate on which the liquid flows is treated with an abrasive M40 for better wetting. The frame from the textolite is attached to textolite plate, and from above is covered with optical glass, thereby creating a minicall with a height of 1.5 mm and a width of 30 mm. Using a liquid nozzle with height 150 microns and with width 30 mm a liquid film is formed. The gas flow enters the minichannel and entrains the film of liquid. A copper rod with a size of 10 x 10 mm^2 was pressed into the substrate. A ceramic heater is attached to the rod. The surface of the rod, which is located in the minichannel, can be smooth (w = 0 mm) or with microgrooves (w = 0.3 and 0.5 mm) (Fig. 2b). The microgrooves are oriented along the flow with the width of the individual structure w = 0.3 or 0.5 mm (Fig. 1b). The temperature of the substrate is measured by 10 K-type thermocouples (Fig. 2a), the distances are in millimeters.

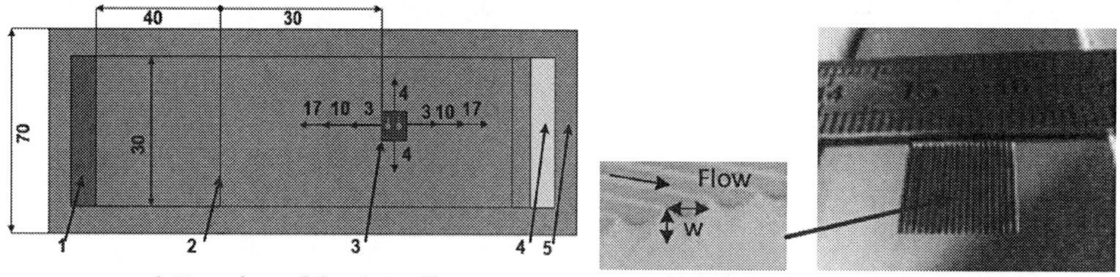

a) Top view of the test cell b) Microgrooves on the heater

Fig. 2. Scheme of the experimental setup.

Visualization of deformations on the film surface is carried out using a high-speed schlieren system, including a high-speed Optronis CL600x2 video camera, IDT 8 LED light source and optical elements.

3 Experimental results

The heat transfer crisis was experimentally studied for FC-72 shear driven liquid film by gas flow in a minichannel with smooth heater and microgrooved heater (the width and height of the micro-finning are equal to w = 0.3 mm) under the action of a nitrogen gas. The dynamics of the flow along a substrate with a smooth heater and on a substrate with microgrooves is significantly different. The critical heat flux on a smooth heater (w = 0) is 3.1 W / cm^2 at Rel = 7.6, Reg = 183.8. The thermocapillary effect has a determining effect. Under the action of thermocapillary forces, liquid is forced out from the heating element, a thickening of the film is formed in the region of the leading edge of the heater with respect to the oncoming flow.

On the heater with microgrooves, the liquid continues to wet the heater due to the capillary effect in the grooves, even for sufficiently high heat fluxes. Comparison of the critical heat fluxes (q_{CHF}) on smooth-surface and micro-finned heaters at different gas and liquid flows shows that the critical heat flux is lower on a smooth heater than on a heater with w = 0.3 mm (Fig. 3). Experimental data show that as the flow of liquids and gases increases, the critical heat flux increases.

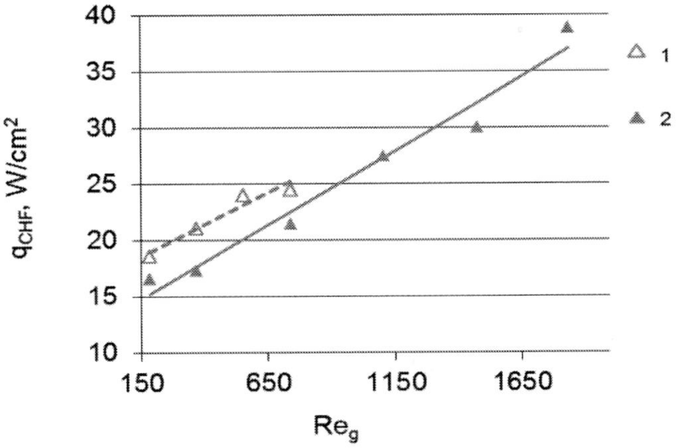

Fig. 3. The critical heat flux for smooth heater and for heater with microgrooves (height 0.3 mm), 1) w=0.3 мм, Re_l=30.2; 2) w=0, Re_l=30.2.

Conclusion

The work shows an increase of the critical heat flux for the FC-72 liquid film flow in a minichannel by a gas flow when using microgrooves on the heating surface.

The work was supported by the RF Ministry of Education and Science (Agreement No. 14.604.21.0053, project ID RFMEFI60414X0053).

References

1. Bar-Cohen A., Wang P., Microgravity Science and Technology, **21**, 351–359 (2009).

2. Chinnov, E. A.; Ron'shin, F. V.; Kabov, O. A., Technical physics letters. **41**, 817-819 (2015)
3. Chinnov, E. A.; Ron'shin, F. V.; Kabov, O. A., Thermophysics and aeromechanics. **22**, 621-629 (2015)
4. Chinnov, E. A.; Ron'shin, F. V.; Kabov, O. A., Thermophysics and aeromechanics. **22**, 265-284 (2015)
5. Kadoura M., Chandra S., Experiments in Fluids, **54,** 1465-1 (2013).
6. Zaitsev D.V., Kabov O.A., Microgravity science and technology, **19**-3/4, 174 (2007).
7. Chinnov E. A., Shatskii E. N, High temperature. **52,** 461-464 (2014)
8. Chinnov E. A., Shatskiy E. N., Technical physics letter. **40,** 7-9 (2014)
9. Kim J.K., Park C.W., Kang Y.T., Int. J. Refrig.-Rev. Int. Froid., **26,** 575–585 (2003).
10. Tiwari N., Davis J. M., Phys. Fluids., **22,** 042106-1 (2010).
11. D. V. Zaitsev, M. Lozano Aviles, H. Auracher, O. A. Kabov, Microgravity science and technology. **19**-3/4, 71 (2007)
12. Yu.O. Kabova, V.V. Kuznetsov, O.A. Kabov, Microgravity Science and Technology, **20**, 187–192 2008

 DOI: 10.1051/epjconf/201715900017

Cooling the vertical surface by conditionally single pulses

Pavel Karpov[1,*], *Alexander* Nazarov[1], *Anatoly* Serov[1,2] and *Victor* Terekhov[1,2]

[1]Kutateladze Institute of Thermophysics SB RAS, Lavrentiev av. 1, Novosibirsk, 630090, Russia
[2]Novosibirsk State Technical University, K. Marx av. 20, Novosibirsk, 630073, Russia

Abstract. You Sprays with periodic supply of the droplet phase have great opportunities to control the heat exchange processes. Varying pulse duration and frequency of their repetition, we can achieve the optimal conditions of evaporative cooling with minimization of the liquid flow rate. The paper presents experimental data on studying local heat transfer on a large subcooled surface, obtained on the original setup with multinozzle controlled system of impact irrigation by the gas-droplet flow. A contribution to intensification of the spray parameters (flow rate, pulse duration, repetition frequency) per a growth of integral heat transfer was studied. Data on instantaneous distribution of the heat flux value helped us to describe the processes occurring on the studied surface. These data could describe the regime of "island" film cooling.

1 Introduction

The current trend of developing miniature electronics dictates the demand for the new ways of removinghigh heat fluxes from small heat exchanging surfaces, and this can be ensured only by maintaining the admissible operating temperatures of equipment and instruments by the new active cooling systems.

Therefore, one of the most effective options for increasing heat transfer from selected power electronic components is controlled local cooling by the single pulses of a gas-droplet flow, working at film and evaporative cooling. The modern electronic components operate in temperature range of up to 100°C, which defines the requirements for the working liquid flows (solid film, island coating, rivulets or drops).

It is known that when cooling the low-temperature surface, the temperature gradient at the liquid-solid surface interface plays a major role in heat transfer due to the Marangoni effect [1] because a significant contribution to heat transfer intensification is made by thermo-capillary convection due to evaporative cooling on a moving film - dry surface boundary.

According to the studies, development of island film cooling is observed on the surface, when tangential forces and local surface tension coefficient arise at insufficient wetting, resulting in shear film flows and dry spot formation, local film areas, rivulets and

*Corresponding author: Flags712008@yandex.ru

DOI: 10.1051/epjconf/201715900017

droplets. The conditions of forming a thin evaporating film with microthickness along the contact line allow maintenanceof a high heat transfer rate.

2 Experimental apparatus and instrumentation

The experimental setup is a multifunctional device, consisting of: programmable multi-jet source of a pulsed gas-droplet flow (pulse injector), automatic calorimeter with removable heat exchanger, automated system of registering the gas-droplet flow parameters, including original capacitive sensors for detecting the thickness and velocity of waves on the liquid film, as well as gradient sensors of the local heat flux [2]. The applied integral method of heat transfer registration allowed us to record the dependence of heat transfer on the regime of gas-droplet flow formation, its intensity, characteristics of roughness the heat exchanging surface and other design features of the cooled body.

The controlled source of the gas-droplet jet is formed via the off-duty factor: droplet flow and continuous gas flow. The programmable multi-jet source of the pulsed gas-droplet flow is designed in the form of a two-chamber unit: for air and water. On the flat part of the source, there are 16 individually controllable liquid injectors in the form of 4 × 4 matrix with a spacing of 30 mm. The liquid nozzle is a sprayer of four nozzles with 125-μm diameter, switched-on by a solenoid valve. On the same surface, there are 25 gas nozzles with the outlet diameter of 0.35 mm around the liquid nozzle, which form the impinging multi-jet air flow. These investigation results relate to the regime, when the heat exchanger surface is normal to the gas-droplet flow and vertical relative to the horizon. The scheme of formation of the wall liquid and air flow is shown in Fig. 1.

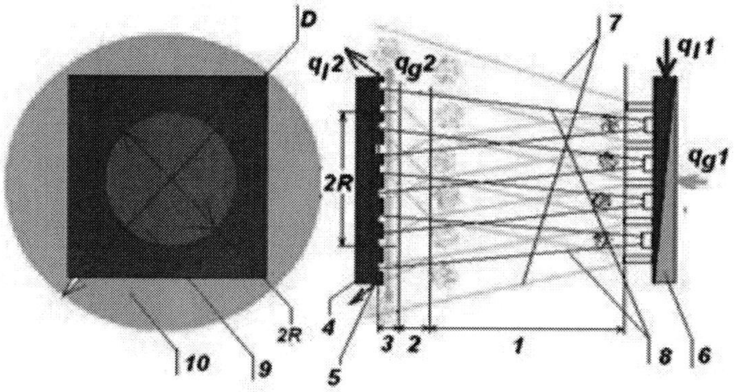

Fig. 1. Scheme of heat exchanger streamlining by the gas-droplet multijet flow.
q_g^1- total mass flow rate of gas, q_g^2- specific mass flow rate of gas on the heat exchanger surface,
q_l^1- total mass flow rate of liquid, q_l^2- specific mass flow rate of liquid on the heat exchanger surface.
1–zone of submerged jets, 2- zone of gas jet turn, 3– zone of near-wall flows, 4–isothermal heat exchanger, 5 – near-wall liquid film, 6 – pulsed droplet and air injector, 7 – air jets, 8 – droplet jets.

Distilled water and water - ethanol mixture are used as the working liquids. To monitor the process of the near-wall film formation on the heat exchanger surface, we have determined the regime, when the impinging multi-jet gas-droplet flow forms a liquid film on the entire surface of heat exchanger. These regime allows observation of hydrodynamics and heat transfer in the central and peripheral areas of the near-wall flow of cooling liquid in the regime of conditionally single pulse, when the film with characteristic currents in specific zones of the surface is formed at the moment of irrigation by the droplet flow. The film evaporates partially or completely in-between the pulses. The above regime relates to cooling by conditionally single pulse.

3 Results and discussion

This work is aimed at studying the processes of intensification of heat transfer from the heat-loaded surface at pulse irrigation by conditionally single multi-jet droplet flows. The short-term operation of all liquid nozzles with the ratio of period between pulses to duration of the droplet flow pulse higher than 20 is considered as a single pulse. The original programmable device forming the multi-jet pulse gas-droplet flows allows us to set the initial conditions of the cooling spray in a wide range.

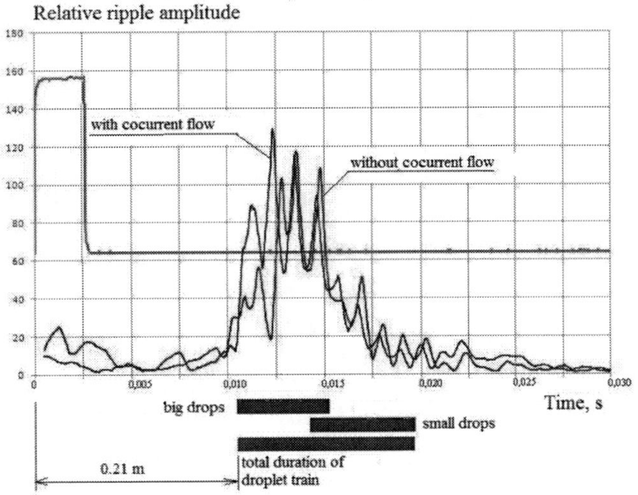

Fig. 2. Instantaneous value of droplet flow density near the heat exchanging surface.

Since formation of the multi-jet flame is studied insufficiently, at first, it was necessary to obtain data on distribution and behavior of the droplet train during target bombing. The effect of the co-current air flow on distribution of the droplet flow densityin the drift space and over the flow cross-section near the cooled surface was studied (Fig.2). The data showed that the droplet mixture flame in the co-current air flow has more uniform distribution of the droplet flow density over the cross-section and droplet grouping along the velocity vector, which result in more effective cooling (~ 3 times). During the movement of a droplet train in the air flow, the large droplets group, forming a denser"head" of a pulse, which provides intensive "turbulization" of the formed film.

The next stage of research was data obtaining on the values of local and integral heat transfer coefficientsat irrigation by conditionally single pulses of the droplet-air mixture. Capabilities ofthe programmable multi-jet source allowed us to form a thin liquid film on the surface of the heat-loaded surface. The main condition was "conditional singlenessof a pulse". Fig.3 shows the dependence of heat transfer obtained with the help of the local heat flux sensor. The peak corresponds to maximal heat transfer. When the bulk liquid is deposited on the surface, formed liquid film heating is observed (horizontal shelf), then there is uniform evaporation of the heated liquid film. The abrupt peak in the diagram corresponds byits time to the "end" of the droplet train. It is also possible to observe contribution of the secondary droplets to the process of heat entrainment from the surface.

DOI: 10.1051/epjconf/201715900017

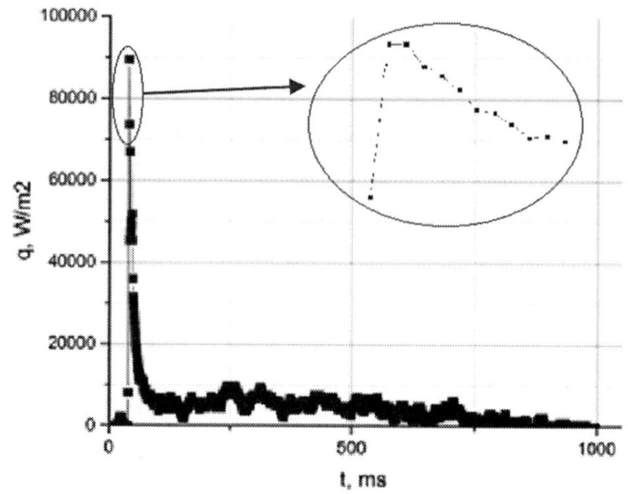

Fig. 3. Heatfluxdensityobtainedbythelocalheatfluxprobe

In all experiments, the pulse duration was constant and equal to 4 ms. Repetition rate varied under the condition that the preceding pulse had no effect on the subsequent one (complete evaporation of all liquid film from the surface). The data on the values of heat transfer coefficients are shown in Table 1.

Table 1. Effectiveness of the spray

F, Hz	r,g/(m2s)	h,W/(m2K)	η
0.25	1.4	269.1	0.96
1	3.6	446.8	0.87
2	5.9	596.4	0.81
4	10.7	795.5	0.75
6	16.2	888.4	0.69

The parameter, which characterizes the efficiency of the spray-system is its thermal efficiency; it equals the ratio of integral heat flux through the heat exchanger to the total value of system enthalpy [3]:

$$\eta = \frac{q}{m \cdot [c_p \cdot (T_w - T_s) + r]} \quad (1)$$

where – is het flux density (W/m^2), m – is specific mass flow rate (kg/m^{2}s), C_p – is specific heat capacity (J/kgK), w, s – are temperatures of the heat-exchanging surface and liquid (K), r – is latent heat of vaporization of liquid phase (J/kg). DataintheTableshowthat at frequencyF=0.25 Hz, the most effective heat removal is observed.

4 Conclusions

The investigations have shown that it is possible to construct the compact efficient cooling systems for electronic devices, which ensure effective cooling at low coolant flow rate via the regime, controlled by coordinate and time, combining film and evaporative cooling.

The work was financially supported by RFBR (project 16-38-00853).
The work was financially supported by the grant of Russian President NSh-8780.2016.8

References

1. E. Gatapova, O. Kabov, I. Marchuk, Technical Physics Letters, **30**, 418—421, (2004)
2. A. Nazarov, A.Serov, M. Bodrov,Technical Physics, **80**,724-727, (2010)
3. C. Pedersen, Int. J. Heat Mass Transfer,**13**, 369-381, (1970)

EPJ Web of Conferences **159**, 00018 (2017)
AVTFG2016
DOI: 10.1051/epjconf/201715900018

Studying the effect of fire retardant coating on the fire hazard characteristics of wood using infrared thermography

Denis Kasymov[1,*] and *Mikhail* Agafontsev[1]

[1]National Research Tomsk State University, 36 Lenin Ave., Tomsk, 634050, Russia

Abstract. This paper represents a study concerning the effect of fire retardant treatment «FUKAM» on the fire hazard characteristics of pine, aspen and larch. The front of model ground fire was investigated to estimate its effect on the surface of wood samples. Infrared thermography was used as a diagnostic method. The surface temperature distribution was obtained for the test wood samples after exposure to a fire front that was modeled using pine needles. The probability of ignition was estimated for the chosen experimental parameters for each kind of wood. The fire hazard characteristics of wood after fire retardant treatment showed a significant reduction in the surface temperature and the resistance to fire for the chosen parameters of the experiment compared to the same untreated samples.

1 Introduction

Wildfires cause significant material and economic damage to the territories. Numerous theoretical [1–3] and experimental investigations [4–7] conducted to study the effect of fires do not explain the mechanism of fire transition to the wooden constructions, their ignition and subsequent combustion.

The development of chemical industry led to the development of means such as fire retardants which prevent partially or completely the ignition of treated surfaces. On the market there are a lot of such means which have the different technical parameters and a different degree of toxicity.

The literature provides a huge amount of theoretical and experimental data on the fire hazard properties of wood [8], the study of burning particles acting on wooden constructions [9–11], and the effect of intense heating through thermal radiation [4,12–13]. It is of interest to study the effect of fire retardants on the fire hazard properties of wood exposed to combustion under laboratory conditions.

In this work the effect of fire front on the surface of wood samples was studied to estimate the effect of fire retardant treatment «FUKAM» [14]. Infrared thermography was used as a diagnostic method. The main advantage of this method is the high spatial and temporal resolution of measurements [15].

* Corresponding author: denkasymov@gmail.com

2 Laboratory equipment and methodology of the experiment

In the infrared region the sample surface characteristics were recorded using a thermal imager JADE J530SB with a 2.5–2.7 micron optical filter that allowed a temperature to be measured in the range of 500–850 K. In order to record a temperature in the range of 293 - 550 K, the recording was conducted without a filter.

The frame frequency was 50 frames/sec. The moisture content of the samples was controlled using a moisture analyzer AND MX-50 with an accuracy of 0.01%, and the mass of the samples was controlled using an electronic balance AND HL-100.

Combustion, simulated by forest fuel (FF), consisted of pine needles (*Pinus Pinaster*) and represented a site with a width that was close to the size of the wood sample. The samples of pine, aspen and larch were used as the samples which imitated the wood used in constructions. The dimensions of the samples in the experiment were (L × W × H): 0.23 × 0.02 × 0.1 m for pine, 0.17 × 0.02 × 0.1 m for aspen, and 0.16 × 0.02 × 0.12 m for larch. The liquid means «FUKAM» was used as a fire-bio wood protection applied to sawn, dressed and log building elements for housing, public, industrial and agricultural purposes [14].

Consumption of the means «FUKAM» that provides the fire retardant properties was approximately 400 g/m^2, which corresponds to the II group of fire retardant resistance. The dipping method was used as a technique of wood treatment.

The wood samples were placed in the sealed plastic containers filled with the «FUKAM» means, so that the wood was completely covered with liquid. The samples were in the containers for 2–3 hours and then removed, giving the unabsorbed solution to flow down. The samples were placed in a force-draft oven FDO–0.5 – 200 with a temperature of 70° C for 12 hours until complete drying.

Fig. 1 shows a scheme of the experiment:

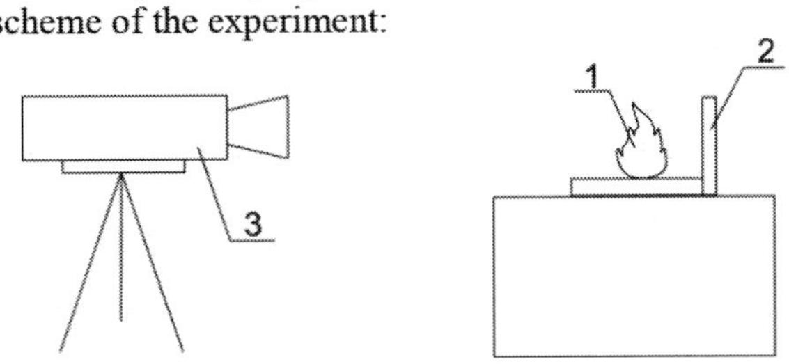

Fig. 1. Experimental setup. 1 – site with FF, 2 – wood sample, 3 – thermal imager JADE J530SB.

During the experiments, the moisture content of the samples was 5.55% for pine, 5.83% for aspen, 3.76% for larch, and 8.03% for pine needles. The FF mass maintained constant and was 50 g in the experiment, thus modeling a ground fire of weak intensity. The model of a ground forest fire of low intensity was used, since this fire was the most widespread in the natural environment, the modeling of which under laboratory conditions did not require significant maintenance costs.

The experiment was conducted as follows. The FF site 1 was located in front of the sample and ignited by a linear source located under the site. The produced fire front acted upon the wood sample 2 that was vertically fixed. The sample surface was recorded using a thermal imager 3. The experiment was conducted in two stages. At the first stage, the wood samples were studied without fire retardant treatment. At the second stage, the samples were covered with a fire-retardant coating. A series of three repetitions were conducted for each stage. The processing of data was carried out using the Altair software.

3 Results

The surface temperature distribution was obtained for the test wood samples (Fig. 2).

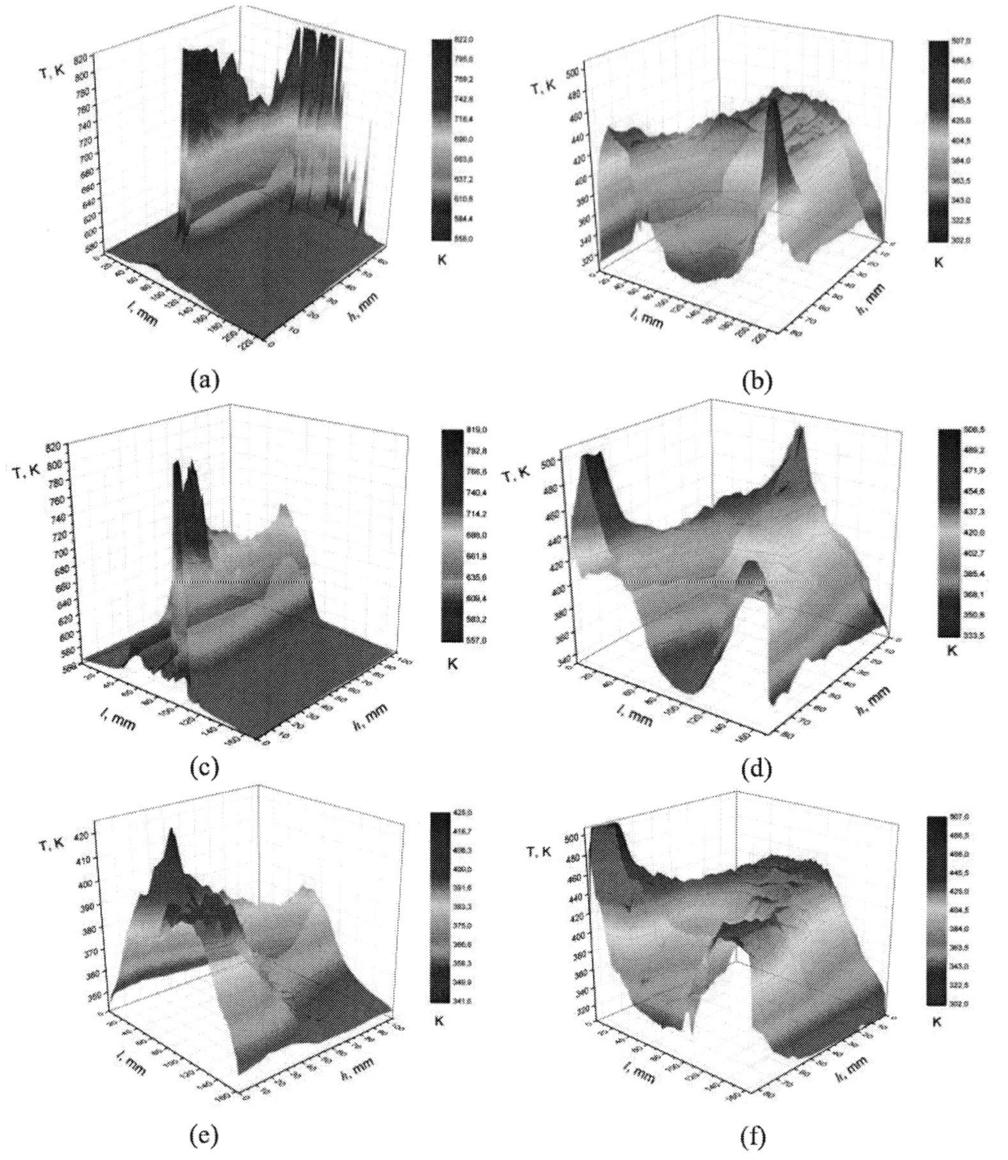

Fig. 2. Surface temperature distribution for the sample of pine ((a) no treatment; (b) treatment with «FUKAM»), aspen ((c) no treatment; (d) treatment with «FUKAM»), and larch ((e) no treatment; (f) with treatment «FUKAM»).

The series of laboratory experiments conducted to investigate the effect of the combustion front initiated by FF showed that charring of the surface was observed for all the wood samples, both with the use of a fire retardant and without it (Fig. 3). However, the further combustion process was not observed for the pine and aspen samples treated with a fire retardant. The fire front had no effect on the samples of larch in both cases.

(a)

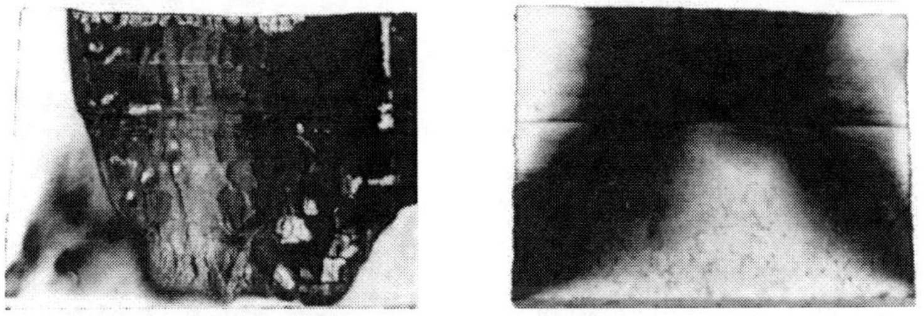

(b)

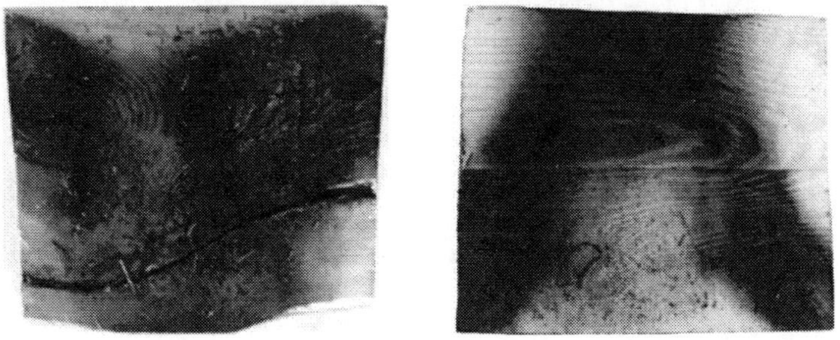

(c)

Fig.3. Wood samples after exposure to the combustion front: (a) pine, (b) aspen, (c) larch (left - no treatment, right - with the fire retardant treatment).

The maximum temperature that was reached on the surface of the samples without a coating was: 820 K for pine, 800 K for aspen, and 410 K for larch and with a coating was 456 K for pine, 506 K for aspen, and 471 K for larch. The nonuniformly laid FF layer was likely to influence on the difference in the temperature values for the larch samples. Fire-retardant treatment «FUKAM» could also influence on the sample heating.

4 Conclusion

For the chosen experimental parameters the surface of the wood samples became charred, but flame combustion was observed only in a few cases, which appears to be connected to the small capacity of the heat flow, as well as the short duration of its effect on wood, which is confirmed, in particular, by the work [16] that analyzes the experimental characteristics of ignition and combustion of wood subjected to the different thermal flows. As a result, it is concluded that the lower the intensity of the heat flow is, the more complete pyrolysis occur at a rate that is low to create a combustible concentration. In addition, as is known, for stable combustion the solid material should obtain the amount of heat that provides a continuous supply of the sufficient amount of gaseous combustible substances in the combustion zone [17].

The use of fire retardant treatment (for example a fire retardant «FUKAM») certainly increases the resistance of wood to fire, which is confirmed by the data on the sample surface temperature after exposure to model combustion.

This study was supported by the Russian Foundation for Basic Research (project No. 16-38-00190) and Tomsk State University, the Fund Program of Academic D.I. Mendeleev, 2016 (grant No. 8.1.27.2016).

References

1. A. M. Grishin, A. S. Yakimov, Thermophys. Aeromech. **20**, 463 (2013).
2. W.E. Mell, S.L. Manzello, A. Maranghides, D. Butry, R.G. Rehm, Int. J. Wild. Fire **19**, 238 (2010).
3. J. D. Cohen, J. Forest. **98**, 3 (2000).
4. V.T. Kuznetsov, A.I. Fil'kov, Combust. Explos. Shock Waves **47**, 65 (2011).
5. A.M. Grishin, A.I. Fil'kov, J. Eng. Phys. Thermophy. **76**, 1139 (2003).
6. A.M. Grishin, A.I. Filkov // Fire Saf. J. **46**, 56 (2011).
7. J.-L. Rossi, K. Chetehouna, A. Collin, B. Moretti, J.-H. Balbi, Combust. Sci. Technol. **182**, 1457 (2010).
8. V. Babrauskas, Fire Saf. J. **40**, 528 (2005).
9. S. D. Tse, A. C. Fernandez-Pello, Fire Saf. J. **30**, 333 (1998).
10. J. D. Cohen, USDA Forest Service Gen.Tech.Rep. PSW-GTR **173**, 189 (2000).
11. S. L. Manzello, S. Suzuki, Y. Hayashi, Fire Saf. J. **54**, 181 (2012).
12. R. Sh. Enaleev, I. V. Krasina, V. S. Gasilov, Yu. S. Chistov, O. A. Tuchkova,Vestnik Kazanskogo tehnologicheskogo universiteta **10**, 99 (2013).
13. L. Lowden, T. Hull, Fire Sci. Rev. **2**, 4 (2013).
14. The special substance for processing different fabrics and building materials «FUKAM»: http://www.fukam.ru/English/
15. V. P. Vavilov, *Thermal/Infrared Nondestructive Testing.-NDT Handbooks series* (Spectrum, Moscow, 2009).
16. I. M. Abduragimov, A. S. Androsov, M. Bartak, Combust. Explos. Shock Waves. **22**: 7 (1986).
17. A. Ya. Korolchenko, *Protsessyi goreniya i vzryiv* (Pozhnauka, Moscow, 2007).

 DOI: 10.1051/epjconf/201715900019

Forced spreading over superhydrophobic and copper surfaces

Dmitry Feoktistov[1,*], *Konstantin* Ponomarev[1] , and *Kseniya* Batishcheva[1]

[1]National Research Tomsk Polytechnic University, Tomsk, Russia

Abstract. Dynamic spreading over superhydrophobic and copper surfaces was studied experimentally under the condition of contact line movement with speed greater than 1 mm/sec. Three modes of spreading of distilled water drop over copper surfaces with sufficient typical roughness (0.591, 5.190 and 6.210 μм) were detected. The first one is drop formation when the contact line speed and dynamic contact angle increase sharply. The second mode is spreading of a drop, which is characterized by a monotonic decrease in the contact line speed and dynamic contact angle. The third one is a formation of an equilibrium contact angle at a constant wetted area (the contact line speed tends to zero, and spreading of a drop occurs as long as the driving force is greater than zero). Some features in spreading were detected on superhydrophobic surface with parameter roughness of 0.751μm compared to other substrates. During drop formation after sharp increase in the contact line speed and dynamic contact angle, there is a mode which is accompanied by a decrease in the contact line speed and monotonic increase in the advancing dynamic contact angle.

1 Introduction

Superhydrophobic surface is not a unique phenomenon in nature, and it is common to many plants and insects. Prospects for the use of superhydrophobic surfaces in everyday life and in industrial processes in virtually all areas of the industry are huge [1-8]. For example, superhydrophobic treatment of buildings allows to reduce dirtying of windows and panes. A similar effect can be obtained in automobile manufacturing at treatment of car body by superhydrophobic coating. In the textile industry hydrophobe processing of clothes will allow to make them waterproof, anti-dirtying in contact with coloring food and drinks. Moreover it can be achieved without change in color, texture properties of clothes. In the power industry there are several examples of using superhydrophobic surfaces: design of filters for cleaning of fuels and oils. The use of such filters allow to separate water-oil emulsions with high efficiency in a wide range of compositions of dispersion systems and the particle size of the dispersed phase. Moreover, such surfaces have a great potential in aviation when creating anti-icing surfaces.

* Corresponding author: fdv@tpu.ru

Currently several methods of creating such surfaces are known: deposition techniques, sublimation, controlled polymerization, nanorod growth, templating and lithographic techniques. However there are no scientific bases [9-12] to control the wetting process. It happens due to the lack of available registration equipment of fast processes of wetting and spreading. In the last decade scientific and technological progress in the field of microelectronics has allowed to create a relatively inexpensive photo and video devices allowing by research groups [13-19] to record the physical mechanisms of fast processes. It should be noted that changes in the dynamic contact angle (DCA) during wetting surfaces (including super hydrophilic) by advancing and receding drops were studied only at slow movement of the three-phase contact line (up to 1 mm/sec). The dynamics of wetting at three-phase contact line speed greater than 1 mm/sec is poorly understood. So the purpose of this work is to study change in the dynamic contact angle of drops spreading over superhydrophobic and copper surfaces at high contact line speeds.

2 Experimental methods

The shadow and Schlieren systems were implemented for conducting experimental studies. The operational principle of the experimental setup is described in detail in [12,19].

According to results of preliminary experiments the main influencing factors were defined (Table 1).

Table 1. Main influencing factors.

Liquid volume	0.3 mL
Flow rate of distilled water	0,005; 0,01; 0,02; 0,04; 0,08; 0,16 mL/sec
Material of substrate	Copper, surface with superhydrophobic coating
Roughness parameter (Ra) of surfaces	flexible copper Ra=0,591 μm; copper Ra=5,190 μm; copper Ra=6,210 μm; surface with superhydrophobic coating Ra=0,751 μm
Wetted liquid	distilled water

3 Results and discussion

According to results of experimental series (Fig. 1 (a-c)) three modes of spreading of the water drop over copper surfaces with typical roughness (0.591, 5.190 и 6.210 μm) were detected [19].

The first mode is characterized by the drop formation; the contact line speed and dynamic contact angle increase sharply. The second one is drop spreading; the contact line speed and DCA decrease monotonously. The third mode is formation of an equilibrium contact angle when the wetted area is constant; the contact line speed tends to zero, and spreading of a drop occurs as long as the driving force is greater than zero.

Duration of modes in the experiments is 1-2% from overall time of spreading for the first mode, 39-50% for the second one, and 48-60% for the third one.

There are some features in spreading modes for superhydrophobic surface with Ra=0.751 μm (Fig. 1 (d)) compared to another surfaces [19]. During drop formation after sharp increase in the contact line speed and dynamic contact angle, there is a mode which is accompanied by a decrease in the contact line speed and monotonic increase in the advancing dynamic contact angle.not allow to use optical methods of processing.

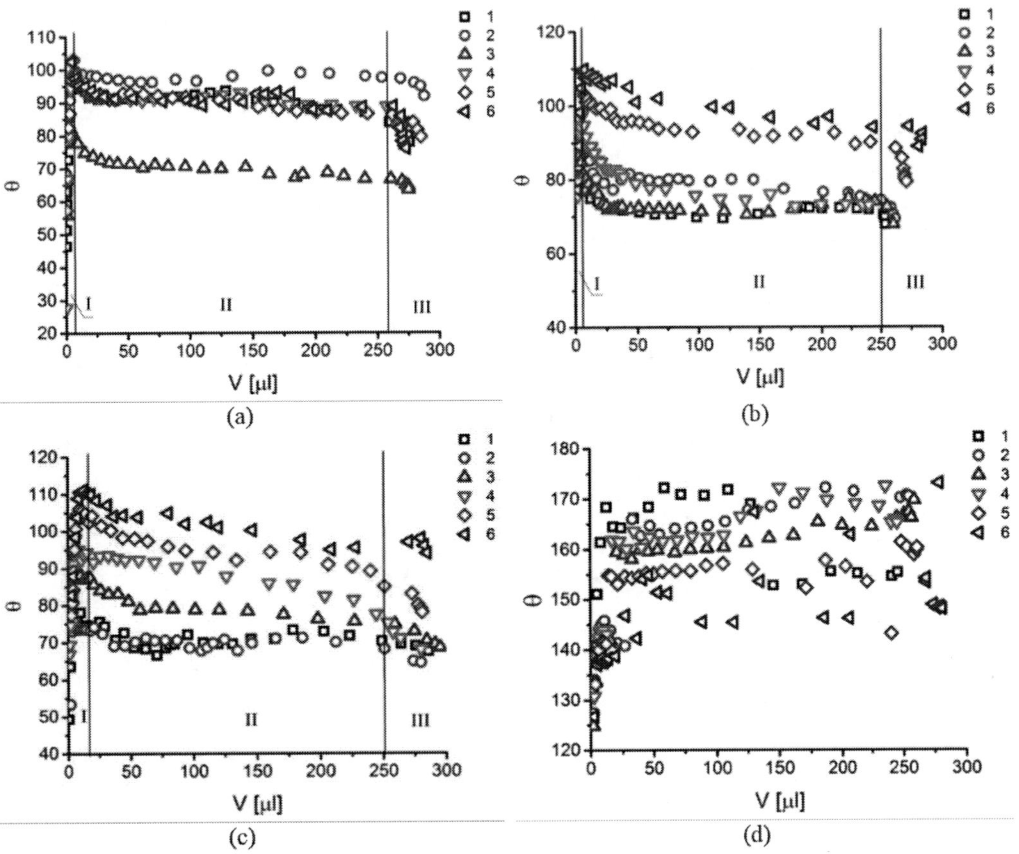

Fig. 1. Dynamic contact angle versus liquid volume during spreading of distilled water drop over surfaces: (a) flexible copper (Ra=0.591 μm); б) copper (Ra=5.190 μm); в) copper (Ra=6.210 μm); г) superhydrophobic surface (Ra=0.751 μm).

It should be noted that analyses of the contact angles values were conducted for the second and third modes. It is found that the maximal value of advancing contact angle increased by 30 % with increasing volumetric flow rate of distilled water from 0.005 mL/sec to 0.16 mL/sec at spreading over copper surface with Ra=5.190 μm, and the angle increases by 37% at spreading on surface with Ra=6.210 μm. Increase in the volumetric flow rate on flexible copper surface has not led to an increase in the value of the advancing contact angle. Advancing DCA in the range of volumetric flow rate from 0.04 mL/sec to 0.16 mL/sec is independent of the drop volume. Wetting replaces by nonwetting during drop spreading on such surfaces. DCA decreases during contact line movement, and after a certain time, which depends on the volumetric flow rate, material of substrate and surface microrelief, becomes less than 90°.

The reported study was supported by the grant of President of Russian Federation for the government support of young Russian scientists (MK-6810.2016.8).

References

1. D. Zaitsev, D. Rodionov, O. Kabov, Tech. Phys. Lett., **35**, 680 (2009)
2. G. Kuznetsov, M. Sheremet, Int. J. Heat Mass Transfer., **54**, 260 (2011)
3. O. Vysokomornaya, G. Kuznetsov, P. Strizhak, Fire Safety J., **70**, 61 (2014)

4. D. Glushkov, G. Kuznetsov, P. Strizhak, R. Volkov, Therm. Sci., **20**, 131 (2016)
5. I. Afanasyev, E. Orlova, D. Feoktistov, EPJ Web of Conferences, **82**, 01054 (2015)
6. I. Shanenkov, A. Sivkov, A. Pak, Y. Kolganova, Adv. Mat. Res., 1040, 813 (2014)
7. G. Kuznetsov, D. Feoktistov, E. Orlova, Thermophys. Aeromech, **23**, 17 (2016)
8. V. Lebedev, V. Lemanov, S. Misyura, V. Terekhov, Fluid Dyn., **28**, 624 (1993)
9. S. Misyura, V. Nakoryakov, Energy Fuels, **27**, 7089 (2013)
10. G. Kuznetsov, N. Kurilenko, V. Maksimov, G. Mamontov, T. Nagornova, J. Eng. Phys. Thermophys., **86**, 519 (2013)
11. Y. Lyulin, D. Feoktistov, I. Afanas'ev, E. Chachilo, O. Kabov, G. Kuznetsov, Tech. Phys. Lett., **41**, 665 (2015)
12. K. Batischeva, E. Orlova, D. Feoktistov, EPJ Web of Conferences, **82**, 01055 (2015)
13. G. Kuznetsov, D. Feoktistov, E. Orlova, J. Eng. Phys. Thermophys., **89**, 317 (2016)
14. D. Zaitsev, M. Aviles, H. Auracher, O. Kabov, Microgravity Sci. Tec., **19**, 71 (2007)
15. R. Volkov, G. Kuznetsov, P. Strizhak, Thermophys. Aeromech., **21**, 255 (2014)
16. D. Zaitsev, D. Kirichenko, O. Kabov, Tech. Phys. Lett., **41**, 551 (2015)
17. I. Shanenkov, A. Pak, A. Sivkov, Y. Shanenkova, MATEC Web of Conferences, **19** (2014)
18. D. Glushkov, J. Legros, P. Strizhak, A. Zakharevich, Fuel, **175**, 105 (2016)
19. D. Feoktistov, E. Orlova, A. Islamova, EPJ Web of Conferences, **110**, 01018 (2016)

EPJ Web of Conferences **159**, 00020 (2017) DOI: 10.1051/epjconf/201715900020
AVTFG2016

Experimental study of geometrical characteristics of distilled water and aqueous ethanol solution droplets evaporating on aluminum surface

Anastasiya Islamova[1,*], *Evgeniya* Orlova[1] and *Ilya* Zykov[1]

[1]National Research Tomsk Polytechnic University, Tomsk, Russia

Abstract. Change of geometrical characteristics of distilled water and aqueous ethanol solution droplets was studied under their evaporation on aluminum surface. According to change in the contact diameter three evaporation modes of distilled water droplet on polished aluminum surface were detected: increase in the contact area, pinning of a droplet (constant contact area), and droplet depinning (decrease in the contact diameter). During evaporation of aqueous ethanol solution droplets, two evaporation modes were detected: increase in the contact area, and droplet depinning.

1 Introduction

The last 20 years the interest for many practical applications has become heat and mass transfer processes in the mini- and microsystems. It is due to the intensive development of microelectronics and medicine, as well as the miniaturization of various kinds of process control and monitoring equipment (aerospace, automobile manufacturing, and transportation field) [1-11]. There is an intensive development of heat exchange systems with micro- and nanoscale. It is known [12-15] that such systems are much more efficient than their analogs with typical dimensions of units and tens of millimeters. For example, in recent years, thermosyphons and heat pipes with the transverse dimensions of 2 millimeters or less are applied for cooling microelectronic equipment [16-17]. But research in this area is being conducted primarily as a search for functional options (using such characteristics of search as refrigerants, materials, regime parameters, size, and configuration items) without analyzing the basic physics of processes occurring in such equipment [18]. Typically, materials for mini heat exchange systems are non-ferrous metals, and the movement of coolant is implemented mainly in the form of rivulet currents, which can disintegrate into separate drops.

To date, theoretical foundations of evaporation of pure liquids and solutions droplets have not been developed to a level that provides a predictive evaluation of the impact of various factors on the physical process.

* Corresponding author: agi2@tpu.ru

There is a lack in the experimental data on heat transfer characteristics in the vicinity of the three-phase contact line (gas/liquid/substrate). Most researchers (R. Hoffman, J. Bachelor, T. Blake, V.M. Starov, R.G. Cox et al.) have been studying the process of evaporation of pure liquids. But the study of the contact angle and the contact line of the drop evaporation of solution on a solid substrate is dedicated to small the number of research.

The intensity of the evaporation process, contact diameter, contact angle are known to be affected by many factors, among them the liquid vapor pressure, thermal conductivity, roughness of the substrate, shape of the surface microstructure, composition of the liquid, impurities in liquids and others [18].

Conducting experimental studies using modern recording equipment will clarify the understanding of the mechanisms of evaporation of pure liquids and solutions droplets. It will develop the physical bases for designing energy-efficient heat exchange systems of small dimensions, in which the heat transfer agent moves in rivulet way or forms droplet irrigation.

The purpose of this work is to study experimentally change in the geometric characteristics of f distilled water and aqueous ethanol solution droplets (41 % and 39 % wt.) evaporating on aluminum substrate. The studies were conducted using the aluminum surface due to the extensive use of non-ferrous metal in the manufacture of heat exchangers.

2 Materials and Methods

Experimental studies were carried out by using the shadow system equipment [16]. A droplet of a predetermined volume was placed on the surface by the electronic dispenser. Droplet volume was ranged from 10 to 50 μL in increments of 10 μL. The geometrical parameters of the drops were obtained after processing the shadow image snapshots in software Drop Shape Analyses by Tangential 1 method.

Fig. 1 shows the obtained dependences of the droplet contact diameter on time at the initial substrate temperature of t = 60 °C.

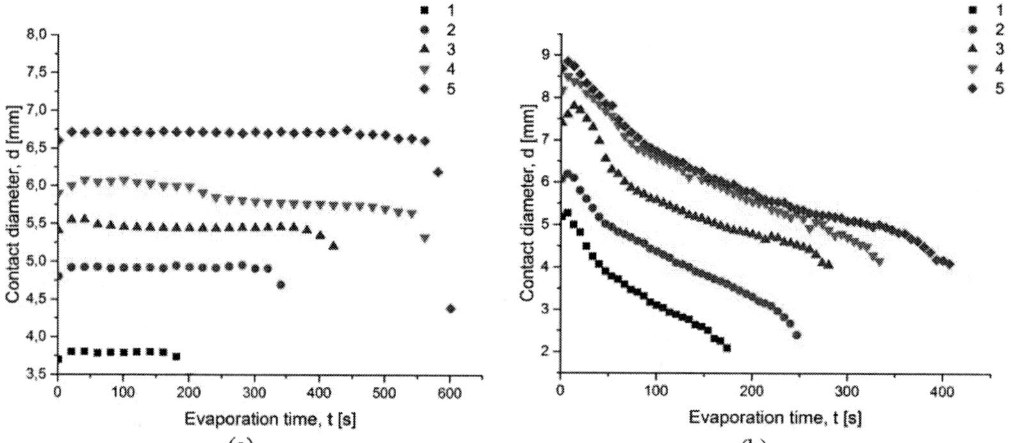

Fig. 1. Time dependences of the contact diameter of evaporating liquids: (a) distilled water; (b) binary liquid on aluminum surface heated up to 60 °C.

According to change in the contact diameter (Fig. 1 (a)) three evaporation modes of distilled water droplets on aluminum surface were detected:

1 – an increase in the contact area;

2 – pinning of a droplet [16] (constant contact diameter);

3 – depinning of a droplet [16] (decrease in the contact diameter).

During evaporation of ethanol aqueous solution droplets, two evaporation modes were detected:

1 – an increase in the contact area;

2 – droplet depinning.

At the initial stage of evaporation an increase in the contact diameter (Fig. 1 (b)) and decrease in the contact angle (Fig. 2 (b)) occurs under the action of the mass forces.

Then, an intensive evaporation of alcohol occurs on the border of the three phases, and under the action of the surface forces a droplet shrinks, the contact area reduces, and the contact angle increases.

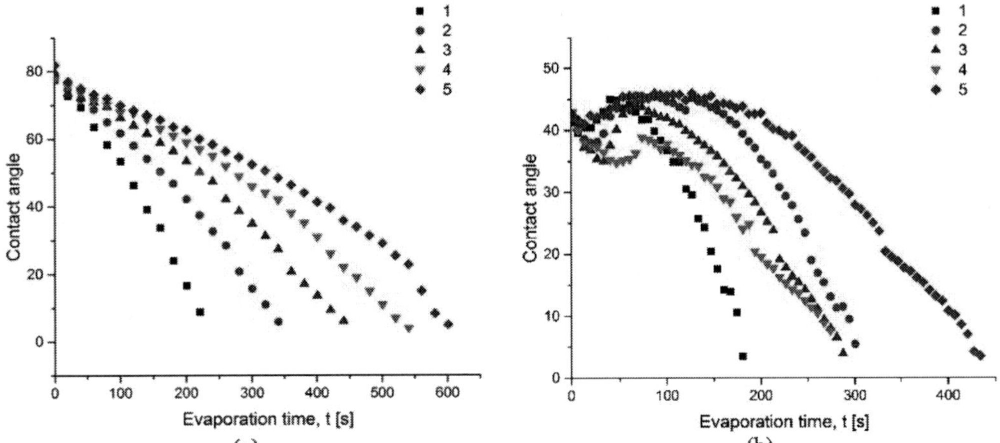

Fig. 2. Time dependences of the contact angle of evaporating droplets: (a) distilled water; (b) binary liquid on aluminum surface heated up to 60 °C.

The errors of image processing by different methods were estimated. The random error of Tangential 1 method does not exceed 3%. The values of the contact angles in the initial period of time were 79.1° ± 2.1° and 42.26° ± 0.57° for water and binary liquid, respectively. They are within the error of the measuring system. It can be concluded that the volume of distilled water, and the binary liquid droplets on polished aluminum surface has no influence on the angle.

The change in volume of distilled water and binary liquid droplet is shown in Figure 3.

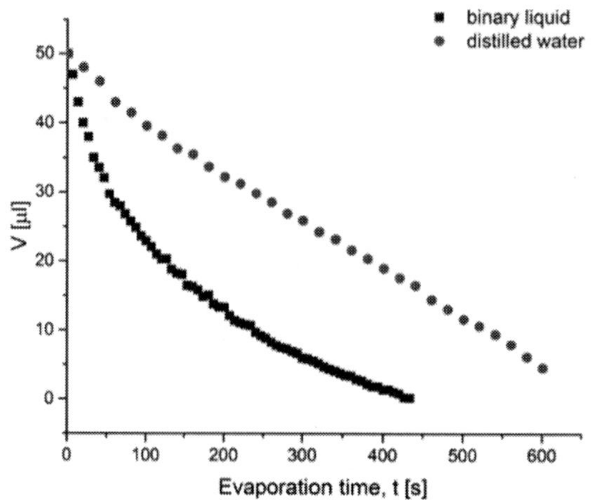

Fig. 3. Change in volume of distilled water and binary liquid droplet in time during evaporation.

It is worth noting that for water evaporation curve has linear nature; more than 70 % of liquid from initial mass of droplet evaporates). For binary liquid in the initial time ((approximately the first 100 seconds) the evaporation rate (dynamics of droplet change) is maximal as ethanol evaporates mostly (heat of vaporization of alcohol is much less than for water). Over time, the alcohol concentration is reduced, and the evaporation rate becomes less intense, water evaporates.

The reported study was supported by the grant of President of Russian Federation for the government support of young Russian scientists (MK-6810.2016.8).

References

1. D. Zaitsev, D. Rodionov, O. Kabov, Tech. Phys. Lett., **35**, 680 (2009)
2. D. Zaitsev, D. Kirichenko, O. Kabov, Tech. Phys. Lett., **41**, 551 (2015)
3. D. Glushkov, J. Legros, P. Strizhak, A. Zakharevich, Fuel, **175**, 105 (2016)
4. D. Glushkov, G. Kuznetsov, P. Strizhak, R. Volkov, Therm. Sci., **20**, 131 (2016)
5. V. Lebedev, V. Lemanov, S. Misyura, V. Terekhov, Fluid Dyn., **28**, 624 (1993)
6. V. Nakoryakov, S. Elistratov, S. Misyura, J. Eng. Phys. Thermophys., **20**, 338 (2011)
7. M. Semenov, A. Manakov, E. Shitz, A. Stoporev, L. Altunina, L. Strelets, S. Misyura, V. Nakoryakov, J. Therm. Anal. Calorim., **119**, 757 (2015)
8. I. Shanenkov, A. Sivkov, A. Pak, Y. Kolganova, Adv. Mat. Res., **1040**, 813 (2014)
9. I. Shanenkov, A. Pak, A. Sivkov, Y. Shanenkova, MATEC Web of Conferences, **19** (2014)
10. G. Kuznetsov, N. Kurilenko, V. Maksimov, G. Mamontov, T. Nagornova, J. Eng. Phys. Thermophys., **86**, 519 (2013)
11. O. Vysokomornaya, G. Kuznetsov, P. Strizhak, Fire Safety J., **70**, 61 (2014)
12. G. Kuznetsov, D. Feoktistov, E. Orlova, Thermophys. Aeromech, **23**, 17 (2016)
13. I. Afanasyev, E. Orlova, D. Feoktistov, EPJ Web of Conferences, **82,** 01054 (2015)
14. G. Kuznetsov, D. Feoktistov, E. Orlova, J. Eng. Phys. Thermophys., **89**, 317 (2016)
15. A. Semenov, D. Feoktistov, D. Zaitsev, G. Kuznetsov, O. Kabov Thermophys. Aeromech., **22**, 771 (2015)
16. G. Kuznetsov, A. Zakharevich, N. Bel'kov, Chem. Pet. Eng., **50**, 1 (2014)
17. K. Batischeva, E. Orlova, D. Feoktistov, MATEC Web of Conferences, **19,** 01001 (2014)
18. G.V. Kuznetsov, D.V. Feoktistov, E.G. Orlova, K.A. Batishcheva, Colloid J. **78**, 335 (2016)

Evolution of liquid bridge rupture in the evaporator model

Elizaveta Gatapova[1,2,*], *Alyona* Novikova[1,3]

[1]Kutateladze Institute of Thermophysics SB RAS, Novosibirsk, Russia
[2]Novosibirsk State University, Novosibirsk, Russia
[3]Novosibirsk State Technical University, Novosibirsk, Russia

Abstract. The liquid bridge rupture confined between two glass plates is experimentally investigated. High–resolution shadow technique is used for video registration. Critical bridge diameters and distance between glass plates are found.

1 Introduction

The development of new technological potential for high-performance cooling systems with manipulation of phase-change at liquid–vapor interfaces for microelectronics and Light Emitting Diodes is highly required by the global miniaturization [1] and by growing up of heat flux densities. Active and passive cooling methods with phase transitions are available, such as microchannel flows [2–4] and heat pipes [5, 6]. Effects near liquid-vapor-solid contact lines [7, 8], as well as temperature jump, which can present near the vapor bubbles [9-11] become important for such systems. The use of different nanocoatings can enhance heat transfer in cooling systems of small size [12].

Investigations of the behavior of the liquid bridges and meniscus occurring in heat pipe are one of the important problems for stable operation of heat pipes. For example, the highest intensity of heat transfer provided by evaporation of a thin liquid film that separates the vapor bubble from the heating surface in the pulsating heat pipe (PHP). The vapor bubbles are formed when the liquid is heated in PHP. The critical temperature difference between the heat source and the heat sink leads to auto-oscillations of the vapor bubbles and liquid plugs present inside the capillary tube [5, 13]. Heat is thus passively transferred, not only by latent heat exchange like in conventional heat pipes, but also by sensible heat transfer between the wall and the fluid. The oscillations are extremely important because they cause a significant increase in heat transfer intensity in comparison with other types of heat pipes [13]. Nowadays, PHP is most reliable and cheap compared to other heat transfer devices.

The processes of liquid transfer between two solid surfaces, formation of liquid bridge and finally formation and evaporation of droplet are important problems for application in PHP. In this paper, we present the results of the experiment on dynamics of liquid bridge confined between two surfaces.

* Corresponding author: gatapova@itp.nsc.ru

2 Experimental setup

The scheme of experimental setup is presented in Fig. 1. To study the dynamics of the liquid bridge the shadow technique providing a resolution of 3 μm per pixel is adopted. The test cell consists of two plates of round shape. Bottom plate has a hole of a diameter less than 1 mm for a liquid injection. The optical glasses BK7 with a diameter of 50 mm are used for experiments. The contact angle of water on BK7 is 40°; contact angle hysteresis is 25°, which are characterized by DSA-100, KRUSS. Ultrapure water (MilliQ) is used as a working fluid. The liquid is injected from a bottom using a high-precision syringe pump (Cole Parmer) and liquid bridge confined between two surfaces is formed (Fig. 2a). The linear micro-positioner (Zaber) travels one of the plates changing and controlling the distance between surfaces. We have put some changes in C # to micro-positioner console. Experiments are conducted at normal atmospheric condition. Air humidity and temperature in vicinity of the test cell are controlled by Testo device.

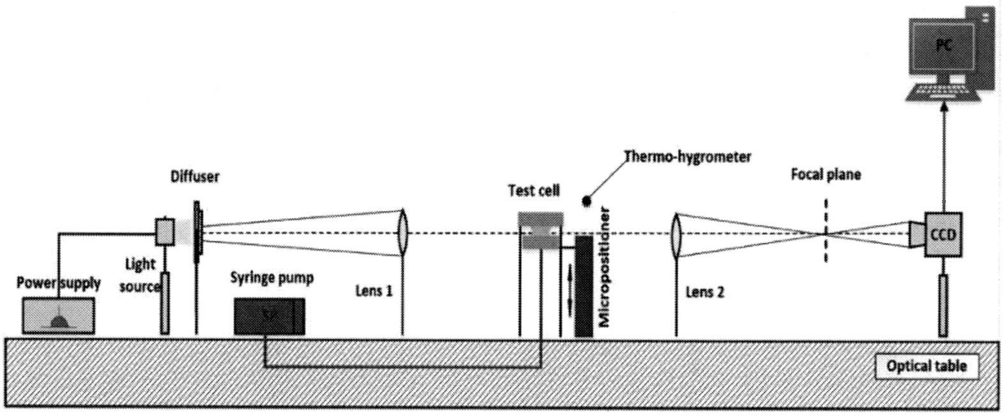

Fig. 1. Scheme of the experimental setup and apparatus.

3 Measurements technique and results

In the present experiments the liquid of volume of 21 μl is injected through a bottom hole with a rate of 0.5 μl/s into the test cell. The liquid bridge confined between two surfaces is formed with a practically the same contact angles at the top and bottom plates. The pressure, temperature and humidity of air were 100.5 kPa, 26.8 °C and 23.8 %, respectively. In order to investigate an evolution of the liquid bridge and contact angles with the changing the distance between two plates the following procedure is organized. The lower plate of the test cell is moved downwards by Zaber micro-positioner with the speed of 40 μm/s until the rupture of the liquid bridge (Fig. 2a, b, c).

Fig. 3 demonstrates evolutions of the two base diameters (D_t, D_b), middle diameter (D_m) and height of liquid bridge (H) in real time. The middle diameter of the liquid bridge is decreasing linearly, while the top and bottom base diameters are decreasing until a critical value. Then the pinning at the top and bottom plates is observed, while the liquid bridge height is increased.

 DOI: 10.1051/epjconf/201715900021

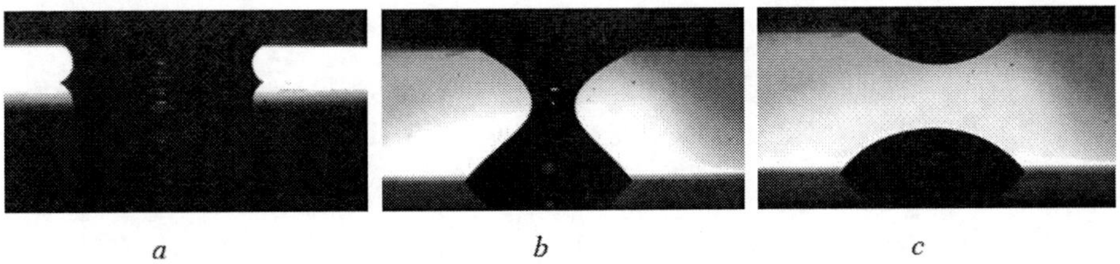

Fig. 2. Evolution of liquid bridge: *a* – initial position; *b* – just before the break; *c* – break and the formation two liquid droplets.

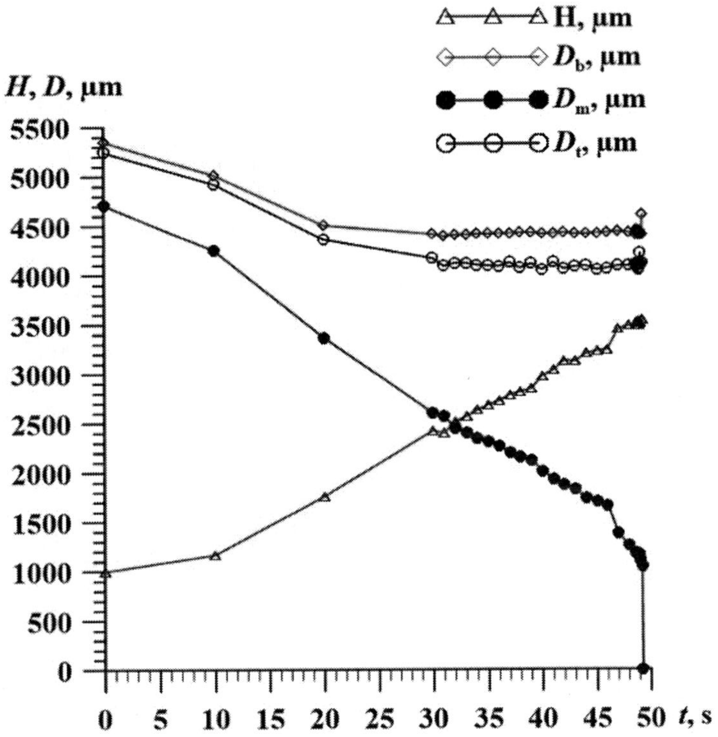

Fig. 3. Evolutions of liquid bridge height and diameters in real time.

In conclusion, the dynamics of liquid bridge rupture between two glass plates is experimentally investigated. High–resolution shadow technique is used for video registration. The pinning at the top and bottom plates is observed for certain distance between the plates until the liquid bridge rupture. While the height and middle diameter change practically linearly. The critical middle diameter to rupture is 1 mm at a moving speed of 40 μm/s and initial liquid volume of 21 μl for particular experiment on BK7 with contact angle hysteresis of 25°.

The authors gratefully acknowledge the support of this work by the Ministry of Education and Science of the Russian Federation (Agreement 14.616.21.0016, project identifier RFMEFI61614X0016).

References

1. T.M. Squires and S.R. Quake, Review of Modern Physics, **77**, 977 (2005)
2. O.A. Kabov, E.Ya. Gatapova, D.V. Zaitsev, 11th IEEE Intersociety Conference on Thermal and Thermomechanical Phenomena in Electronic Systems, I-THERM 2008, Article number 4544312, 520 (2008)
3. E.Ya. Gatapova, O.A. Kabov, International Journal of Heat and Mass Transfer, **51**, 4797 (2008)
4. A. Bar-Cohen and G.A. Holloway, Interfacial Phenomena and Heat Transfer, **3**, 393 (2015)
5. H. Akachi, U.S. Patent, Patent Number 4921041, (1990)
6. M.A. Chernysheva, S.I. Yushakova, Y.F. Maydanik, International Journal of Heat and Mass Transfer, **81**, 297 (2015)
7. E.Ya. Gatapova, O.A. Kabov, Microgravity Science and Technology, **19**, 132 (2007)
8. S.S. Panchamgam, A. Chatterjee, J.L. Plawsky, P.C. Wayner, Jr., International Journal of Heat and Mass Transfer, **51**, 5368 (2008)
9. E.Y. Gatapova, I.A. Graur, F. Sharipov, O.A. Kabov, International Journal of Heat and Mass Transfer, **83**, 235 (2015)
10. E.Y. Gatapova, R.A. Filipenko, Y.V. Lyulin, I.A. Graur, I.V. Marchuk, O.A. Kabov, Thermophysics and Aeromechanics, **22**, 701 (2015)
11. E.Y. Gatapova, I.A. Graur, O.A. Kabov, V.M. Aniskin, M.A. Filipenko, F. Sharipov, L. Tadrist, International Journal of Heat and Mass Transfer, **104**, 800 (2017)
12. N. Miljkovic, R. Enright, Y. Nam, K. Lopez, N. Dou, J. Sack, E.N. Wang, Nano Letters, **13**, 179 (2013)
13. L.L. Vasiliev, Applied Thermal Engineering, **25**, 1 (2005)

DOI: 10.1051/epjconf/201715900022

Interaction of levitating liquid microdroplets with intensive vapor flow near the contact line

Dmitry Kirichenko[1,2], *Dmitry* Zaitsev[1,3], and *Oleg* Kabov[1,3]

[1]Kutateladze Institute of Thermophysics SB RAS, Novosibirsk, Russia
[2]National Research Novosibirsk State University, Novosibirsk, Russia
[3]National Research Tomsk Polytechnic University, Tomsk, Russia

Abstract. In this research, interaction of liquid microdroplets with the air-vapor flow near the contact line (the three-phase solid - liquid - gas boundary) is studied. It is shown experimentally that the flight of levitating microdroplets from the wetted surface to the dry one is accompanied by a significant change in the levitation height, which proves the existence of a zone near the contact line with intense evaporation.

1 Introduction

Two-phase cooling systems represent the effective solution for removing high heat flux densities. Some applications, such as spray cooling [1], heat pipes [2], and film cooling systems [3-5], require a study of the processes taking place near the contact line, i.e., the line, where liquid contacts with the gas phase and heated surface. Recent studies have been dealing with droplet evaporation on the heated substrates [6-8]. It was found out that with decreasing droplet size, the specific evaporation rate increases, which can be associated with increased contribution of evaporation near the contact line to the total evaporation rate. The study of heat and mass transfer near the contact line is also important for the applied research in various fields, such as the study of heat transfer at fuel ignition [9-10]. In this paper, we study interaction of levitating liquid microdroplets with the air-vapor flow in the contact line region. It is shown that in this region, there is an area of intense evaporation, where the flight of microdroplets is accompanied by a significant change in the height of droplet levitation.

2 Experimental setup and research methods

In the experiments, the stainless steel working section with embedded copper rod of 10x10 mm^2 was used. The rod was heated using a nichrome spiral, with the current supplied from the power source. The substrate surface temperature was controlled by means of thermocouples installed at the surface of the copper rod. The scheme of experimental setup is shown in Fig. 1. The experiment was performed in the horizontal working section open to the atmosphere at the air temperature of around 25°C. The working surface of the test section was mechanically polished. The working surface morphology was studied using scanning electron microscope JEOL JSM6700F and atomic-force microscope Solver Pro

NT MDT. An image of the surface obtained using the electron microscope is shown in Fig. 2. The value of root-mean-square roughness of the working surface, obtained with the help of the atomic-force microscope, is RMS = 500 nm.

Distilled deionized nano-filtered water at the room temperature was used as the working liquid. The droplet diameter and height of droplet levitation were measured using the images obtained by the high-speed camera FASTCAM SA1.1 (5600 frames per second at resolution of 1024x1024 pixels). Optical resolution of the system was 0.78 μm/pixel.

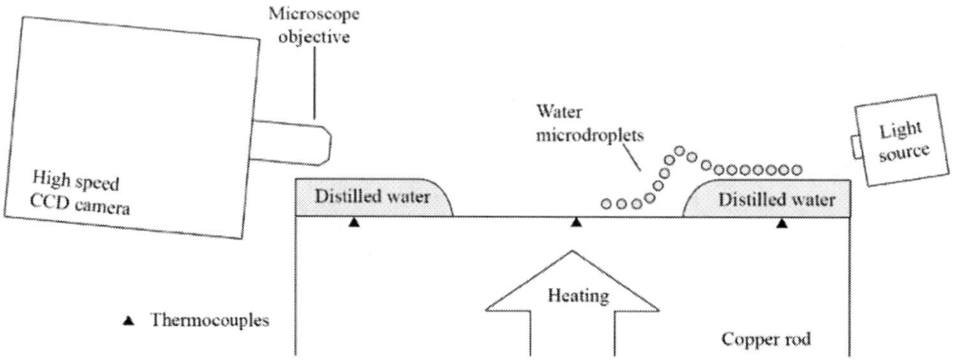

Fig. 1. Schematic of the experimental setup.

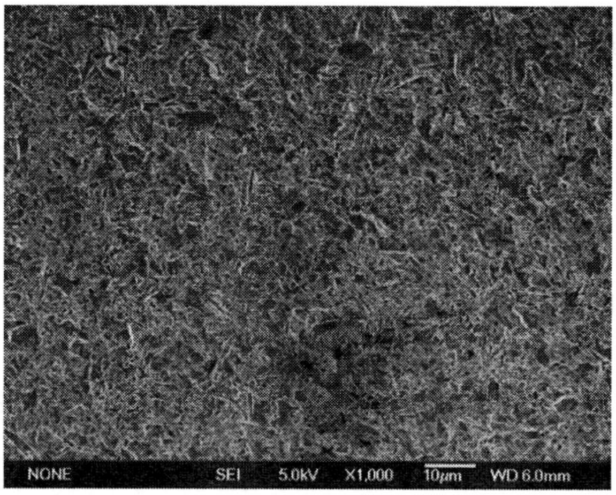

Fig. 2. The working surface image obtained using the scanning electron microscope JEOL JSM6700F.

3 Investigation results

Under intensive heating of a horizontal liquid film, a monolayer of microdroplets with ordered hexagonal structure is formed, levitating at a certain height above the free surface of liquid [11, 12]. If a dry spot is formed in the film, the droplets will "roll down" to the area of the dry spot. The picture of droplets moving towards the dry spot area and flying over the contact line is shown in Fig. 3. Possible scenarios of microdroplet flight are shown in Fig. 4. With the help of high-speed shooting, three possible scenarios of droplet flight over the contact line were distinguished: 1 – flight with a change in the height near the contact line; 2 – flight with a circular trajectory; 3 – levitating at a fixed location. All three cases are observed within the temperature range from 75°C to 100°C, however, the scenario No 1 is much more common. In the case of "normal" flight across the contact line, the droplet changes the height of levitation several times, when approaching the contact line. This means that in this area there is more intense evaporation, which throws up a droplet. Thus, we can conclude that evaporation intensity in the contact line area can be several times higher than evaporation intensity of liquid at a distance from the contact line. Droplet circular trajectory near the contact line (scenario No 2) can be connected with the vortex flow of the air-vapor mixture formed near the contact line. The most unlikely scenario is droplet hanging near the contact line (scenario No 3); in this case, the force of dynamic vapor pressure from the contact line balances the gravity force acting on the droplet.

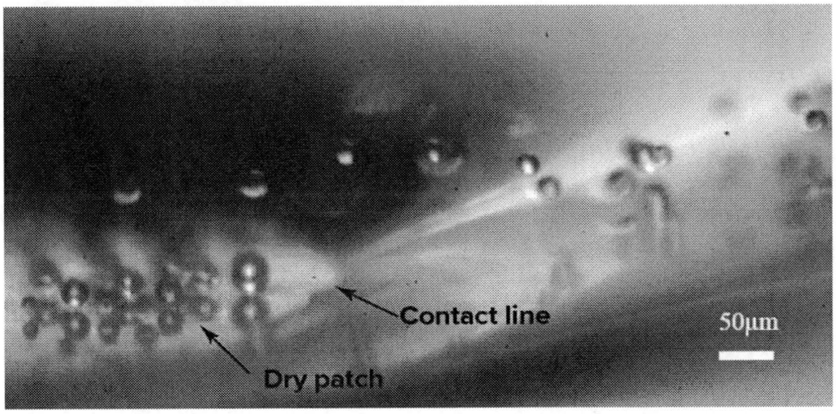

Fig. 3. The flight of liquid microdroplets across the contact line to the dry spot zone. Substrate temperature is 79°C.

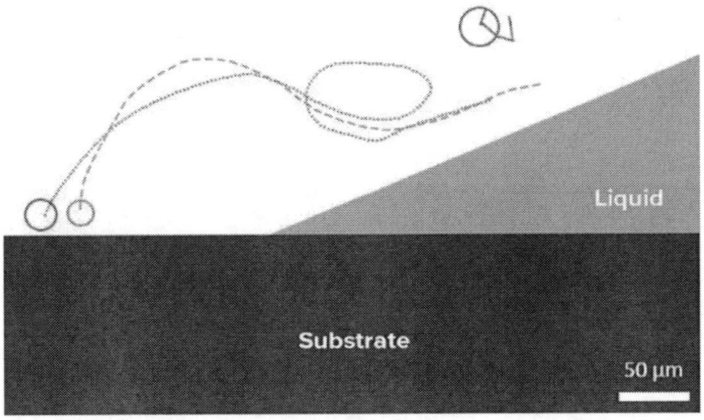

Fig. 4. Scenarios of microdroplet flight across the contact line: blue dashed line – with a change in the height near the contact line (substrate temperature is 79°C); red dotted line – with circular trajectory (substrate temperature is 81°C); green line – with levitation at a fixed location (substrate temperature is 97°C). Droplet size is in the scale.

4 Conclusions

Interaction of liquid microdroplets with the size of about 10 μm with air-vapor flow near the contact line was studied experimentally. It is shown that:

- transition of microdroplets from the wetted to the dry surface is accompanied by a change in the height of droplet levitation near the contact line, which proves the existence of the zone with intense evaporation near the contact line;
- three scenarios of microdroplet flight across the contact line are possible.

The study was financially supported by the grant of Russian Science Foundation (agreement No. 14-19-01755).

References

1. J. Kim, Int. J. Heat Fluid Flow **28**, 753 (2007)
2. A. Chatterjee, J. L. Plawsky, and P. Wayner Jr., Adv. Colloid Interface Sci. **168**, 40 (2011)
3. Gatapova E.Ya., Kabov O.A., Microgravity Science and Technology **XIX**-3/4, 132 (2007)
4. Zaitsev D.V., Rodionov D.A. and Kabov O.A., Technical Physics letters **35**, 680 (2009)
5. Kabov O.A. and Zaitsev D.V., Multiphase Science and Technology **21**, 249 (2009)
6. E.Ya. Gatapova, A.A. Semenov, D.V. Zaitsev, O.A. Kabov, Colloids and Surfaces A: Physicochemical and Engineering Aspects **441**, 776 (2014)
7. A.A. Semenov, D.V. Feoktistov, D.V. Zaitsev, G.V. Kuznetsov, O.A. Kabov, Thermophysics and Aeromechanics **22**, 771 (2015)
8. G. V. Kuznetsov, D. V. Feoktistov, and E. G. Orlova. Journal of Engineering Physics and Thermophysics **89**, 317 (2016)
9. Glushkov D.O., Syrodoy S.V., Zakharevich A.V., Strizhak P.A., Fuel Processing Technology **148**, 224 (2016)
10. Glushkov D.O., Legros J.-C., Strizhak P.A. Zakharevich A.V., Fuel **175**, 105 (2016)
11. A.A. Fedorets, I.V. Marchuk, O.A. Kabov, Tech. Phys. Lett. **37**, 2 (2011)
12. A.A. Fedorets, I.V. Marchuk, P.A. Stryzhak, O.A. Kabov, Thermophysics and Aeromechanics **22**, 4 (2015)

EPJ Web of Conferences **159**, 00023 (2017) DOI: 10.1051/epjconf/201715900023
AVTFG2016

Effect of heat conductivity and thickness of substrate on thermocapillary rupture of locally heated horizontal liquid film

Dmitry Kochkin[1,2,*], *Veronica* Sulyaeva[3]

[1]Kutateladze Institute of Thermophysics SB RAS, 630090 Novosibirsk, Russia
[2]Novosibirsk State Technical University, 630073 Novosibirsk, Russia
[3]Nikolaev Institute of Inorganic Chemistry SB RAS, 630090 Novosibirsk, Russia

Abstract. Experimental results on thermocapillary rupture of a horizontal locally heated water film on the removable substrates of different thicknesses, made of materials with different heat conductivity, are presented in this paper. The studies have shown that with an increase in substrate heat conductivity and thickness, the critical heat flux, required for rupture of the film, increases. The heat flux redistributes over the copper and brass substrates; at that, the local character of heat flux is lost, and rupture does not occur.

1 Introduction

The devices, which use liquid film flow, are widely used in industry. Thin liquid films are promising for cooling the equipment with high local heat generation, in particular, high-performance electronics [1-3]. Film cooling systems are effective because they provide high intensity of heat and mass transfer at relatively low coolant flow rates. Reducing the film thickness allows intensification of heat and mass transfer, but thin films are inclined to rupture, which causes a rapid increase in temperature of the cooled section and its breakdown.

The thermocapillary rupture of the locally heated liquid film flowing down under the action of gravity for the first time was studied in [4-6]. The influence of substrate wettability and liquid properties on thermocapillary rupture of the falling films was investigated in [7-9]. The effect of substrate structuring on the break of the falling liquid films was studied in [10]. The thermocapillary rupture of a horizontal liquid layer for the first time was investigated in [11, 12]. Studies of [7, 13] have shown that the critical heat flux for the horizontal liquid films can be several times higher than the heat flux for the films falling under the action of gravity. This explains the relevance of our research.

The main aim of our study is to determine the dependence of the critical heat flux on the thickness of horizontal liquid film on the substrates of different materials and different thicknesses.

* Corresponding author: kochkin1995@mail.ru

DOI: 10.1051/epjconf/201715900023

2 Experimental setup

The experiments were carried out on locally heated substrates of copper, brass, stainless steel, and glass with the diameter of 50.8 mm and thickness of 1 and 3 mm. The substrate was placed on the textolite base with the copper rod of 12.7 mm in diameter (Fig. 1). A ceramic heater was connected to the rod from the bottom. The heat flux was determined by electrical power released on the ceramic heater. Distilled water with the initial temperature of 25°C was used as the working liquid. The liquid was fed to the substrate using a syringe pump. The substrate was cooled along the edges with the help of a cooling circuit (Fig. 1). The water temperature in the cooling circuit was either 10 or 25°C.

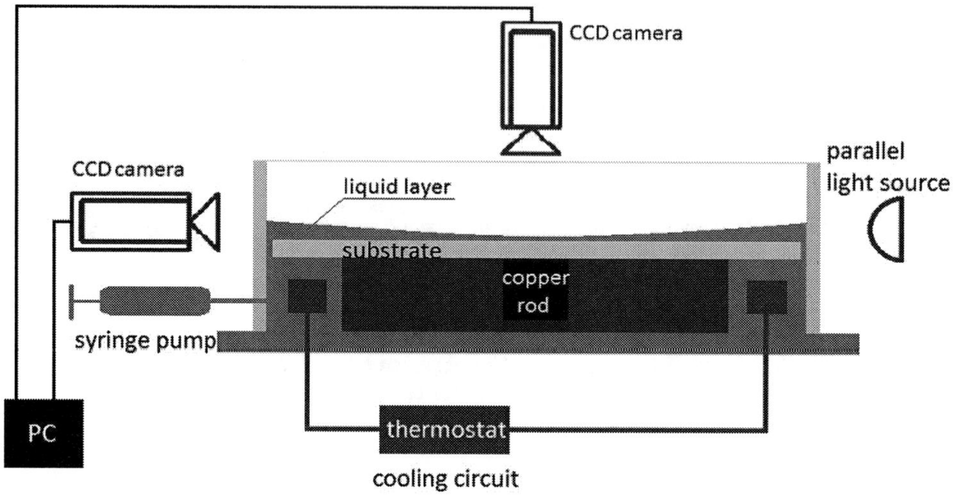

Fig. 1. Layout of setup (power source of the heater is not shown).

To determine the film thickness, we applied the optical method using the CCD DMK 23GP031 digital camera and source of parallel light MI-150 (Fig. 1). Another camera was placed above the test section to observe film rupture dynamics.

The morphology of the surface of the glass substrate was analyzed using a scanning electron microscope (HITACHI S3400N) and atomic force microscope (Solver Pro NT MDT), Fig. 2. The root mean square (RMS) surface roughness was found to be 5.417 nm, with the average roughness of 4.057 nm.

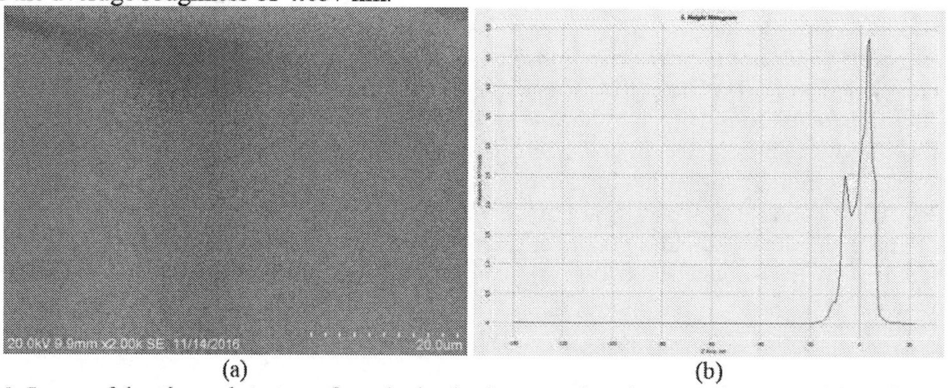

Fig. 2. Image of the glass substrate surface obtained using scanning electron microscope (a), and height histogram obtained using atomic force microscope (b).

3 Results of experiment

Dependence of the critical heat flux on initial liquid film thickness on stainless steel substrates with the thicknesses of 1 and 3 mm, and glass substrate of 3-mm thickness is shown in Fig. 3. It is seen that on the stainless steel substrates of 1-mm thickness, the critical heat flux is several times lower than on the substrates of the same material with the thickness of 3 mm. This suggests that increasing the substrate thickness, we increase its thermal resistance, therefore, the temperature on the substrate surface becomes lower, the local heat flux density also decreases due to increased heat spreading into the substrate, and the critical heat flux increases due to the effect of these two factors. According to comparison of dependencies for the 3-mm substrates of stainless steel and glass, the critical heat flux on the stainless steel substrate is several times higher than on the glass substrate. Apparently, this is caused by the fact that heat conductivity of stainless steel is ~ 20 times higher than heat conductivity of glass, and heat flow spreading into the substrate is much larger; at that, heating localization decreases and the critical heat flux becomes higher.

In Fig. 3, our data are compared with results of [7] on rupture of a horizontal water layer on the stainless steel substrate with an in-built copper heater of 1x1 cm, set into a steel plate (without removable substrates). The difference between our data and data of [7] can be, probably, caused by the fact that due to heat spreading, the heat flux density on the working surface in experiment of [7] can be either lower than in our experiment (for the 3-mm glass substrate and 1-mm steel substrate) or higher (for the 3-mm steel substrate).

In the experiments on the copper and brass substrates of 1-mm and 3-mm thickness, the film rupture in the range of heat fluxes of up to 60 W was not observed. Due to high heat conductivity of these materials, the local character of heating was lost because the heat flux was redistributed over the substrate and discharged into the cooling circuit.

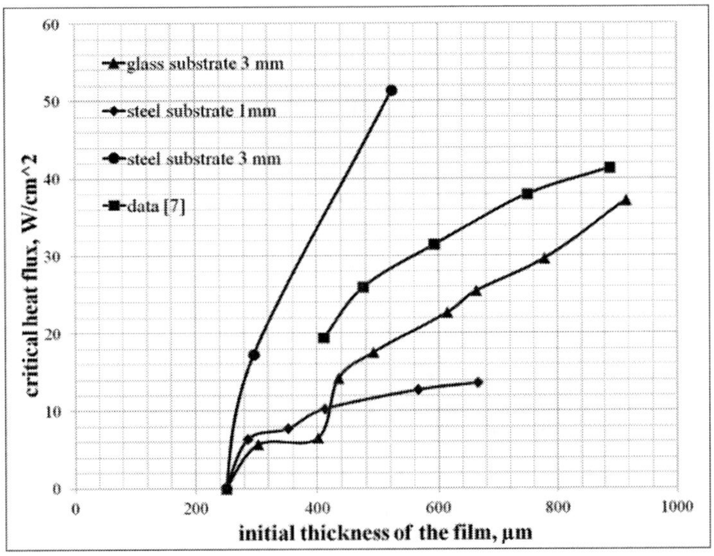

Fig. 3. Dependence of the critical heat flux on initial film thickness.

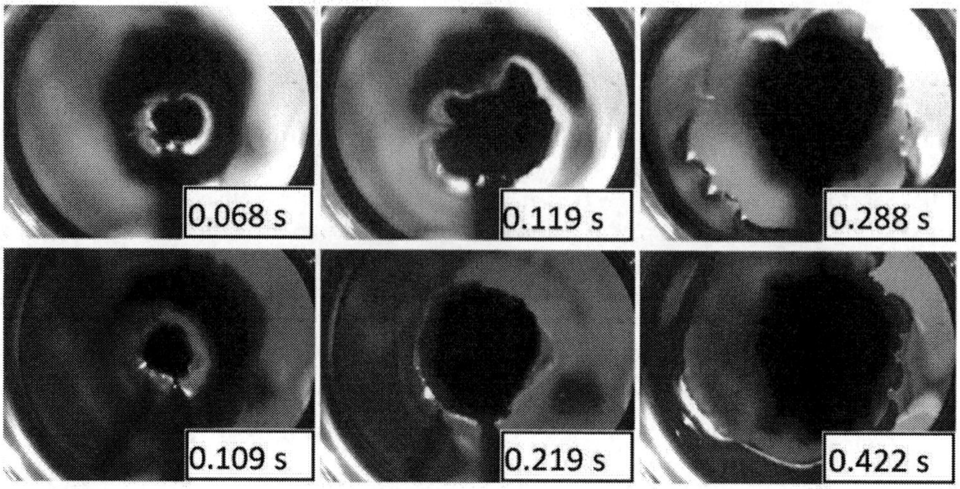

Fig. 4. Dynamics of dry spot spreading. The upper and lower series of pictures – film with the thickness of 0.4 mm (q = 6.86 W/cm^2) and 0.83 mm (q = 18.6 W/cm^2), respectively. Time from the moment of film rupture onset is indicated.

Dynamics of dry spot spreading (on the glass substrate of 3-mm thickness) on the film with a thickness of 0.4 mm at q = 6.86 W/cm^2 and film with thickness of 0.83 mm at q = 18.6 W/cm^2 are shown in Fig. 4. It is seen that a dry spot in both cases is formed in the substrate center above the heater and it spreads throughout the substrate. As can be seen according to times, indicated in Fig. 4, the velocity of dry spot spreading for the film thickness of 0.4 mm is slightly higher than for the film with thickness of 0.83 mm.

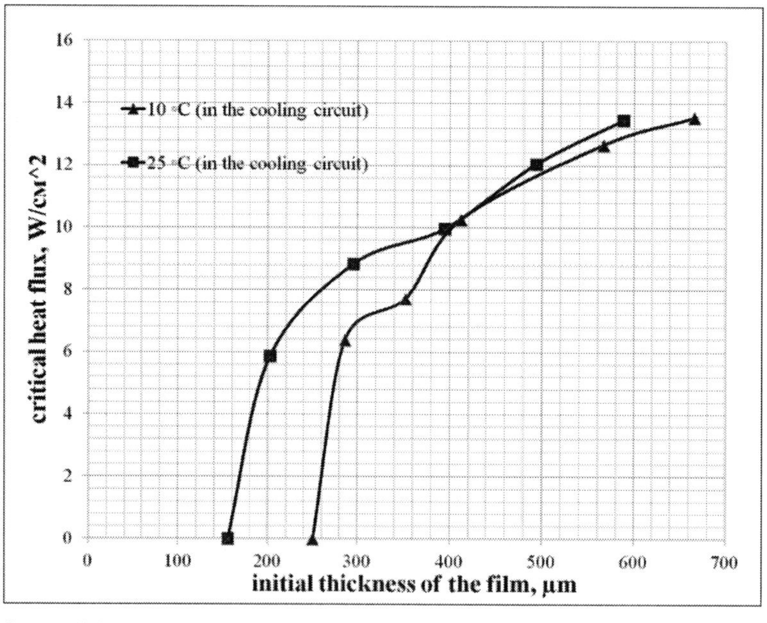

Fig. 5. Dependence of the critical heat flux on initial film thickness at different temperatures in the cooling circuit.

Dependence of the critical heat flux on initial film thickness on the stainless steel substrates of 1-mm thickness is shown in Fig. 5 for different water temperatures in the cooling circuit (10°C and 25°C). At water temperature of 10°C in the cooling circuit, the film is broken at the thickness of ~ 250 μm without heater switching on. This is due to the influence of thermocapillary forces caused by the difference between the temperature of water in the cooling circuit and air in the room. Increasing the temperature of water in the cooling circuit to the temperature of air in the room (25°C) eliminates thermocapillary effect with breakdown of films with the thickness higher than ~200 μm. As seen from Fig. 5, for liquid films with initial thickness higher than ~ 400 μm the water temperature in the cooling circuit stops to significantly influence on the critical heat flux.

4 Conclusion

It was found out that with an increase in heat conductivity and substrate thickness, the critical heat flux required for film rupture increases. When using the substrates made of materials with high heat conductivity (copper, brass), the heat flux is redistributed over the substrate, and heat is removed to the cooling circuit; at that, there is no film rupture.

This work was supported by the Ministry of Education and Science of Russia (Agreement No 14.604.21.0053, project identifier RFMEFI60414X0053).

References

1. A. Bar-Cohen, P. Wang. ASME. J. Heat Transfer, **134**, 051017 (2012)
2. E.Ya. Gatapova, O.A. Kabov, Microgravity Science and Technology, **19**-3/4, 132, (2007)
3. D.V. Zaitsev, D.A. Rodionov, O.A. Kabov, Technical Physics L etters, **35**, 680 (2009)
4. O.A. Kabov, Thermophysics and Aerodynamics, **7**, 513 (2000)
5. E.A. Chinnov, O.A. Kabov, A.V. Muzykantov and D.V. Zaitsev, Intern. Journal of Heat and Technology, **19**, 31 (2001)
6. D.V. Zaitsev, D.A. Rodionov, O.A. Kabov, Microgravity science and technology, **19**-3/4, 100 (2007)
7. D. V. Zaitsev, D. P. Kirichenko, and O. A. Kabov, Technical Physics Letters, **41**, 551 (2015)
8. D. V. Zaitsev, O. A. Kabov, V. V. Cheverda, and N. S. Bufetov, High Temperature, **42**, 450 (2004)
9. D.V. Zaitsev, A.A. Semenov, and O.A. Kabov, Thermophysics and Aeromechanics, **23**, 625 (2016)
10. D.V. Zaitsev, A.M. Lozano, H. Auracher, O.A. Kabov, Microgravity science and technology, **19**-3/4, 71 (2007)
11. A. Orell, S.G. Bankoff, Int. J. Heat and Mass Transfer, **14**, 1835 (1971)
12. J. P. Burelbach, S.G. Bankoff, S.H. Davis, Phys. Fluids **A 2**, 321 (1990)
13. D.V. Zaitsev, O.A. Kabov, Microgravity science and technology, **19**-3/4, 174 (2007)

On the development of controlled stationary and travelling disturbances in the supersonic boundary layer of a swept wing

Gleb Kolosov[1,2,*], *Alexandra* Panina[1] , and *Alexander* Kosinov[1,2]

[1]Khristianovich Institute of Theoretical and Applied Mechanics, SB RAS, Institutskaya Str. 4/1, Novosibirsk 630090, Russia
[2]Novosibirsk State University, Novosibirsk 630090, Russia

Abstract. Experimental data on the evolution of controlled stationary and travelling disturbances in a 3D supersonic boundary layer, over a 45° swept-wing at Mach number 2.0, is presented. Travelling artificial disturbances were introduced in the boundary layer by periodical glow discharge, at a frequency of 20 kHz. Stationary disturbances were acquired by setting the roughness elements on the surface of the model. Spatial-temporal and spectral-wave characteristics of the wave train at the frequency 20 kHz, in the linear region of development, were obtained. It was found that the periodic modulation of mean flow can lead to the stabilization of unstable travelling disturbances in the supersonic boundary layer of a swept wing. These experiments have investigated the viability of using roughness elements to control laminar-turbulent transition.

1 Introduction

The transition to turbulence in swept-wing boundary layers has been the focus of considerable attention by researchers for more than 30 years [1, 2]. Such research is of great practical relevance: an ability to control the transition process on airplane wings would lead to significant fuel savings in future supersonic transports. Unfortunately, a number of difficulties arise in studying the process of laminar–turbulent transition. An inherent property of swept-wing boundary layers is the presence of crossflow, which leads to the appearance of a new type of instability. Modern theory on transition mechanisms in three-dimensional (3D) boundary layers gives four basic types of flow instabilities: (1) attachment line boundary layer instability on the swept leading edge; (2) crossflow instability; (3) centrifugal instability; and (4) possible instability of Tollmien–Schlichting waves. Their relative roles in laminar-turbulent transition strongly depends upon a variety of factors, such as the level of free stream disturbances, surface quality, Mach number, model geometry etc.

Most of the theoretical and experimental results on stability and transition of the 3D boundary layers have been obtained for subsonic flow. An important feature in such boundary layers is that flow becomes primarily unstable to both stationary vortices and

* Corresponding author: kolosov@itam.nsc.ru

oblique travelling waves. Depending on the environmental conditions, two main scenarios for transition are possible [1]. One scenario concerns the domination of stationary disturbances (crossflow vortices), and can occur if the free-stream turbulence level is small [3]. In this case, stationary vortices, excited by surface roughness, are developed linearly and eventually saturate and become subject to high-frequency secondary instability. This results in laminar-flow breakdown [4, 5]. An alternative scenario is that involving the domination of travelling disturbances, excited by the vortical content of the free stream turbulence. Observations of travelling instability modes are usually performed under controlled disturbance conditions, when wave packets are generated by a localized pulse [6]. It is important to note that a paper authored by Wasserman and Kloker [7], is the only paper to date to scrutinize crossflow breakdown by travelling modes.

A supersonic boundary layer is far more complex than an incompressible boundary layer and has been thoroughly investigated both experimentally and numerically. Several studies have been devoted to the evolution of travelling disturbances in 3D supersonic boundary layers, and it has been suggested that travelling disturbances are what dominates the transition [8, 9]. In order to excite controlled travelling disturbances in a supersonic boundary layer, a glow discharge technique was developed at ITAM SB RAS [10]. Such a technique allows one to observe the linear development of a wave train with frequencies of 10 KHz and 20 kHz, on a swept wing at Mach number M = 2.0. [11, 12]. Moreover, at ITAM SB RAS, the technique of pulsed excitation of a wave packet with a broad disturbance spectrum has been implemented to obtain wave characteristics for a wide range of frequencies [13].

Later stages of the transition are characterized by nonlinearity of the disturbance flow field, for example the nonlinear interaction between stationary and travelling waves. Using roughness elements on the model allows us to obtain stationary waves with specified characteristics. The effect of distributed roughness on transition control in supersonic boundary layers on swept-wing models was studied in [14, 15]. It was shown that the periodic surface roughness near the leading edge may stabilize the flow or move the laminar-turbulent transition closer to the leading edge of the model. Note that the flow stabilization was caused by the influence of distributed roughness on the stationary crossflow instability mode.

The goal of this work is (1) to use periodical discharge to obtain wave characteristics of the most unstable disturbances in the swept-wing supersonic boundary layer at M = 2.0 and (2) to use distributed roughness elements to investigate the effects of transverse mean flow modulation on the generation and development of a wave packet with the same income flow parameters.

2 Experimental setup and data processing

The experiments were performed in a T-325 low-noise supersonic wind tunnel of ITAM SB RAS, at Mach number M = 2.0 and unit Reynolds number $Re_1 = 5\times10^6$ m^{-1}.

Disturbances in the flow were measured with a constant-temperature hot-wire anemometer. Probes with tungsten wire, 10 microns in diameter and 1.5 mm long, were used. The value of overheat loading of the wire was set equal to 0.8, and this resulted in 90% of the measured disturbances corresponding to mass-flow fluctuations. To move the probe along the x, y, and z coordinates, traverse gear was used. The accuracy of determination of the probe position in the x- and z-coordinates was 0.1 mm, and 0.01 mm for y-coordinate.

A swept-wing model was used in the experiments. The model had a sweep angle of 45° and a slightly blunted leading edge of radius 0.2 mm. The swept wing had a 2.6% lens-shape airfoil on the upper side and a flat surface on the bottom side, with a maximum thickness of 12 mm. The curvature radius of the upper side of the model was approximately 4000 mm. A source of localized artificial disturbances was built into the model, similar to that described by Kosinov et al. [10]. Controlled pulsations were generated by using high-frequency glow discharge in a chamber and were injected into the boundary layer through an aperture 0.4 mm in diameter, in the upper surface of the model. The source aperture was located at $x_0 = 56.6 \pm 0.3$ mm from the leading edge on the line of symmetry of the model. The pulsation measurements were synchronized using glow discharge, which was ignited with a fundamental frequency of 20 kHz. AC and DC signals from CTA were written to the PC using a 12-bit ADC with sampling rate 750 kHz and by DC voltmeter correspondingly. Four time traces, each of length 65536 points, were measured and written to file for each position of the hot wire.

Transverse modulation of the mean flow in the boundary layer was created using adhesive tape stickers. Roughness elements were located on the top surface of the model, parallel to the leading edge. The stickers were diamond in shape and their size was 3×3 mm. Two variants of the experiment were conducted, corresponding to spacing between roughness elements of 6 mm and 12 mm respectively. The roughness elements each had a thickness of 60 mm. The experimental set-up is presented in Figure 1. The spanwise measurements were made at a fixed normal distance from the model surface and at $y/\delta \approx$ const for each tranverse.

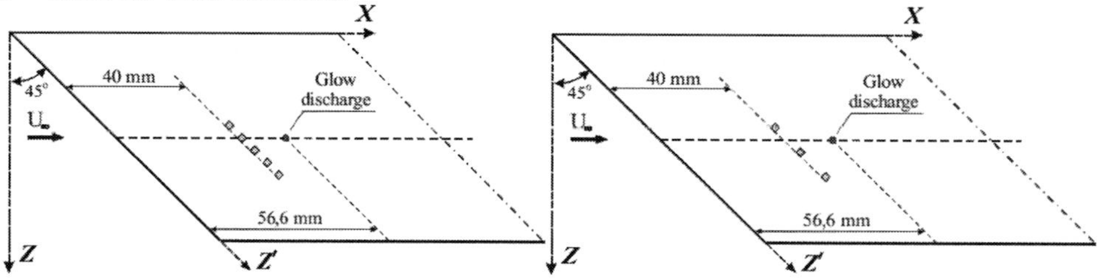

Fig. 1. Swept wing and set-up of the roughness elements; spacing of 6mm between roughness elements (left), spacing of 12mm between roughness elements (right)

The fluctuation time traces obtained, normalized by the mean voltage CTA output at each spatial measurement point, were transformed to mass flow fluctuations according to the relation $m(x',z',t) = e'(x',z',t)/(Q{\cdot}E(x',z'))$. Here $e'(x',z',t)$ is the fluctuation signal, $E(x', z')$ is the mean voltage from the CTA output, and Q is the sensitivity of the probe to the mass flow fluctuations. Note that the probe measures the x-component of the mass flow fluctuation. The frequency-wave disturbance spectra at a fixed x'-coordinate were calculated using the discrete Fourier transform of the following form:

$$m_{f\beta'} = \frac{\sqrt{2}}{T\delta_n} \sum_{j,k} m(z'_j, t_k) \exp[i(\beta' z'_j - \omega t_k)] \Delta t \Delta z'_j, \tag{1}$$

where T is the time-length of realization, β' is the spanwise component of the wave vector, and δ_n (= 1 mm) is the boundary layer thickness scale for normalizing the spectra in β'. The Fourier transform in time was performed using the Hanning window. After performing the Fourier transform, the amplitude and phase of the controlled disturbances were calculating using:

$$A_{f\beta'} = \sqrt{\mathrm{Re}^2(m_{f\beta'}) + \mathrm{Im}^2(m_{f\beta'})}, \tag{2}$$

$$\Phi_{f\beta'} = \arctan[-\mathrm{Im}(m_{f\beta'}) / \mathrm{Re}(m_{f\beta'})]. \tag{3}$$

3 Results

Results concerning the development of natural disturbances are presented in detail in [11, 12]. Note that the source of the controlled disturbances is located in the linear region of the natural fluctuation development and, moreover, the frequency of the artificial pulsations (f = 20 kHz) falls within the frequency range for most unstable natural disturbances.

The controlled disturbance development on a smooth swept wing was measured in four different sections by means of displacing the anemometer probe along the z'-coordinate, that is parallel to the leading edge of the model at x = 20, 30, 40 and 50 mm from the discharge. The amplitude distributions of perturbations with frequency f = 20 kHz in the spanwise direction are shown in Figure 2. One can see that in some places inside the wave packet, the amplitude grows downstream, whereas the main amplitude maximum is slowly displaced in the positive direction of the z'-coordinate. Recall that the asymmetry of the amplitude distributions along the spanwise z'-coordinate, relative to the position of the discharge (z' = 0 mm), are associated with the presence of crossflow on the swept wing. The wave train is spread along the leading edge of the model in the direction opposite to the crossflow, and this is consistent with previously obtained results for the same Mach number (2.0) [11, 12].

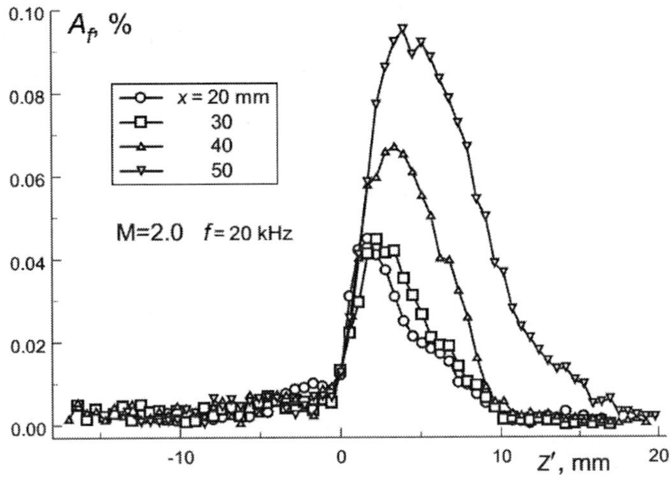

Fig. 2. Amplitude distributions of disturbances in the spanwise direction

Figure 3 shows the amplitude β'-spectra of the controlled disturbances at frequency 20 kHz. Unlike in the flat plate case, the swept-wing β'-spectra are asymmetric. The amplitude growth is observed in a wide range of panwise wavenumbers, but the amplitude maximum is located at β' = 1.1 rad/mm. This value corresponds to the linear peak in downstream wave train development. It should be noted that the traverses being considered don't belong to the nonlinear region of disturbance development. Thus one cannot expect the emergence of additional peaks in β'-spectra other than the linear one. Growth over a wider range of β' is only observed for the cross section x = 50 mm.

The results in the transverse-modulated supersonic boundary layer on a swept wing will now be discussed. Roughness element thickness was constant in all experiments and was 60 microns. Spacing between roughness elements was varied. All measurements were taken at the location of the maximum pulsations through the boundary layer, and at one of the edges of the tranverses where there was no mean flow modulation from the stickers.

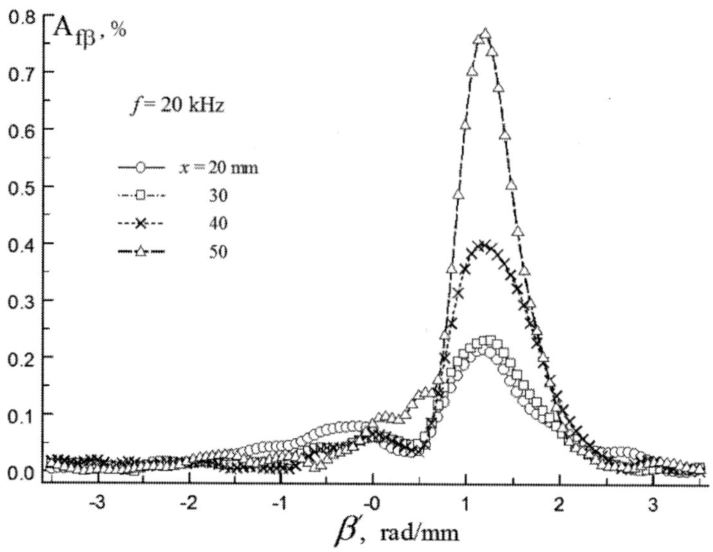

Fig. 3. Amplitude β'-spectra of disturbances

Figure 4a shows the comparison of the mean flow modulation across the model in the initial section ($x = 14.14$ mm from the discharge) for the cases of a smooth wing and swept wing with periodic stickers, with spacing between roughness elements of 6 mm and 12 mm respectively. It is clear from Figure 4a that the period of mean flow modulation corresponds to the period of roughness on the model surface. For the case of 6 mm spacing of roughness elements it was possible, in some region, to produce flow modulation resembling the sine function. Note that the disturbance source aperture on the model surface introduces an additional mean flow distortion, which can be observed near $z' \approx 2$ mm (see Figure 4a).

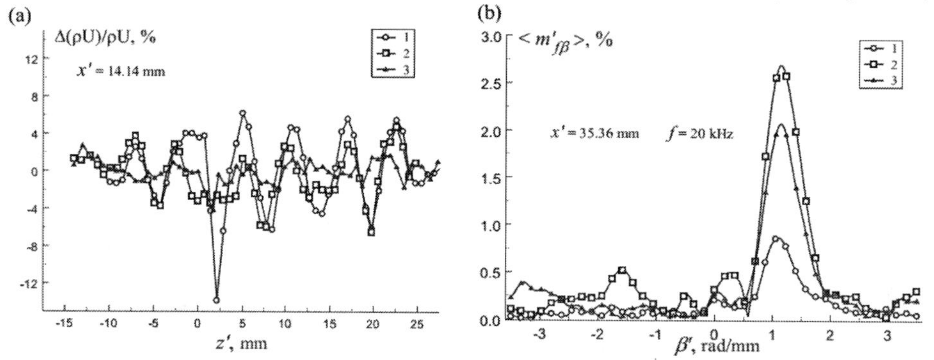

Fig. 4. (a) Mass flow variation; (b) amplitude β'-spectra. 1 – spacing of 6mm between roughness elements, 2 – spacing of 12mm between roughness elements, 3 – smooth surface

The amplitude β'-spectra of disturbances for the fundamental frequency ($f = 20$ kHz) at $x = 35.4$ mm is shown in Figure 4b. The results show that, in all cross sections, the perturbations showing most growth correspond to $\beta' \approx 1$ rad/mm wave number. This corresponds to the linear peak and it is in agreement with previously obtained results (see Figure 3). Note that the electrical power of the disturbance source was the same for all experiments. However, the data obtained shows that the relative receptivity of the boundary layer to nonstationary disturbances were significantly different. Also note that, in addition to the main linear peak, we can see a number of additional peaks. Their appearance cannot be explained by the linear theory, since they are the result of nonlinear wave interactions caused by the introduction of artificial stationary disturbances.

Figure 5 shows the relationship between transfer factor dependency and transverse wave number β', for the frequency 20 kHz. The results presented in Figure 5a show that at the initial stage (for the case of a swept wing with roughness elements separated by 12 mm) the growth rate is greatest in the wide region, near $\beta' = 1$ rad/mm. However, at $x = 35.36$ mm from the disturbance source aperture, the growth rate of disturbances for the smooth wing is greater by a factor of approximately 1.5 times, compared to the case of the swept wing (for both values of roughness element separation) (see Figure 5b). Due to this experimental conclusion, it can be concluded that using microroughness enables the possibility of stabilizing disturbance development at the fundamental frequency in the boundary layer of a swept wing.

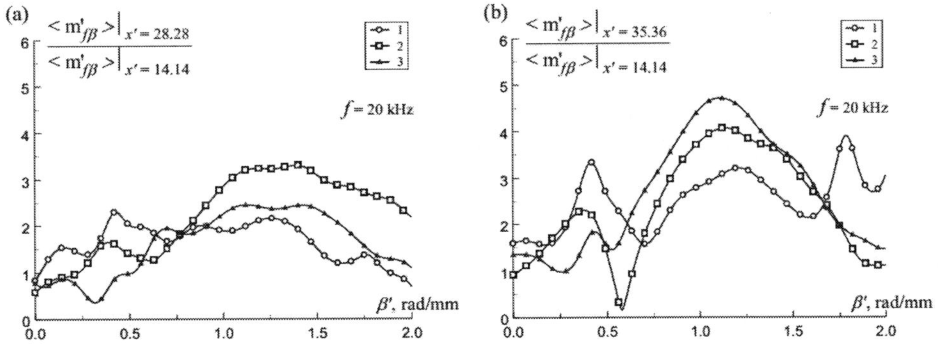

Fig. 5. Comparison of the transfer dependences over β' at the 20 kHz frequency. 1 – spacing of 6 mm between roughness elements, 2 – spacing of 12mm between roughness elements, 3 – smooth surface

4 Conclusion

An experimental study was conducted on the development of controlled travelling and stationary disturbances in a supersonic boundary layer of swept wing. This was performed at Mach 2.0 and unit Reynolds number $Re_1 = 5\times10^6$ m^{-1}. On the smooth swept wing, the region corresponding to linear stage of nonstationary disturbance evolution was examined. The amplitude distributions were shown to be spatially asymmetric, which can be attributed to the presence of a crossflow in 3D boundary layer. It was found that the amplitude maximum is located at $\beta' = 1.1$ rad/mm for 20 kHz.

In experiments with roughness elements on swept wing it was shown that the variation of mean flow inside the boundary layer emerged that corresponds to the introduced stationary wave. It was determined that the roughness elements and their relative position can affect the downstream development of travelling pulsations. The experiments showed that the periodic modulation of mean flow by roughness elements can lead to the stabilization of travelling disturbances in the supersonic boundary layer of a swept wing. Therefore it is possible to utilize weak flow inhomogeneity, by using surface microroughness, to control supersonic boundary layer transition.

The work was supported by the Russian Foundation for Basic Research (Grant No. 16-31-00290).

References

1. H. Bippes, Prog. Aerospace Sci. **35**, 363 (1999)
2. W. S. Saric, H. L. Reed, E. B. White, Annu. Rev. Fluid Mech. **35**, 413 (2003)

3. M.R. Malik, F. Li, C.-L. Chang, J. Fluid Mech. **268**, 1 (1994)
4. D.I.A. Poll, J. Fluid Mech. **150**, 329 (1985)
5. E.B. White, W.S. Saric, J. Fluid Mech. **525**, 275 (2005)
6. V.R. Gaponenko, A.V. Ivanov, Y.S. Kachanov, Thermophys. Aeromechanics **2**, 312 (1995)
7. P. Wasserman, M. Kloker, J. Fluid Mech. **483**, 67 (2003)
8. L.N. Cattafesta III, V. Iyer, J.A. Masad, R.A. King, J.R. Dagenhart, AIAA J. **33**, 2032 (1995)
9. F. Li, M. Choudhari, Theor. Comput. Fluid Dyn. **25**, 65 (2011)
10. A.D. Kosinov, A.A. Maslov, S.G. Shevelkov, J. Fluid Mech. **219**, 621 (1990)
11. Yu.G. Ermolaev, G.L. Kolosov , A.D. Kosinov, N.V. Semenov, Fluid Dynamics **49**, 188 (2014)
12. A.D. Kosinov, G.L. Kolosov, N.V. Semionov, Yu.G. Yermolaev, Phys. Fluids **28**, 064101 (2016)
13. A.A. Yatskikh, Yu.G. Ermolaev, A.D. Kosinov, N.V. Semionov, Thermophys. Aeromechanics **22**, 17 (2015)
14. W.S. Saric, H.L. Reed, AIAA Paper **2002-0147**, 10 (2002)
15. N.V. Semionov, A.D. Kosinov, Thermophys. Aeromechanics **14**, 337 (2007)

EPJ Web of Conferences **159**, 00025 (2017) DOI: 10.1051/epjconf/201715900025
AVTFG2016

Experimental study of metastable liquids decay during depressurization in cylindrical channel

Igor Kozulin[1,2,*], *German* Bartkus[1,2] , and *Sergey* Dimov[1]

[1]SB RAS, Institute of Thermophysics, 630090 pr. Akademika Lavrentjeva 1, Novosibirsk, Russia
[2]Novosibirsk State University, 630090 st. Pirogova 1, Novosibirsk, Russia

Abstract. Patterns of the nonequilibrium phase transitions during propagation of an adiabatic wave phase transition in metastable refrigerants R141b, R113 and pentane C_5H_{12} were experimentally established. Experiments were carried out in a wide range of initial temperatures and pressures in a transparent cylindrical channel. Visualization of liquids decay was performed using high-speed video camera.

1 Introduction

A research of fluid destruction at high specific energy impacts is necessary for development of the nonequilibrium phase transitions theory, as well as for applications related to developing the security systems, the energy systems, and solving the security problems during storage of liquefied gases.

A distribution of phase transition at depressurization of the cylindrical channel was studied in [1]. The phase transition in channels with different shapes was studied in the paper [2] and it was shown that the wave speed is independent from the channel diameter and its shape. This paper continues the experimental study of [3] and [4]. In these articles, the adiabatic evaporation waves were investigated during the rapid transition of Freon in the metastable state after pressure relief. An analysis of the published studies shows that the phase transition in the metastable liquid was not completely studied. A liquid vaporization mechanism was not defined on the wave front. The aim of this work is to determine the patterns of surface decay during depressurization of metastable Freon R141b, R113 and pentane C_5H_{12}. The fluids selection is caused by a significant difference in their thermal properties, including vapor density, liquid heat capacity and latent heat of vaporization. A phase transition visualization was performed using a high-speed video camera.

2 Experimental setup

The study of propagation of wave phase transition in metastable liquid was made using the experimental setup, Fig. 1.

* Corresponding author: igornt@yandex.ru

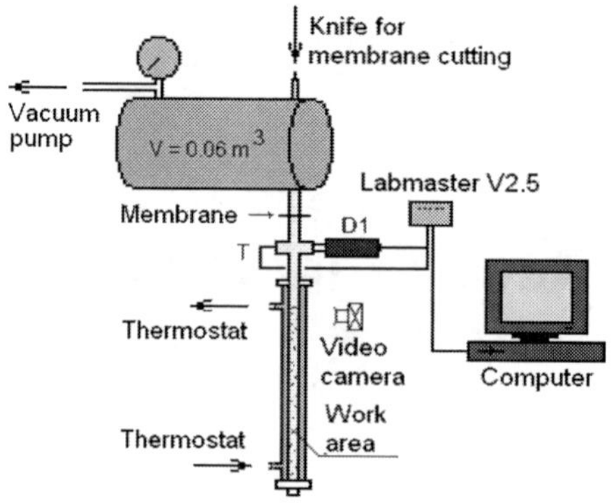

Fig. 1. The experimental setup.

The experimental setup includes a low-pressure chamber volume of 60 liters and a test section with a superheated liquid separated from chamber by a membrane. The membrane was cut by a special knife at a specified time. Refrigerants R141b, R113 and pentane C_5H_{12} were used as working fluids. The cylindrical glass channel was used as a test section. The inner diameter is 16 mm, the cannel length is 300 mm. The test section is located vertically and it has a thermostatic jacket in which water is pumped from the thermostat. The thermostat heated the test section to a determined temperature T_0.

The originality of this set-up is that that the test section is made from glass. It allows us to visualize the boiling liquid and study the detailed structure of the adiabatic wave front in the superheated liquid. The initial temperature range was varied from 35 to 80 ^{0}C and measured with an accuracy of 0.3 ^{0}C. Prior to the experiment in a low-pressure chamber with vacuum, the pump created a vacuum in a low-pressure chamber measured with an accuracy of 0.005 bar. The initial pressure range produced in the low-pressure chamber was P_{res}=0.04-0.07 bar. The pressure sensor DMP 330L with a measuring range from 0 to 16 bar is located at the top of the test section. The copper - constantan thermocouple was mounted at the top of the test section. During the experiment, the data from sensors handled from ADC Labmaster V2.5 in PC. Using high-speed video camera Optronis CR600x2, the registration of metastable liquids decay was made.

3 Results and discussion

The adiabatic phase transition initiated by the pressure relief was caused by the membrane rupture. The bubbles' microbursts are triggered on the surface layer of the superheated liquid. The metastable liquid formed pulsating waves of intense phase transition with the cellular structure interphase border after the establishment of pressure.

Fig. 2 (a) shows the dynamics of boiling refrigerant R141b at the initial temperature and pressure: T_0 = 35.4 ^{0}C, P_{res} = 0.04 bar. A video of the phase transition was carried out using a high-speed video camera (500 fps) with transmitted light. It is seen from the photos that there are the end stage of emission and stop boundaries (t=0 ms) and the stage of intensive evaporation and moving the boundaries (t=6 ms) of supersaturated vapor. This process is periodic; as a result of which, it is possible to observe the development of the interfacial waves of surface boiling. Velocity of boiling refrigerant R141b is shown in Fig. 2 (b) at initial temperature T_0=35.4 ^{0}C.

The local interface velocity of boiling was determined with the help of a video image processing program which allows determining the coordinates of the current boiling front position along the entire border between the two phases. These coordinates are averaged to obtain the current boiling front position for a given image.

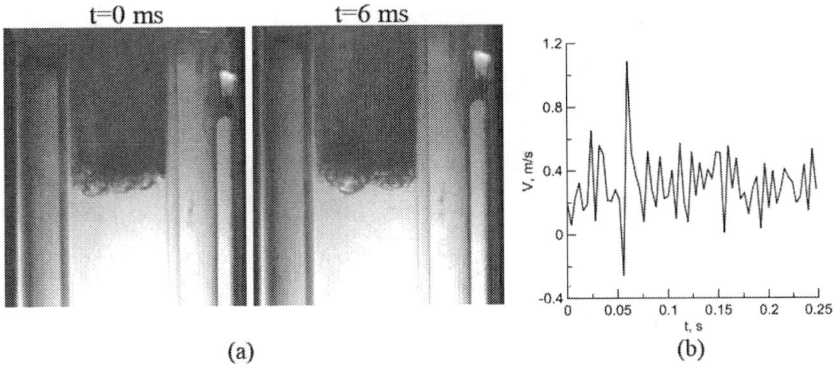

Fig. 2. The dynamics of refrigerant R141b boiling at initial temperature and pressure: $T_0 = 35.4$ °C, P_{res}=0.04 bar. (a) Visualization at 500 fps. (b) Velocity of boiling, depending from the time.

The average speed of the boiling front at T_0=35.4 °C for refrigerant R141b was 0.31 m/s. There are significant pulsations of velocities while moving an interphase surface. These pulsations are greatly exceeded its mean value. The stage of active vaporization and moving of the front changes to the stage of stopping and collapsing the front, but the average speed of the wave is constant during the whole phase transition process. The average velocity of the wave depends slightly on the pressure in the low pressure chamber and is determined by initial temperature of the liquid.

The experimental study of the metastable decay of refrigerant R113 was performed. Fig. 3 (a) shows the dynamic of the surface phase transition for refrigerant R113 at temperature and pressure: T_0 = 65.2 °C, P_{res}=0.036 bar at boiling nucleation in a surface layer. The figure shows that the evaporation front moves from top to bottom, the boiling is accompanied by high-intensity fluctuations at the interface surface between the liquid and gas phases. Fig. 3 (b) shows the velocity of interface surface of boiling refrigerant R113 at temperature T_0=65.2 °C. The average speed of boiling front is 0.32 m/s.

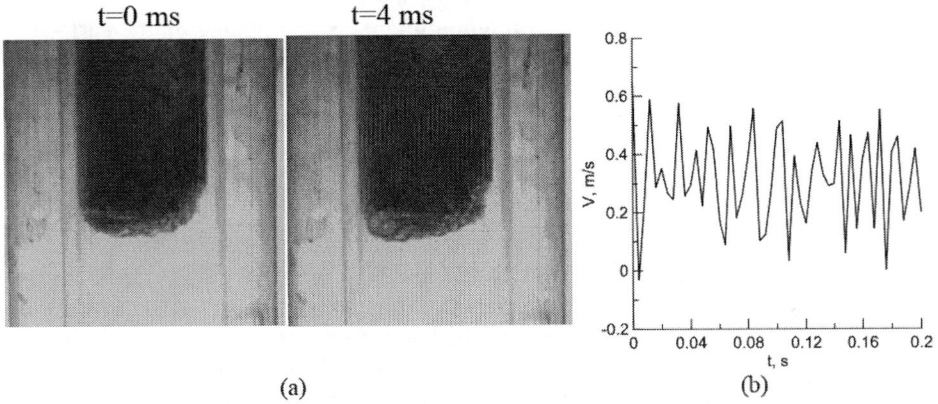

Fig. 3. The dynamics of refrigerant R113 boiling at initial temperature and pressure: $T_0 = 65.2$ °C, P_{res}=0.036 bar. (a) Visualization at 500 fps. (b) Velocity of boiling, depending from the time.

The experimental study of the metastable decay of pentane was performed. Fig. 4 (a) shows the dynamic of the surface phase transition for refrigerant R113 at temperature and pressure: T_0 = 48.55 °C, P_{res}=0.04 bar at boiling nucleation in a surface layer. Fig. 4 (b) shows velocity of the pentane wave boiling at temperature T_0=48.55 °C. The average speed of boiling is 0.52 m/s.

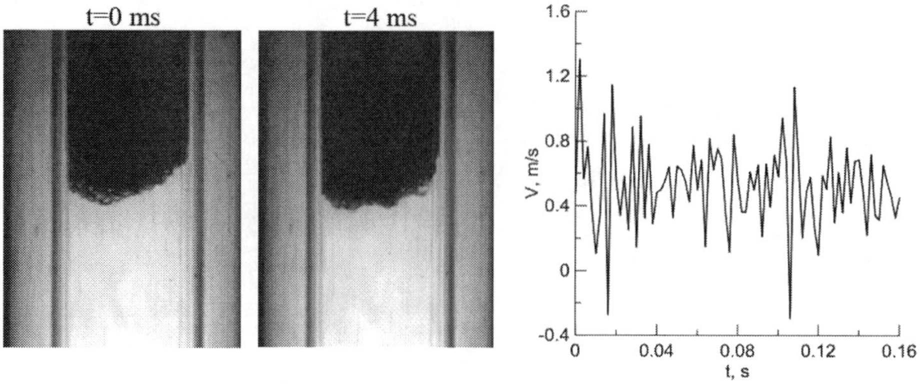

(a) (b)

Fig. 4. The dynamics of pentane boiling at initial temperature and pressure: $T_0 = 48.55\ ^0C$, $P_{res}=0.04$ bar. (a) Visualization at 500 fps. (b) Velocity of boiling, depending from the time.

A visualization of the first nucleation wave on the interfacial meniscus was made for pentane with the help of video camera Optronis CR600x2. The camera settings were 15 000 fps and the picture resolution 128x128 px, Fig. 5.

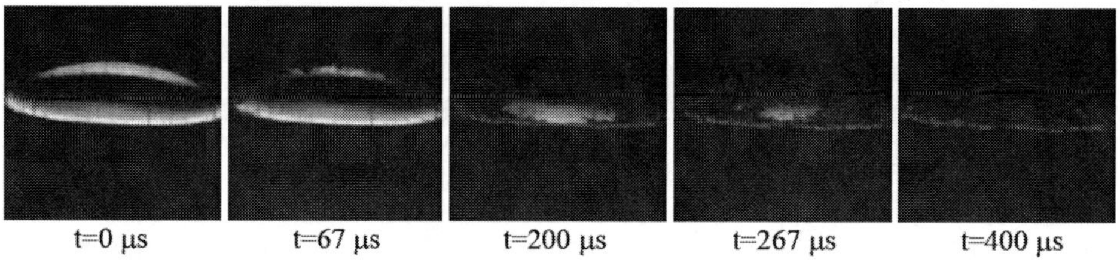

Fig. 5. The initial stage of a phase transition of pentane at temperature and pressure: $T_0 = 50.07\ ^0C$, $P_{res} = 0.05$ bar. Video frame rate - 15 000 fps.

Figure 5 shows the nucleation of the liquid boiling centers initiated on the surface layer in the meniscus interfacial region. The propagation of the wave nucleation of boiling centers is most often from the channel walls to its center. The bubble's growth is triggered by the rupture of the membrane on the surface layer of the superheated liquid. Then the formation of liquid pulsating decay comes with the cellular structure in the inter phase surface. The wave begins on the fluid surface layer near the wall in the area where the work for generation of critical bubble is reduced.

Fig. 6 (a) shows the average speed of the interphase boundary depending on the initial temperature during boiling Freon R141b, R 113 and pentane. The graph shows that at the same heating rate the velocity of interface motion for pentane is greater than the average velocity of the inter phase boundary for Freon. The average speed of the initial stage of the phase transition for pentane was determined, Fig 6 (b). Figure 6 shows that the rate of initial wave formation at the interface surface is significantly greater than the rate of liquid evaporation during the developed boiling.

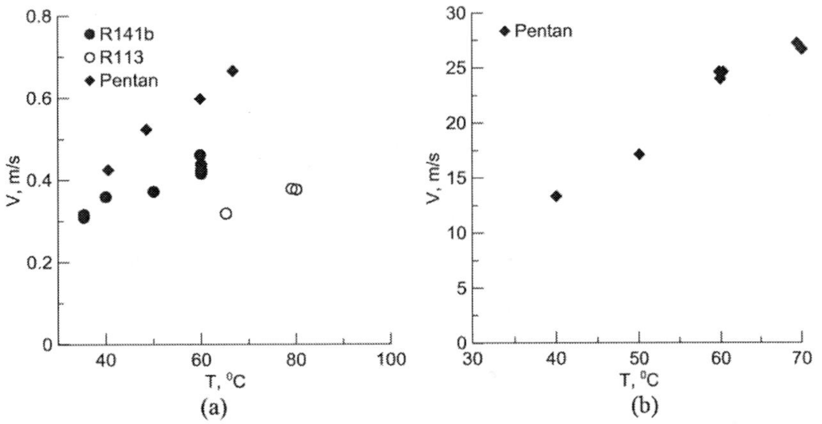

Fig. 6. (a) The average speed of the interphase boundary during boiling depending on the initial temperature; (b) The average speed of the initial stage of the phase transition for pentane depending on the initial temperature.

4 Conclusion

The experimental studies were made for investigation of the adiabatic wave evaporation with rapid transition refrigerant R141b, R113, pentane in a metastable state by the pressure reduction. It was found that the phase transition in the metastable liquid passes under development of multiscale turbulence in the liquid and vapor phases. Using the high-speed video camera, the images of surface phase transition were taken, the pulsating character was established. The rate of formation of the wave phase transition for the metastable pentane was established. Velocity pulsations of the phase transition were determined for the metastable pentane and Freon during the boiling depending on the initial temperature difference.

The study was performed in the Kutateladze Institute of Thermophysics SB RAS and was supported by grant RFBR No. 15-08-06870 A.

References

1. V.V. Kuznetsov, I.A. Kozulin, O.V. Vitovsky, J. Eng.Therm. **21**, 136-143 (2012)
2. P. Reinke, G. Yadigaroglu, Int. J. of Multiphase Flow. **27**, 1487–1516 (2001)
3. L.G. Hill, B. Sturtevant, Int. Union of Theor. and Applied Mechanics, 25-37 (1990)
4. P. Reinke, G. Yadigaroglu, Int. J. of Multiphase Flow. **27**, 1487–1516 (2001)

EPJ Web of Conferences **159**, 00026 (2017) DOI: 10.1051/epjconf/201715900026
AVTFG2016

Effect of a fire retardant on the ignition of pine wood exposed to smoldering particles of pine bark

Denis Kasymov[1,*] *and Alexander Paletsky*[2]

[1]National Research Tomsk State University, 36 Lenin Ave., Tomsk, 634050, Russia

[2]Institute of Chemical Kinetics and Combustion, Novosibirsk, 630090, Russia

Abstract. This paper represents the laboratory research results on the interaction of smoldering pine bark particles with pine wood. Fire retardant treatment "FUKAM" was studied to estimate its effect on the probability of wood ignition. The wind speed and the number and geometric sizes of particles were varied in the experiments. The results have shown that the increase in the wind speed leads to the increase in the probability of wood ignition by the particles of the same size, and fire-retardant treatment significantly increases the protective properties of wood exposed to smoldering pine bark particles.

1 Introduction

The main factors influencing on the ignition of construction materials and the propagation of fires in natural and urban areas are the radiative and convective heat transfer from fire, as well as the smoldering particles which can accumulate on the roof, in the corners of buildings, near fences or indoors, which may result in the ignition of buildings [1].

The works [2, 3] report that the production of smoldering particles is a process in which the combustible materials, such as shrubs, trees and building materials are heated and separated into smaller smoldering particles during a fire. After that the particles are transported from the fire by a convective column, initiate spot fires and cause the ignition of residential and industrial buildings.

One of the ways to reduce the flammability of synthetic and natural polymer materials (including wood) is the use of fire retardants which can be applied to a polymer, both throughout the volume during the production of materials and in the surface treatment of finished products.

Thus, the goal of this work is to study the ignition of preheated wood samples exposed to smoldering natural wood particles, as well as the effect of a fire-retardant coating on the fire hazard characteristics of wood.

2 Experimental

* Corresponding author: denkasymov@gmail.com

The effect of the fire retardant coating was investigated using the "FUKAM" means [4].

To study the probable ignition of wood samples exposed to smoldering particles, the setup was modernized to carry out investigations concerning the ignition of fuel bed by smoldering particles [5]. The scheme of the setup is shown in Figure 1.

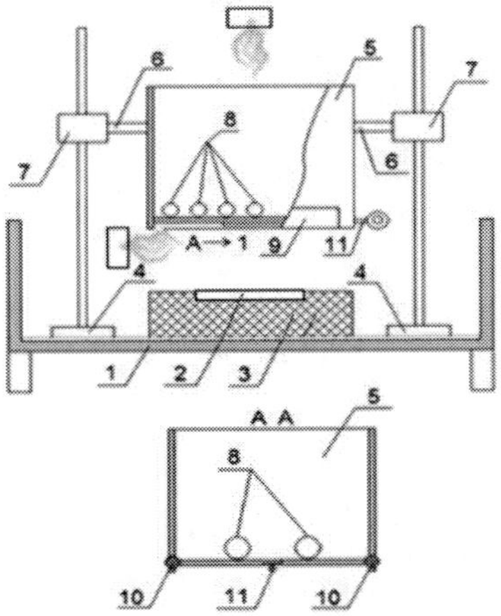

Fig. 1. Experimental setup: 1 is the – pan; 2 – wood sample; 3 – heat-insulating material; 4 – support; 5 – box; 6 – bracket; 7 – holder; 8 – particle samples; 9 – separated bottom; 10 – balls; 11 – stopper.

The imitators of smoldering particles were placed in the box (5) fixed to the supports (4) and having the holes in the bottom for the uniform heating of particles, excluding the probability of particle loss during the experiments. The particles were ignited and started smoldering using two gas burners (Energy GT–1–100). The gas flow rate of the burner was 80 g/hour. The composition of the gaseous mixture was as follows: 9% of propane, 25% of butane, 60% of isobutane, and 6% of impurities. For uniform heating the one burner was placed under the box (5) and the other one was over it. Before conducting the experiment, the wood sample was heated up to a temperature of 200 ÷ 220 °C for 4 minutes [6, 7]. It allowed the conditions to be modeled when the wooden building was exposed to the heat flow from the forest fire front [8, 9]. After that, the substrate with the sample was moved under the box with the particles which were preheated by the gas burners and then thrown down.

The standard solid pine board with the size of (0.23×0.095×0.022) m^3 was used as the sample in the experiments. The size measurement accuracy was ±0.002 m^3. The samples were insulated from the environment by an insulating material in such a way that only the largest part of the surface (S = 218.5 cm^2) remained open. Figure 2 shows the photographs of wood before the experiment, as well as the heating element.

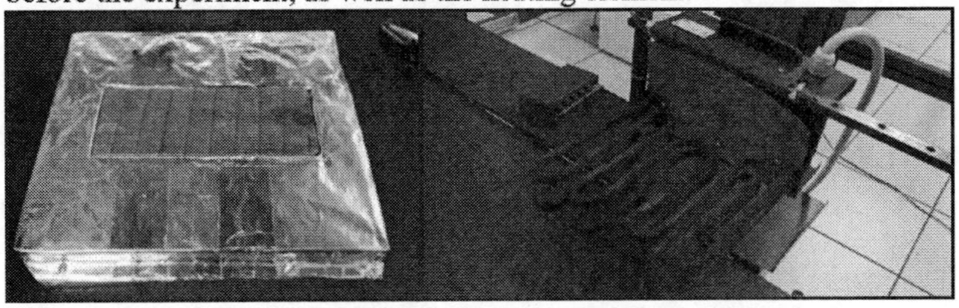

Fig. 2. Wood prepared for the experiment (left) and the heating element (right).

The dipping treatment of the sample was used as a method to add a fire retardant solution to the wood sample. The wood samples were in the liquid solution "FUKAM" for 3 hours, so that the wood was completely covered with liquid. Then the samples were placed in a force-draft oven FDO-0.5 - 200 with a temperature of 70° C until complete drying. Consumption of the means "FUKAM" corresponds to the ASTM E119–00a [10] that provides the II group of fire retardant resistance.

Fresh-gathered pine bark with the characteristic linear sizes $L = (10, 15; 20, 25, 30) \cdot 10^{-3}$ m and a thickness $h = (4 \div 5) \cdot 10^{-3}$ m was used as the imitators of smoldering particles in all experiments. The particle size was chosen based on the data obtained during the field experiments [1].

In a series of experiments a one smoldering particle was first used, then two particles and so on up to 10 particles to model the ignition of wood beginning from one particle and finishing with "fire rain". Three repetitions were performed for each experiment. The wood sample was considered to be ignited if the ignition occurred at least in one of the three experiments.

The weight of the wood samples was 290 g, the wood density was 520 kg/m^3. The initial moisture content of the wood samples was 7.8% and that of the pine bark particles was 2.9%.

The exposure time of the burners was determined depending on the particle size and was: 12 seconds for the particles with a size $L = (10, 15; 20) \cdot 10^{-3}$ m and 15 seconds for the particles with a size $L = (25, 30) \cdot 10^{-3}$ m. According to the data obtained earlier [11], the minimum temperature of smoldering pine bark was 190 °C.

Since in real conditions the exposure of smoldering particles to fuel bed and different wooden buildings is accompanied by a number of natural factors, in particular, by the heated air flow from the front of forest fire, so for this reason in the experiments the smoldering particles which were thrown down on the test wood samples were blown with a flow of heated air supplied by a drying fan with a rate and a temperature of 1, 1.5, 2, 2.5 m/s and 40, 50, 60, 110 °C, respectively. A further increase in the flow rate led to the entrainment of particles outside the site. The rate of the flow directed perpendicular to the falling smoldering particles was determined using an anemometer CFM Master 8901 with a measurement error of 2%. The air flow was directed to the surface of the wood sample in the area of throwing the particles by using a nozzle.

3 Results

Figure 3 shows the comparison of the experimental results to estimate the ignition of the preheated pine wood surface depending on the size and number of the smoldering particles interacting with the surface at various air flow rates. In the graphs, the areas I and II represent the values at which the wood sample ignition is possible (I) or not (II).

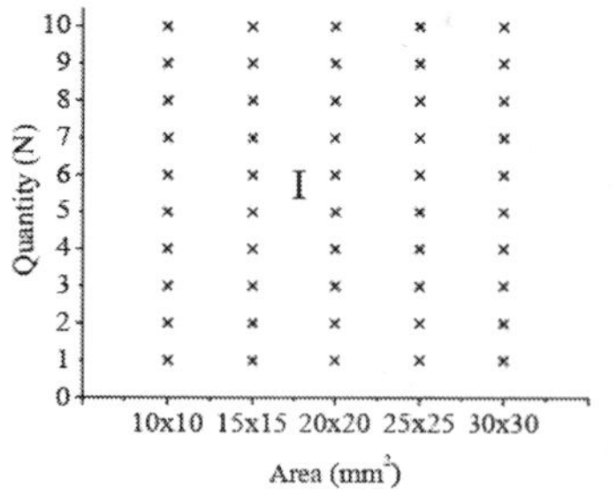

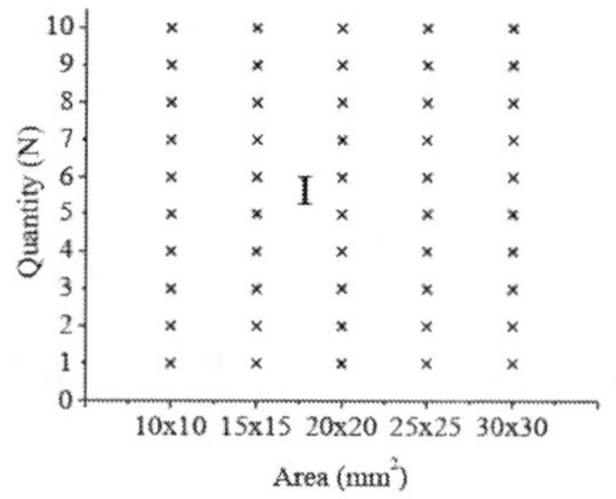

(a)

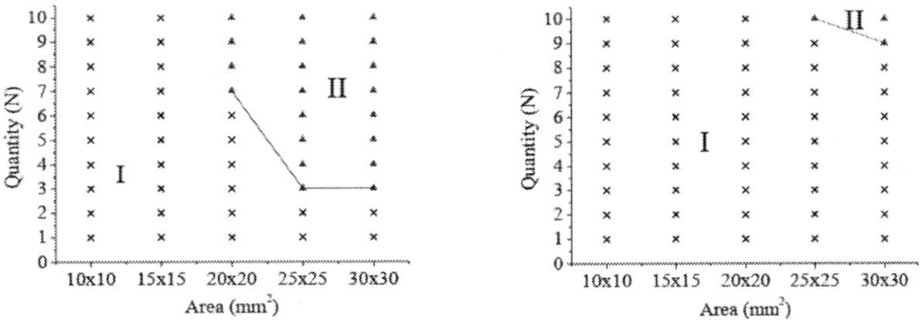

(b)

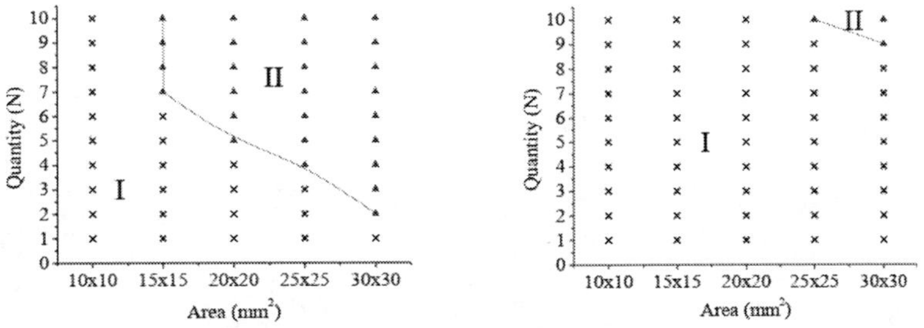

(c)

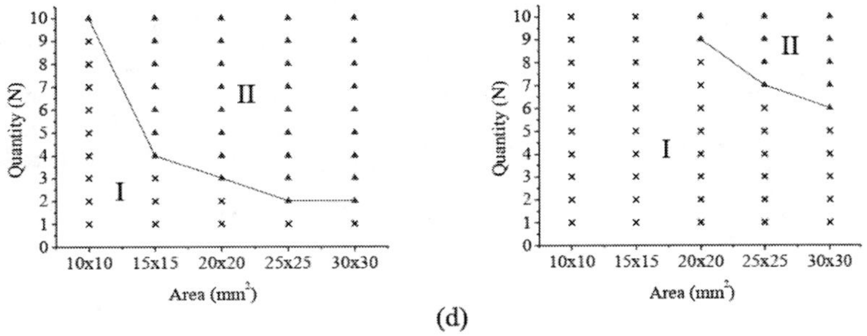

(d)

Fig. 3. Ignition of preheated pine wood with fire retardant treatment (right) and without (left) at different wind speeds (a) 1 m/s, T = 40 °C; (b) 1.5 m/s, T = 50 °C; (c) 2 m/s, T = 60 °C; (d) 2.5 m/s, T = 110 °C.

Figure 3 demonstrates that ignition is not observed for all sizes of particles (Fig. 3a) at a wind speed of 1 m/s, both for the untreated wood and the wood with fire retardant treatment. For the speed V = 1.5 m/s, there is a region II in which the wood is ignited. The surface was found to be ignited by the particles, the geometric dimensions of which exceeded $20\cdot10^{-3}$ m for the untreated wood and $25\cdot10^{-3}$ for the treated one. In addition, as is seen from the graph, the increase in the size of the particles leads to the decrease in their number.

The analysis of the graphs shows that the increase in the wind speed results in the increase in the probability of the ignition by the particles of the same size for the wood that is not treated with a flame retardant. For example, increasing the wind speed from 1.5 to 2.5 m/s leads to the fact that the minimum number of the particles with a size $L = (20, 30)\cdot10^{-3}$ m, sufficient to ignite the wood, decreases from 7 and 3, up to 3 and 2, respectively. The number of particles also influences on the wood ignition process. The minimum total area of smoldering particles, which is equal to the product of the particle number, sufficient for

 DOI: 10.1051/epjconf/201715900026

ignition, by the characteristic area, significantly increases when the wood is treated with a fire retardant. The average value of this quantity calculated for the different characteristic sizes of particles at a speed of 1.5 m/s is 24±5 cm^2 for the untreated wood and 72±9 cm^2 for the wood treated with "FUKAM", respectively. This value is inversely proportional to the wind speed. In addition, the surface occupied by the smoldering particles during ignition was about 11% for the untreated wood and ~ 33% for the wood treated with "FUKAM".

The observations have shown that the wood treated with a fire retardant is ignited when the particles dropped on it start the flaming combustion due to the air flow, and then the combustion spreads to the wood.

Analyzing the results on the exposure of the smoldering pine bark particles to the wood treated with a fire retardant, it is noted that only in a few cases the prolonged exposure of flame caused by the particles blown by the air flow at the speeds of 1.5 ÷ 2.5 m/s leads to the ignition of wood.

Figure 4 shows a typical picture that demonstrates the exposure of smoldering particles to the wood treated with a fire retardant (5 particles with a size of $30\cdot10^{-3}$ m at a wind speed of 2.5 m/s).

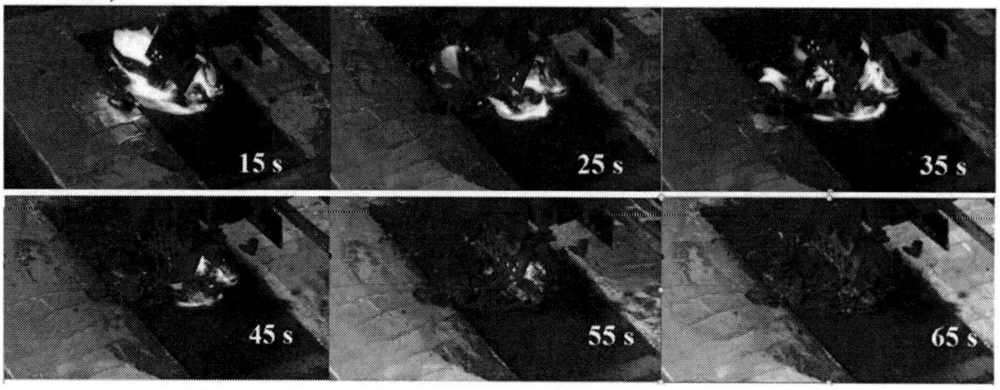

Fig.4. Exposure of smoldering particles to the wood sample treated with a fire retardant (5 particles with a size of $30\cdot10^{-3}$ m were used at a wind speed of 2.5 m/s).

It is seen that the large inflow of oxidizer into the combustion zone results in the flaming particle combustion that lasts about 1 minute. However, this does not lead to the ignition of wood, but there is local charring in the contact area of particles with the wood sample, which is likely to be connected to the inhibitory properties of a fire retardant.

4 Conclusion

For the chosen parameters of the experiment the pine wood samples treated with the means "FUKAM" were resistant to the exposure of smoldering pine bark particles (up to 10 smoldering particle with a size of $25\cdot10^{-3}$ m) and a wind speed (up to 2 m/s). The empirical value was determined within the frameworks of the conditions chosen to conduct the experiment and was equal to the total area of smoldering pine bark particles, normalized to the wind speed sufficient to initiate the ignition (1.5 m/s), namely 24±5 cm^2 for the preheated pine wood and 72±9 cm^2 for the board treated with a fire retardant "FUKAM". The results of the experiments show that the effect of fire retardant treatment "FUKAM" significantly increases the protective properties of wood exposed to smoldering pine bark particles.

DOI: 10.1051/epjconf/201715900026

This study was supported by the Russian Foundation for Basic Research (project No. 16-33-50133).

References

1. M. El Houssami, E. Mueller, A. Filkov, J. C. Thomas, N. Skowronski, M. R. Gallagher, K. Clark, R. Kremens, A. Simeoni, Fire Technol. **52**, 731 (2015).
2. S. L. Manzello, S. Suzuki, Y. Hayashi, Fire Saf, J. **52**, 181 (2012).
3. A. C. Fernandez-Pello, C. Lautenberger, D. Rich, C. Zak, J. Urban, R. Hadden, S. Scott, S. Fereres, Combust. Sci. Technol. **187**, 269 (2015)
4. The special substance for processing different fabrics and building materials "FUKAM": http://www.fukam.ru/English.
5. A. Filkov, D. Kasymov, V. Zima, O. Matvienko, AIP Conf. Proc. **1698**, 060004 (2016).
6. R. M. Aseeva, B. B. Serkov, A. B. Sivenkov, *Gorenie drevesiny i ee pozharoopasnye svojstva*, (EMERCOM of Russia, Moscow, 2010).
7. V. Babrauskas, Fire Saf. J. **40**, 528 (2005).
8. J. D. Cohen, J. Forest. **98**, 15 (2000).
9. R. G. Rehm, A. Hamins, H. R. Baum NISTIR, **6891** (2001).
10. ASTM E119–00a, Fire Tests of Building Construction and Materials, (2004).
11. A. Filkov, S. Prohanov, E. Mueller, D. Kasymov, P. Martynov, M. El Houssami, Jan Thomas, N. Skowronski, B. Butler, M. Gallagher, K. Clark, W. Mell, R. Kremens, R. M. Hadden, A. Simeoni, Proc. Combust. Inst. (to be published).

EPJ Web of Conferences **159**, 00027 (2017) DOI: 10.1051/epjconf/201715900027
AVTFG2016

Modelling of conjugate heat transfer in barium titanate plates heated by the air flow

Nikolay Kozyulin[1, *], *Maxim* Bobrov[1] , *and Michael* Hrebtov[1]

[1]Institute of Thermophysics of SB RAS, 630090 Ac. Lavrentiev ave. 1, Novosibirsk, Russia

Abstract. We present the results of simulation of conjugate heat transfer between the grid of barium titanate plates and the hot air flow. The air temperature undergoes rapid change and the thermal front propagation and heat exchange with the solid plates have been studied for several plate configurations. The results show that the air heat could be effectively absorbed by the plates during the time of thermal front propagation, making such configuration attractive for pyroelectric energy harvesting applications.

1 Introduction

The problem of conjugate heat transfer, when a solid is heated or cooled by the flow of gas or liquid has many practical applications, from environmental flows to engineering devices and power plants. At present, because of ecological needs and technical progress the development of environmental energy converters becomes more and more relevant. One way of utilizing thermal energy of the environment is by using pyroelectric effect [1,2].

The pyroelectric effect is the change in electrical polarization of material with temperature. It can be used for generation of electric current and producing some electric power. For efficient power generation by this process the rapid change in temperature of the material is required. In practice this can be achieved by passing a volume of hot air through the grid of colder pyroelectric material [3].

In the present paper we concentrate on the numerical estimate of effective parameters of barium titanate ($BaTiO_3$) grid being heated by a hot air flow. The grid is constructed of flat plates of fixed longitudinal length placed on top of each other with some regular spacing between them. The grid spacing and plate thickness were varied in numerical simulations to find out the best parameters for the fastest and fullest heat absorption from the air.

2 Computational details

The single "cell" of the plate-grid was modelled. The computational domain contains two "half-plates" and the air channel in between (Fig 1). Each half-plate has symmetry conditions specified on the outer boundary and wall (i.e., no-slip) boundary conditions on

* Corresponding author: nikkozyulin@gmail.com

 DOI: 10.1051/epjconf/201715900027

the inner boundary where it contacts with the air flow. At the inner boundary the conjugate heat transfer conditions were set up, by balancing the heat flux between the solid and the fluid. At the inlet the constant air velocity profile was specified with the value of 15 m/s. This value was selected as practically achievable in experimental setups.

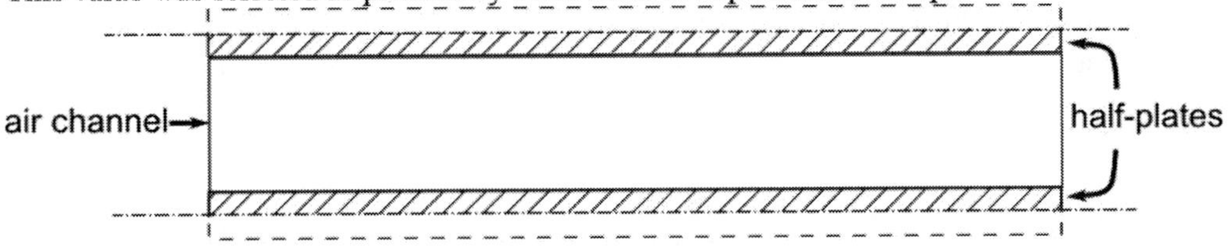

Fig. 1. The scheme of computational domain.

Air channel height (the plate-grid spacing) was varied between 0.1 – 2 mm. The Reynolds numbers for this setup were between 75 – 1500. Taking into account the prescribed length of the channel (2 cm) for this small Reynolds numbers the three-dimensional turbulent effects in a spatially developing boundary layer are weak [4], thus we decided to perform a two dimensional simulation of the flow.

The simulation starts with the flow and the plates having the same temperature (273 K), then at time t_0 the inlet air temperature rises fast by 100 K and the plates start to heat up. The inlet air temperature changes from 273 K to 373 K in 0.1 ms, and the hot air front passes through the plates in 1.33 ms.

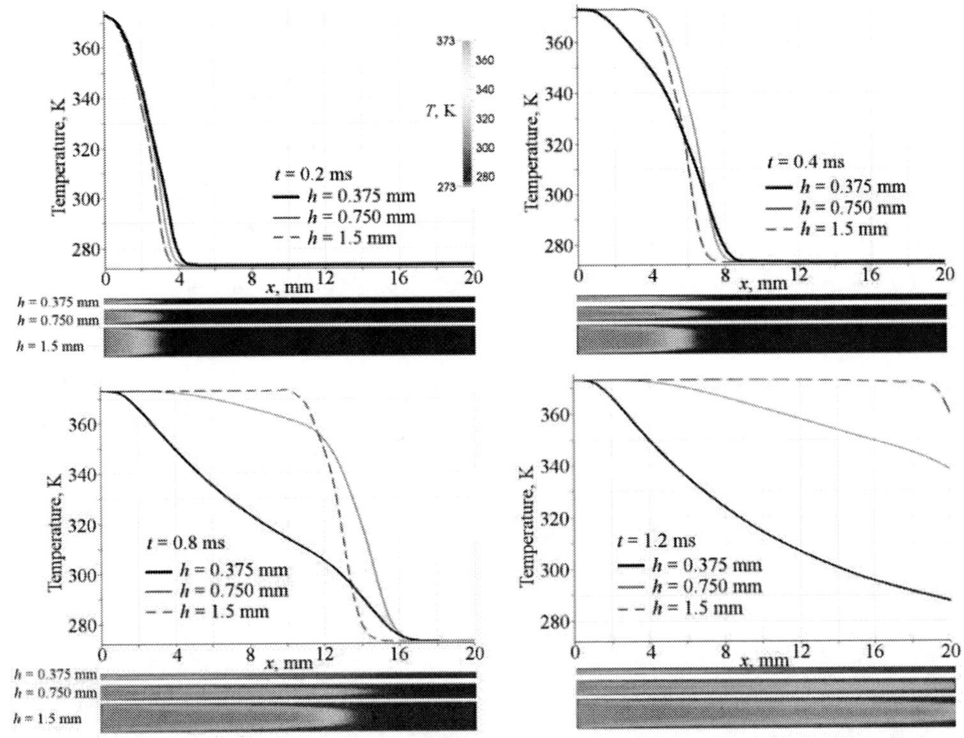

Fig. 2. The longitudinal temperature profiles through the central cross-section of the air channel for three channel heights at the different stages of passing of the hot temperature front. The bottom parts of the subfigures show the temperature distribution in the air channel. The plate thickness is 200 nm.

 DOI: 10.1051/epjconf/201715900027

Three plate thickness values were tested: 20 μm, 2 μm and 200 nm.

The domain has 220 × 120 computational points with clustering towards the inner walls in the air channel. In the solid part of the domain the grid spacing was uniform in both directions.

The $BaTiO_3$ physical properties was taken from [5]. The computational code has finite-volume discretization with second order in space and time. No turbulence model was used. The buoyancy and compressibility effects were taken into account but were found to be negligible in the current simulations.

3 Results and discussion

3.1 The effect of different plate spacing

Figure 2 shows temperature field evolution in the air channel while the hot temperature front is proceeding through the channel. It can be seen that for the channel height (h) of 1.5 mm the shape of the profile at the front is virtually not changing, reflecting that only a small fraction of heat is transferred to the solid plates in that case, thus rendering this spacing value inefficient. For smaller h the shape of the profile starts to change with the hot front propagation progress. This is especially noticeable for h = 0.375 mm, where the profile has two distinct parts, one (after the thermal front boundary) with evidently exponential behaviour, reflecting the constant heat loss to the plates, and the other with steeper slope at the thermal front boundary.

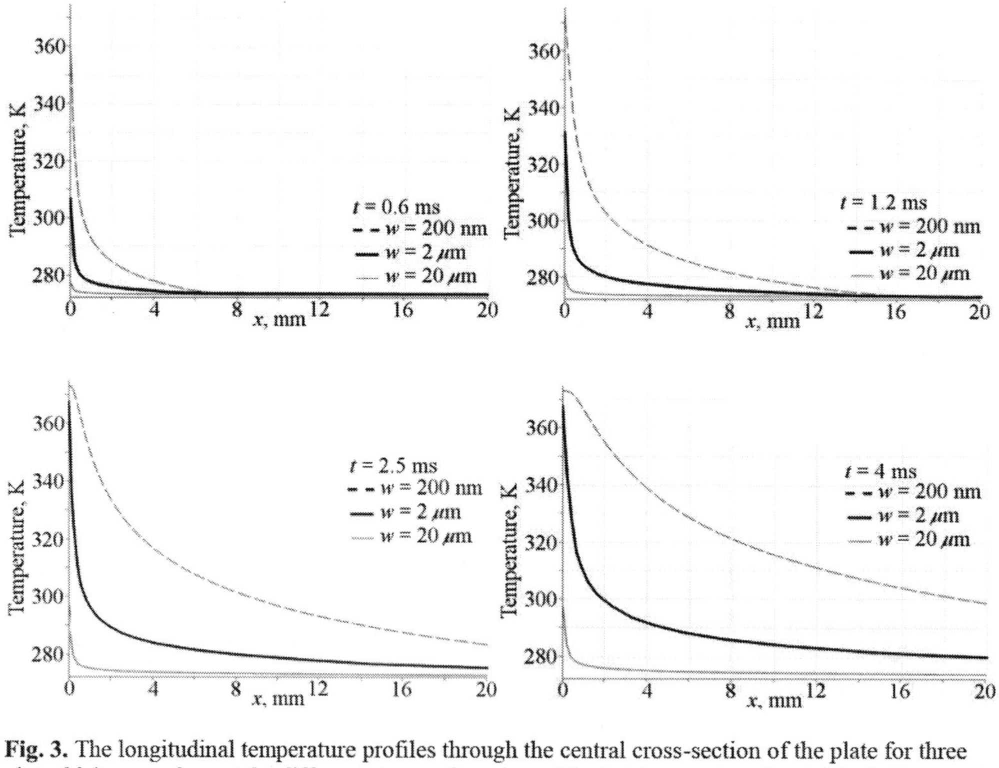

Fig. 3. The longitudinal temperature profiles through the central cross-section of the plate for three plate thickness values at the different stages of passing of the hot temperature front. The bottom part shows temperature distributions for times after the thermal front propagation. The air channel height is 0.75 mm.

After the front passes through the channel, the equilibrium temperature profiles are set up throughout the channel, varying slowly as the plates get more and more heated up. Another interesting feature is that smaller values of spacing lead to faster temperature propagation in the air. This is may be due to the mutual influence of the wall boundary layers, which is negligible for large h (1.5 mm), and becomes more pronounced for smaller h (0.375 mm).

As a result it can be seen that choosing the spacing of 0.375 mm leads to the transfer of major part of air thermal energy to the plates during the hot temperature front propagation which makes this setup promising for thermal energy harvesting.

3.2 The effect of different plate thickness

Figure 3 shows the temperature distribution in the plates during and after the heat front propagation. It can be seen that only for the thinnest plates (thickness w=200 nm) significant change in temperature is observed during the thermal front propagation time. The longitudinal distribution shows that the outlet parts of the plates are getting much less heated then the inlet parts for the prescribed channel length (2 cm).

In general, all of the tested plate configurations are not allowing the prescribed air flow to heat up the whole plate sufficiently during the time of thermal front propagation. Three or four times larger heating period is needed (Fig 3, bottom) to heat up even the thinnest (200 nm) of tested plates.

The thermal inertia of the plate leads to a very little sensitivity of the solid temperature field to the thermal front boundary position in the air.

4.Conclusion

The present study shows that the fast rise in the inlet air temperature may be effectively absorbed by the ferroelectric walls during the time of thermal front propagation if the plate spacing is small enough. In our case the plate spacing should be about 0.375 mm.

The thermal response of barium titanate plates to the hot air flow with a steep thermal front was estimated numerically. The results show that for the fast temperature response in the plates which is needed for effective thermal power harvesting one is forced to use very thin plates (possibly coatings) of ferroelectric material in air heated setups. For reasonable air velocities (of the order of tens of m/s) the optimal plate sizes might be calculed. Our results show that the chosen plate length of 2 cm is too high, and should be shortened that will lead to a faster growth of the plate temperature and less pressure drop in the air channel.

The optimal parameters for the specified air velocity of 15 m/s and the temperature difference of 100 K could be the following: the plate length of about 1 cm, the plates spacing of 0.375 mm and the plate thickness of 200 nm or less.

This work was supported by Russian Science Foundation (grant №16-19-00119).

References

1. S.R. Etesami, J.Berakdar, Appl. Phys. Lett., **108** (2016)
2. M. Sharma, A. Chauhan, R. Vaish, V. S. Chauhan, Solar Energy, **115** (2015)
3. D. Vanderpool, J. H. Yoon, L. Pilon, Int. J. Heat Mass Tran, **51** (2008)
4. M. Lesieur, *Turbulence in fluids* (Springer Science & Business Media, (2012)
5. Y. He, Thermochimica acta, **419** (2004)

EPJ Web of Conferences **159**, 00028 (2017) DOI: 10.1051/epjconf/201715900028
AVTFG2016

Influence of perturbance frequency on intensity of mixing of fluids in microchannel devices

Aleksandra Kravtsova[1,2,*], and *Yuri* Meshalkin[1,2]

[1]Kutateladze Institute of Thermophysics, Siberian Branch of the Russian Academy of Sciences, 630090 Novosibirsk, Russia
[2]Novosibirsk State University, Department of Physics, 630090 Novosibirsk, Russia

Abstract. Influence of perturbation frequency on mixing flow velocity and the varying flow structure in T-shape microchannels are investigated using laser induced fluorescence. The external disturbances of flow are created excitation system based on the piezoelectric actuator. Several increases in mixing efficiency were common to all regimes, but also some different features, were observed. For the stationary vortex flow regime, we obtained a decrease in mixing efficiency of 10% at a frequency of 650Hz. For the stationary asymmetric vortex flow regime, the mixing efficiency increases by 33% and 23% for the frequencies 500Hz and 800 Hz, respectively. For Re=400 and Re=300, the flow structures without disturbance are virtually identical; however, in the case of external influence for the quasiperiodic unsteady flow regime, the mixing efficiency decreases by 22% and 29% at distances of 1 and 5 calibres, respectively, for frequency of 1000 Hz.

1 Introduction

The development of precision technology, both in medical and biological industries, is impossible without the miniaturization of devices. Effective and resource-efficient systems, such as fuel cells, provide a long autonomous operation in different devices (i.e. tablets, laptops, electric cars, microreactors, etc.) and are becoming more important due to their growing use in the manufacture of drugs and regenerative medicine. The performance of these devices depends on the velocity of mixing liquids inside them. Therefore, the development of microfluidic devices with high mixing velocity, throughout the application of external perturbations, is a necessary area of research of hydrodynamic processes in microchannels.

The increase of mixing velocity can be achieved by applying either passive or active perturbations to the flow. The conventional passive methods of increasing the mixing velocity involve decreasing the thickness of the fluid layers by changing the microchannel size ([1], [2]) and extending the contact area with adjacent liquids. To increase the mixing velocity in the microchannel, an excitation of circulations in the flow and a curvature of the flow lines can be used. Engel et al. [3] and Kockmann et al. [4] have discovered that flow in

* Corresponding author: Kravtsova.Alya@gmail.com

microchannels has three flow regimes: layered laminar flow, vortex flow, and stationary asymmetric vortex flow. In the first case, lines of flow move through the channel with minimal distortion. Small vortices form throughout the flow in the case of vortex regime. Appearance of instability in flow leads to destruction of axial vortex symmetry, and we can obtain stationary asymmetric vortex flow. T-microchannels are eminently suitable for fundamental investigations of mixing processes in the convective micromixers. Hoffman et al. [5] were the first researchers to investigate the different flow regimes in T-shape microchannels for various Reynolds numbers. A detailed study of the concentration profile in a microchannel T-type was obtained by Gobert et al. [6]. They used the computational modeling and technique Particle Tracking Velocimetry in their work. Later, Wong et al. [7] presented a detailed study of flow behavior and properties of mixing in microchannel T-type, but this simulation did not provide a complete understanding of convective effects. Dreher et al. [8] have conducted a numerical simulation in the T-channel with rectangular cross section and showed that the best mixing occurs at Reynolds number from 240 to 700. N. Nama et al. [9] have studied the effect of piezoelectric transducer to the flow in the microchannel, in which the liquid has been supplied at different angles. They showed that the propagating waves from the transducer change the structure formed in the liquid and thereby can influence the mixing velocity. Claudio P. Fonte et al. [10], using Computational Fluid Dynamics (CFD) simulation, studied the effect of flow rate pulsations on the dynamics of the flow and mixing efficiency in the T-shape microchannel; however, an experimental confirmation has still not been obtained. Thus, a significant number of studies of flows in microchannels presented in the literature have been made by mathematical methods of numerical simulation. For examination and validation of these models, an experimental investigation of mixing in microchannels is necessary, and we present our results in the following sections.

2 Experimental conditions and measurement techniques

To obtain the characteristics of the flow the method of laser induced fluorescence (micro-LIF) inverted in microscope Carl Zeiss Axio Observer.Z1 was used. A T-shaped microchannel made of optically transparent material SU-8 with dimensions of $120\times120\times240$ μm (height, width of the input channel and the width of the output channel) was mounted on the microscope stage. Distilled water and a solution of distilled water with the dissolving on a molecular level Rhodamine 6G dye were fed in microcannel. The concentration of the solution is 15mg/l. A flow rate of fluid in the T-microchannel is defined by a syringe pump KD Scientific company with two outputs. A mercury lamp with light filter is used for illumination of the flow in channel. Filter allocates the light with a wavelength of 532 nm, which corresponds to the maximum of spectrum absorption for Rhodamine 6G dye. The dye emitted light in the red wavelength range via the dichroic mirror radiation is directed to a CCD camera with space resolution of 2048×2048 pixels. The resulting images are operated via a computer using "Actualflow" software. The experimental facility is shown in Figure 1.

To create the external disturbances of flow with different frequency, unique excitation system of the flow based on the piezoelectric actuator was mounted (Figure 1). The free running of the piezoactuator with a minimum 1kN preload and the maximum voltage 150V is 150 μm. The system creates an in-phase variable flow rate to both inputs of the T-shaped microchannel. The piezoelectric actuator deforming with predetermined frequency and amplitude passes the perturbation in the flow through the rod by direct mechanical contact with the actuator.

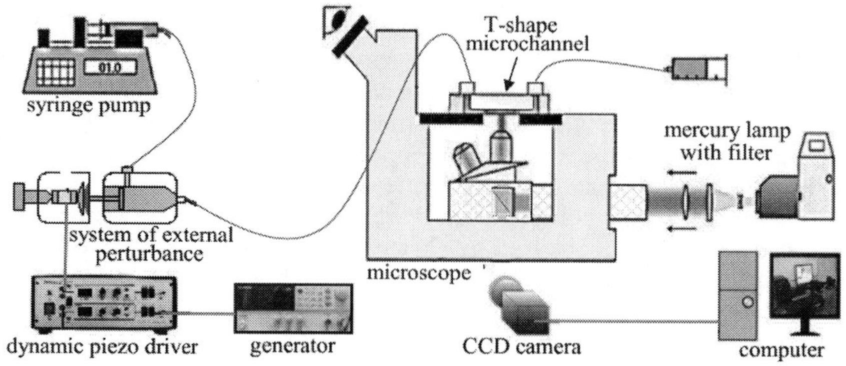

Fig. 1. Sketch of experimental setup

3 Results

Experimental study of influence of the perturbation frequency on the process of mixing flow velocity in a microchannel T-type and the flow structure were examined. The experiments were carried out for Reynolds numbers Re = 120, 186, 300, 400. The system of external perturbance was generated the flow pulsation in the same phase on both inputs of microchannel with the frequency, *f*, range from 0 Hz to 1000 Hz with step 100 Hz. The sampling frequency of measurements was increased four times (step of the measurements was 25 Hz) in region of greatest change of the mixing efficiency (400 Hz to 800 Hz).

The mixing efficiency was calculated for each frequency of disturbance. To calculate the mixing efficiency, M, in T-channels the Dankwerts segregation intensity was used. Mixing efficiency was calculated on basis of concentration fields obtained the micro-LIF method at a distance of 1 and 5 calibers from the entrance of the mixing channel for each flow regime with and without superimposed pulsations.

The first series of experiments are presented for the stationary vortex flow regime Re = 120 (Figure 2a). We can see two symmetrical vortices on this regime which are called the Dean vortices. In Figure 2a the mixing efficiency increases in diapason of frequencies from 0 Hz to 500 Hz at a distance of 1 caliber from the entrance at 6% and at a distance of 5 calibers at 12%. At a further increase of the frequency of the external influence of up to 650 Hz the mixing efficiency decrease at 6% at a distance of 1 caliber and 10% 5 calibers relative to the undisturbed flow. Further increasing the frequency leads to a continuous growth of the mixing efficiency. It becomes maximum by a frequency 1000 Hz and increase by 13% and 16% at a distance of 1 and 5 calibers, respectively. According to the visualization of flow mixing efficiency increases significantly.

For stationary asymmetric vortex flow regime Re = 186 (Figure 2b) the mixing efficiency increase by 33% with raise frequency from 0 Hz to 500 Hz at a distance of 1 caliber from the entrance in working channel. A further increase of the frequency of the external perturbation up to 650 Hz leads to a decrease in the mixing efficiency however it is greater by 10% then in case of undisturbed stream. The mixing efficiency increases by 23% for the growth frequency up to 800 Hz. In the frequency range from 800 to 900 Hz the mixing efficiency decreases to the level of the frequency value for 650 Hz. Next the mixing efficiency increases by 10% up to 1000 Hz. We can observe monotonic increase of the mixing efficiency by 15% for rising frequency of excitation at a distance of 5 calibers from the entrance in working channel.

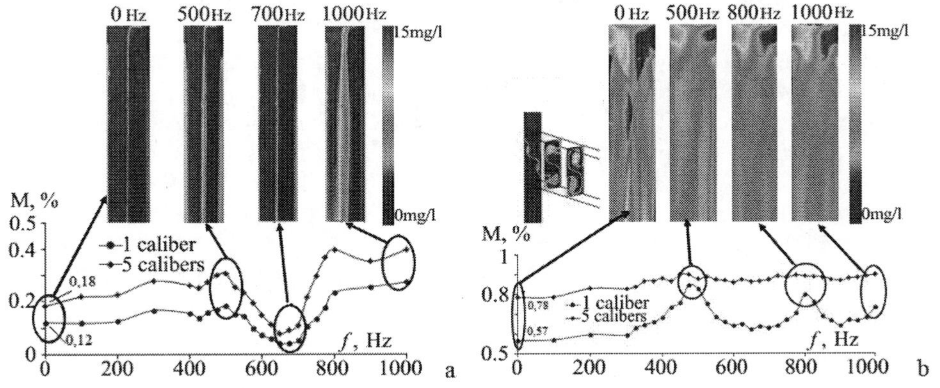

Fig. 2. The mixing efficiency against perturbation frequency for: a - Re = 120, b - Re = 186.

For the periodic unsteady flow regime Re = 300 (Figure 3a) significant influence on the flow structure and mixing efficiency are not detected. With increasing of perturbance frequency the mixing efficiency is virtually unchanged at a distance of 1 and 5 calibers due to availability of the turbulent agitation mechanism in flow, which in turn allows mixing fluids well enough.

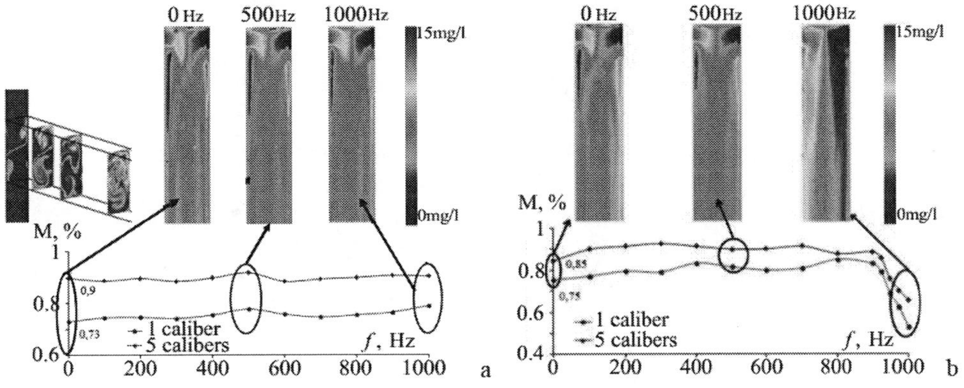

Fig. 3. The mixing efficiency against perturbation frequency for: a - Re = 300, b - Re = 400.

For Re = 400 and Re = 300 the flow structure without disturbance are virtually identical. However, in case of external influence for the quasiperiodic unsteady flow regime (Figure 3b) the mixing efficiency decreases by 22% and 29% at a distance of 1 and 5 calibres, respectively, for frequency of 1000 Hz.

Based on the obtained data we can conclude that in the investigated range of Reynolds numbers the mixing efficiency in T-microchannels1 can be significantly increased by superimposing on the flow of external disturbances of a certain frequency.

The reported study was funded by RFBR according to the research project No. 16-31-00149 mol_a.

References

1. J. Marschewski, S. Jung, P. Ruch, N. Prasad, S. Mazzotti, B. Michelb, D. Poulikakos, Lab Chip, 1923-1933 (2015)
2. E. Kjeang, R. Michel, D.A. Harrington, N. Djilali, D. Sinton, J. Am. Chem. Soc. **130,** 4000-4006 (2008)

3. M. Engler, N. Kockmann, T. Kiefer, P. Woias, Chem. Eng. J. **101,** 315-322 (2004)
4. N. Kockmann, C. Foll, P. Woias, SPIE, **4982**, 319-329 (2003)
5. M. Hoffmann, N. Rabiger, M. Schluter, S. Blazy, D. Bothe, C. Stemich, H.-J. *11th Europ. Conf. on Mixing* (Bamberg, Germany, 2003)
6. C. Gobert, F. Schwertfirm, M. Manhart, *2nd Joint U.S.-European Fluid Engineering Summer Meeting,* (Mi-ami, Florida, USA, 2006)
7. S. H. Wong, M.C.L. Ward, C.W. Wharton, Sens. & Act. B. **100**, 359-379 (2004)
8. S. Dreher, N. Kockmann, P. Woias, Heat Transfer Eng. **30**, 91-100 (2009)
9. N. Nama, P-H. Huang, T.J. Huang, F. Costanzo, Lab Chip. **14**, 2824-2836 (2014)
10. E. Erko, C.P. Fonte, M.M. Dias, J.C.B. Lopes, R.J. Santos, Chem. Eng. Res. Des. **106**, 74-91 (2016)

Infrared study of surface of the HFE-7100 liquid evaporating into gas flow

Aleksei Kreta[1,2*]

[1]Novosibirsk State University, Novosibirsk, Russia
[2]Kutateladze Institute of Thermophysics SB RAS, Novosibirsk, Russia

Abstract. An experimental study of the temperature field on the surface of horizontal liquid layer (HFE-7100) evaporating into gas flow (Air) has been performed. Temperature gradient of the gas-liquid interface has been measured with the help of Titanium 570M IR camera. Shear stresses on gas-liquid interface induced by thermocapillary effect and inert gas flow have been defined.

1 Introduction

The fluid motion driven by thermal, concentration or charge gradients is a common and important phenomenon in nature. In many cases, the concentration and the charge gradients are absent or could be neglected (pure liquids, the absence of electric or magnetic field) but the presence of a temperature gradient in a fluid layer can nonetheless trigger the important convective motions. In these cases, two mechanisms have been spotted to be responsible for the convection. One of them is the gravity that acts essentially on density differences due to temperature gradients, the other one is the thermo-capillarity originated by the temperature dependence of the surface tension at the liquid interface. The convective motions caused by buoyancy are known as the Rayleigh-Bénard convection. When the convection is induced by the surface tension, it is referred to as Marangoni-Bénard convection. In the most of investigations the fluid layers are steadily heated from below with a static temperature gradient and without evaporation. The evaporating fluid layers with nonlinear temperature gradient have much less been studied due to the complexity of the problem. The works [1] report the experimental results concerning the effects of evaporation on the thermocapillary convection in a thin liquid layer. It was found that evaporation is not only a means of cooling the layer surface, but also a driving power for the development of the convective instabilities.

The problem of convection of the evaporating liquid layer under the action of the inert gas flow is much more complicated than described above, because includes also the shear-stress effect induced by gas flow [2]. The coupling between evaporation and convection cased by a shear-stress, thermocapillary and buoyancy effects in the liquid layer, has a direct effect on the evaporation flow rate, being relevant, for example, for heat-pipe grooves and thin-film evaporators. The increased interest to the problem is caused by experiments under microgravity conditions [3–5], some experiments under normal gravity

* Corresponding author: alexy.slav@yandex.ru

and the preparation of experiments on the International Space Station in the framework of the scientific project "Convection and Interfacial Mass Exchange" (CIMEX) of the European Space Agency [6]. These experiments are aimed to investigate the features of convective flows of a fluid in an open cavity and in an open horizontal layer under conditions of a co-current gas flow and evaporation. The processes of convection, accompanied by evaporation at the interface, are actively studied experimentally [7], [12-14], numerically [8, 9] and theoretically in the present time [9, 10], [15, 16].

The aim of this work is to experimentally study the structure of convective flows and to measure temperature distribution on the surface of horizontal liquid layer under the action of the gas flow using IR camera

2 Experimental setup

The experimental investigations were conducted on the setup shown in Fig. 1. The setup consists of the following components: test section, gas and liquid loops, data acquisition system, thermal stabilization system, and optical techniques.

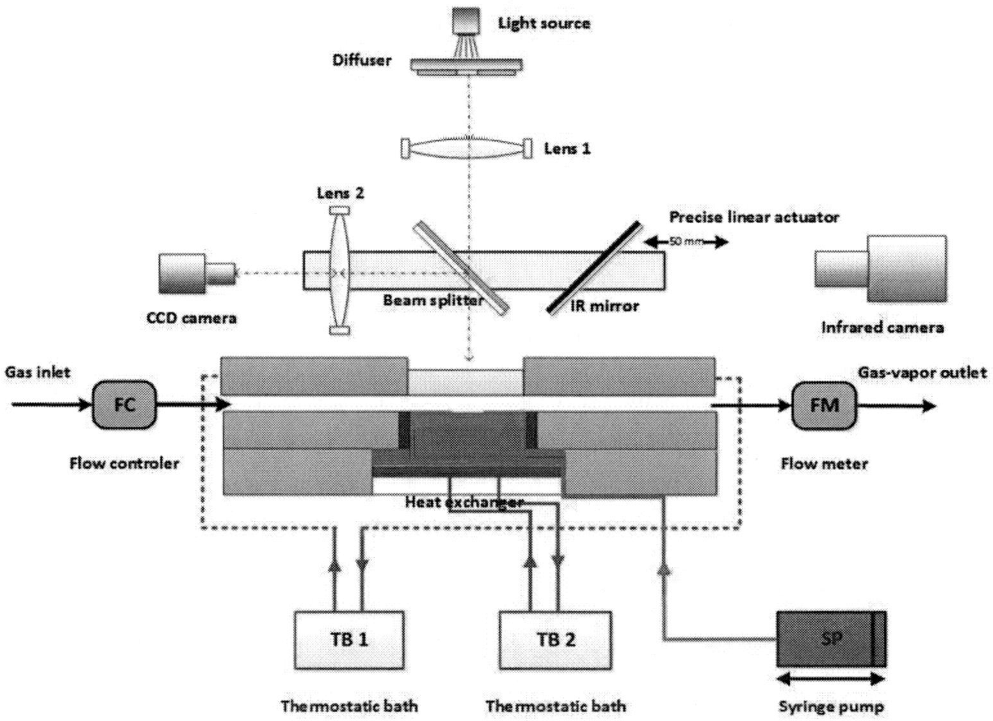

Fig. 1. Scheme of the experimental setup.

Pure gas enters to the test section from the compressor. The mass flow controller sets flow rate of the gas at the inlet of the test section. The liquid is injected into the test section by a precision syringe pump. The working liquid evaporates under the gas flow and the vapor-gas mixture flows to the outlet of the gas channel in test section. The vapor-gas mixture flow at the outlet is measured by a mass flow meter. The liquid temperature in the test section is determined by the temperature of the substrate base, which controlled with the help of thermoelectric module. The position of the interface level is controlled with an accuracy of 10 micrometres using a Schlieren technique and the syringe pump [7]. The micro-groove as barrier for liquid spreading outside to the evaporation zone is used, [17].

Measurements of the temperature distribution on gas-liquid interface were conducted with the help with of Titanium 570M IR camera. An infrared wavelength region is 3-5 micrometres. Infrared imaging of liquid surface was performing through quartz optical window that implemented in the top cover of the test cell. IR camera was placed on the optical tabletop horizontally and measuring of the surface temperature of the liquid was contacted with the help of the IR mirror. The IR mirror was inclined to 45^0 to infrared camera and liquid surface. Optical components of the Schlieren technique and IR mirror were mounted on an optical platform of the linear actuator. Linear actuator is controlled from personal computer and by special software. The optical platform moved in the range of 50 mm and speed of 104 mm/s in horizontal direction. Initially, the platform was set in position where optical axes of Schlieren technique coincide and the interface visualized. When a flatness of interface and the stationary regime of the experiment were established the linear actuator moved the platform in position where optical axes of the IR camera and mirror also coincide and measurement of surface temperature was running.

3 Results and discussion

The experiments were conducted at atmospheric pressure in a test section, and a 5 mm thick liquid layer. HFE 7100 liquid was used as the working liquid. The evaporation surface area was 100 mm^2 with the corresponding 10x10 mm cut-out in the plate. The gas flow rate was in the range 100-1000 ml/min that corresponds to gas average velocities for 0,0138 to 0.138 m/s (Re=2,8-28), respectively. The temperature of the "liquid-gas" system was 20 °C. The difference between the temperature of the gas and liquid does not exceed 0.1 °C.

Fig. 2 shows a thermogram of the liquid surface at the intensive evaporation. The IR image snapshot does not explain all dynamic of the process. Analysing of the video the movement of gas-liquid interface in counter-current direction to the gas flow is observed. In Fig. 3 a typical temperature distribution of the interface in the direction of gas flow is plotted. It is shown the differences in the interfacial temperature. Intensive evaporation induces temperature decreasing in initial of the evaporation surface and temperature gradient is appeared. In this case the thermocapillary effect on the gas-liquid interface takes place. The shear stress on the liquid surface τ_{sur} caused by the thermocapillary effect can be estimated as:

$$\tau_{sur} = \frac{\delta\sigma}{\delta T}\frac{\delta T}{\delta x} \tag{1}$$

where, $\sigma_T = \dfrac{\delta\sigma}{\delta T}$ - temperature coefficient of surface tension is equal 1,14 x 10^{-4} N/mK.

The temperature gradient is defined as: $\dfrac{\delta T}{\delta x} = \dfrac{T_2 - T_1}{\Delta x}$. Points of the temperature measuring and length x=0,008 m are shown in Fig. 2.

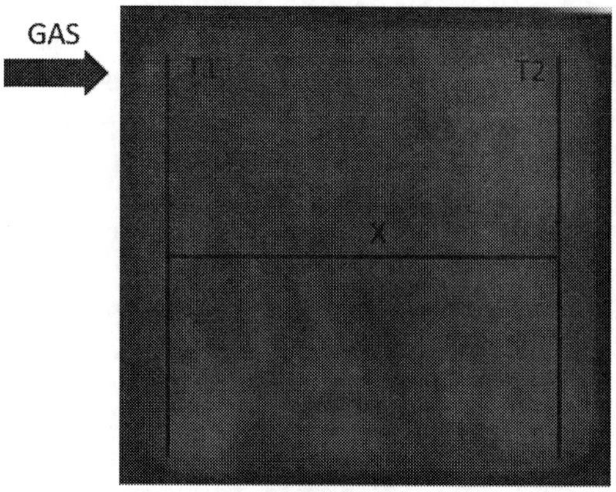

Fig. 2. The thermogram of the evaporating liquid layer

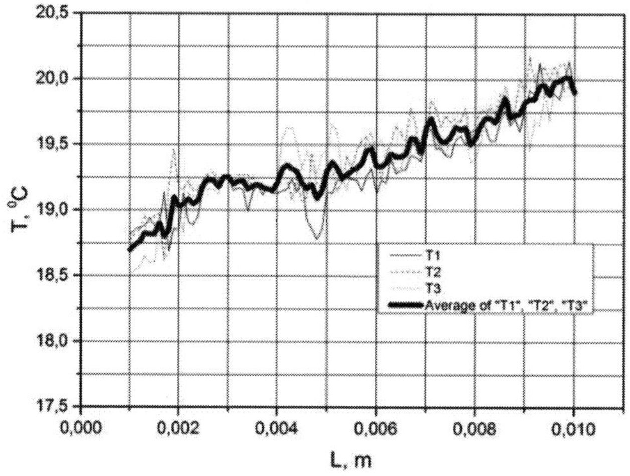

Fig. 3. The temperature distribution on the surface of the evaporating liquid layer

Shear stresses on the liquid surface induced by gas flow can be estimated as [11]:

$$\tau_g = \eta \frac{dV}{dy} = -\frac{H}{2}\frac{\delta p}{\delta x} \tag{2}$$

where, H= 0,005 m is height of gas channel; η – dynamic viscosity.

Pressure gradient is calculated from the formula: $V_g = -\frac{H^2}{12\eta}\frac{\delta P}{\delta x}$, where average gas velocity V_g, m/c is known value. It is assumed that in the gas channel the Poiseuille flow is realised. The gas-liquid interface is considered as a rigid surface.

In Fig. 4 the calculations of the shear stresses caused by the thermocapillary effect and gas flow in depends on the gas Reynolds number are presented. With the growth of the gas flow the thermocapillary and gas shear stresses are increased. The plot shows that the thermocapillary shear stresses are bigger than gas shear stresses in several times for all gas Reynolds number. It means the thermocapillary forces dominate the gas shear stresses.

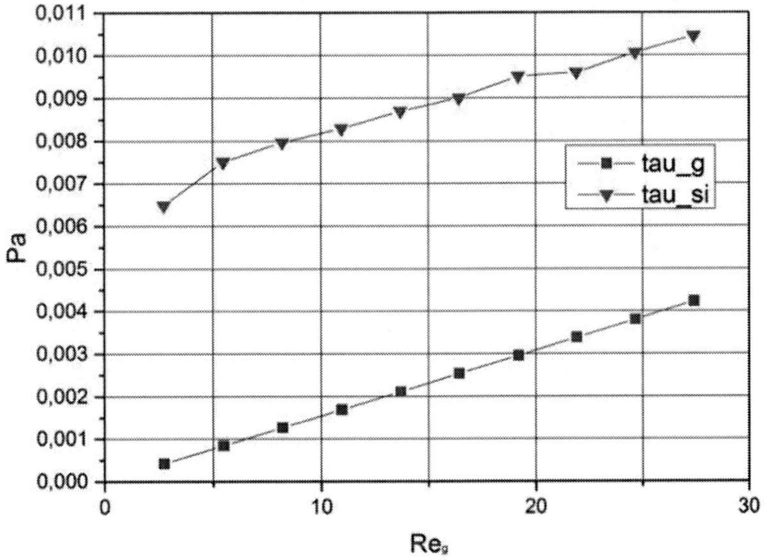

Fig. 4. Shear stresses

4 Conclusion

Infrared study and measurements of the temperature field on the surface of horizontal liquid layer evaporating into gas flow has been performed. Temperature gradient of the gas-liquid interface has been measured with the help of Titanium 570M IR camera. Shear stresses on gas-liquid interface induced by thermocapillary effect and inert gas flow have been defined. It is shown that thermocapillary stresses acts opposite to the gas flow and moves liquid at the gas-liquid interface in the direction of the attack border of the evaporation surface.

The work was financially supported by the Russian Ministry of Education and Science (Project identifier RFMEFI61314X0011).

References

1. N. Zhang, D. F. Chao, Int. Comm. Heat and Mass Transfer, **26**, 1069 (1999)
2. O.A. Kabov, International Journal of Emerging Multidisciplinary Fluid Sciences, **2,** 161 (2010)
3. L. Joannes, O. Dupont, S. Gorza, P. Colinet, J. Legros, T. Lanen, H. Oksuzoglu, *Proceedings of Fringe 2001, 4th international workshop on automatic processing of fringe patterns*, 506 (2002)
4. K. Löth, B. Larsson, H. Schneider, O. Janson, Y. Houltz, P. Colinet, C. Iorio, L. Joannes, O. Dupont, *Proceedings of the 53rd international astronautical congress, Houston, Texas (The World Space Congress)*, (2002)
5. P. Colinet, L. Joannes, C.S. Iorio, B. Haut, M. Bestehorn, G. Lebon, J.C. Legros, Advances in Space Research, **32**, 119 (2003)
6. T. Balázs, Microgravity Sci. Technol., **24**, 189 (2012)
7. Yu.V. Lyulin, O.A. Kabov, Int. J. Heat and Mass Transfer, **70**, 599 (2014)
8. H. Machrafi, C.S. Iorio, P.C. Dauby, Interfacial Phenomena and Heat Transfer, **2,** 1 (2014)

9. O.N. Goncharova, O.A. Kabov, Int. Journal of Heat and Mass Transfer, **53**, 2795 (2010)
10. O.N. Goncharova, M. Hennenberg, E.V. Rezanova, O.A. Kabov, Interfacial phenomena and heat transfer, **1**, 317 (2013)
11. L. Landau, E. Lifshitz, *Hydrodynamics* (Science, Moscow, 1986)
12. Yu.V. Lyulin, D.V. Feoktistov, I.A. Afanasjev, E.S. Chachilo, O.A. Kabov and G.V. Kuznetsov Technical Physics Letters **41**, 665 (2015)
13. Yu.V. Lyulin, O. A. Kabov, Technical Physics Letters **39**, 795 (2013)
14. E.Y. Gatapova, R.A. Filipenko, Y.V. Lyulin, I.A. Graur, I.V. Marchuk, O.A. Kabov, Thermophysics and Aeromechanics **22**, 701 (2015)
15. O.N. Goncharova, E.V. Rezanova, Yu.V. Lyulin, and O.A. Kabov, Thermophysics and Aeromechanics **22**, 631 (2015)
16. Y.O. Kabova, V.V. Kuznetsov, O.A. Kabov, International Journal of Heat and Mass Transfer, **55**, 1271 (2012)
17. V. Grishaev, A. Amirfazl, S. Chikov, Yu. Lyulin, O. Kabov, Microgravity Science and Technology **1**, 27 (2013).

Adiabatic wall temperature and heat transfer coefficient influenced by separated supersonic flow

Alexander Leontiev[1], *Sergey* Popovich[1,*], *Mark* Strongin[1] and *Yurii* Vinogradov[1]

[1]Institute of Mechanics, Lomonosov Moscow State University, 1, Michurinsky Pr., Moscow, 119192, Russia

Abstract. Investigations of supersonic air flow around plane surface behind a rib perpendicular to the flow direction are performed. Research was carried out for free stream Mach number 2.25 and turbulent flow regime – $Re_x > 2 \cdot 10^7$. Rib height was varied in range from 2 to 8 mm while boundary layer thickness at the nozzle exit section was about 6 mm. As a result adiabatic wall temperature and heat transfer coefficient are obtained for flow around plane surface behind a rib in contrast with the flow around plane surface without any disturbances.

1 Introduction

One of the major problems in high speed vehicles engineering is the accuracy of experimental data obtained in supersonic wind tunnels while its extrapolation to the real flight conditions. Flight speed increasing over 0.4·M results in the requirement for accounting of compressibility effect [1, 2]. Flow deceleration from supersonic speed to the zero-order meaning on the wall causes increased flow kinetic energy transforming into heat in comparison with subsonic flows.

Heat transfer factor h for supersonic gas flow around the wall is evaluated by means of specific heat flow rate q divided by the difference between adiabatic wall temperature T_{aw} and the measured wall temperature T_w:

$$h = \frac{q}{T_{aw} - T_w} \tag{1}$$

One of the main problems in the equation (1) is definition of adiabatic wall temperature T_{aw} [3, 4]. In engineering applications it is defined by means of temperature recovery factor r, total temperature T_0, Mach number M and the ratio of specific heats γ:

* Corresponding author: pss1@mail.ru

$$T_{aw} = \frac{T_0 \cdot \left(1 + r \cdot \frac{\gamma - 1}{2} \cdot \mathrm{M}^2\right)}{1 + \frac{\gamma - 1}{2} \cdot \mathrm{M}^2} \tag{2}$$

The physical meaning of temperature recovery factor r is that it shows a part of the flow kinetic energy that is transformed to the heat on the wall [1-4]. It is defined by means of adiabatic wall temperature T_{aw}, total temperature T_0 and static temperature T:

$$r = \frac{T_{aw} - T}{T_0 - T} \tag{3}$$

There is a theoretical relation for recovery factor – it depends on the working medium Prandtl number, i.e. its viscosity, specific heat and thermal conductivity: $r = \mathrm{Pr}^n$. Numerous experimental research for air flows [5] show that for a turbulent boundary layer in supersonic flow around the plate and along the cylinder recovery factor r lies in the range 0.885±0.01. Theoretical relation for recovery factor in turbulent conditions is: $r = \sqrt[3]{\mathrm{Pr}}$. This relation was experimentally proven for air in and around Pr=0.7. At the same time some researchers consider influence of recovery factor from Reynolds, Mach and Prandtl numbers, injection/suction into/from the boundary layer, shape and relief of streamlined surface [5-11]. These results show that recovery factor can significantly differ from the mentioned approximation.

High influence of heat transfer coefficient from temperature recovery factor is also intensified by low gas temperature in most supersonic wind tunnels. In addition if researched Mach number is not high (less than M=3) then temperature difference between adiabatic wall temperature and wall temperature is not high [4]. Therefore, mistakes in temperature recovery factor cause high uncertainties in heat transfer data [12].

For flow around simple bodies (plane surfaces, cylinders, cones) adiabatic wall temperature is defined with the use of theoretical meaning for recovery factor as a function of Prandlt number. However, when the flow with a boundary layer separation is considered (behind ribs, dimples, holes) then it is difficult to define adiabatic wall temperature and it is often changed for the flow stagnation temperature. Such approach results in high errors in defining heat transfer coefficient – up to 50%. The goal of this research was applying of an unsteady state methodology for adiabatic wall temperature acquisition in aerodynamic experiment together with heat transfer coefficient definition.

2 Experimental apparatus, instrumentation and technique

Experiments have been conducted in supersonic wind tunnel during its launching [12, 13]. Flow velocity in the working area was changed with the use of plane variable supersonic nozzle in range of Mach numbers 2.0÷3.0. Stagnation pressure was up to 7 atm. Stagnation temperature changed from 278 to 298K. Maximum gas flow rate was about 5 kg/sec.

Working area has a rectangular cross-section 70 mm width and 90 mm height (Fig. 1). For the purpose of experimental observation optical glass windows were installed on side walls of the working area. Flow visualization was made with schlieren photography thus providing an opportunity of shock waves observation for separation flow behind the rib. As an experimental model a plate made of plexiglass was taken. This material has a low thermal conductivity coefficient k = 0.19 W/(m·K) thus providing an assumption of heat insulated surface. The model was mounted on the lower side of the working area in parallel with the main flow (Fig. 1). Width of the model fitted the working area – 70 mm, its length was 200 mm. On the lower side of the channel in front of the model a rib was placed perpendicularly to the flow direction. Rib height was changed from 2 to 8 mm. Boundary layer thickness in the region of rib placement was about 6 mm (for flow around plane surface without any disturbances). Reynolds number was calculated for boundary layer length from the nozzle throat and its meaning was $Re_x \geq 2\cdot10^7$, which corresponds to turbulent flow regime.

In order to measure static pressure experimental model was drained along the central line. Through this perforation air bleed was taken to the static pressure probe. Stagnation parameters were measured in settling chamber of the wind tunnel: temperature – with the use of two insulated chromel-alumel thermocouples, static pressure – with the use of probes Honeywell ML-300PS2PC. Wall temperature of the experimental model was recorded by 6 thermocouples pressurized on the wall surface.

Automation program was made in LabView. Data from all the probes was taken to the signal conditioning unit NI SCXI-1303, then it goes through the instrumentation amplifier SCXI-1102 (thermocouple) and SCXI-1102B (pressure probe) and finally comes into the analog-digital converter NI PCI-6220.

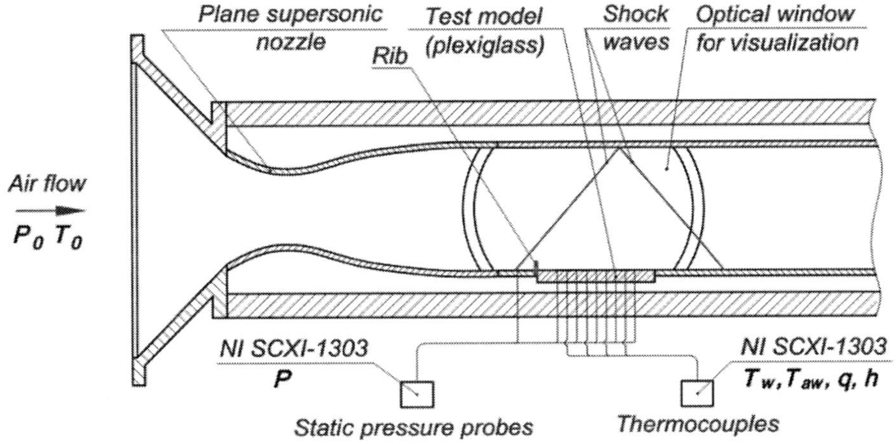

Fig. 1. Experimental apparatus and instrumentation schematic model

The methodology for adiabatic wall temperature acquisition implies registering wall temperature during launching of the wind tunnel. Then the heat flux rate is calculated from the registered temperature time history T_w, plexiglass properties: density ρ, thermal conductivity k and specific heat C (4). Further on the heat flux rate can be presented in dimensionless form so that it could be linearly extrapolated to the zero heat flux meaning [3, 4, 15]. The calculated wall temperature at the zero heat flux rate is equal to the adiabatic wall temperature. Finally, Stanton number can be derived from heat flux rate and difference between measured wall temperature and calculated adiabatic wall temperature.

$$q = \frac{\sqrt{\rho k C}}{\sqrt{\pi}} \int_0^t \frac{\frac{dT_w(\tau)}{d\tau}}{(t-\tau)^{1/2}} d\tau \quad (4)$$

Measurement uncertainties of the main parameters were calculated for 95% confidence interval [12]: ±1.2% for Mach number, ±2.7% for Reynolds number, ±0.8% for temperature recovery factor, ±9.6% for heat transfer coefficient.

3 Results and discussion

The results show that in the separated region behind the rib adiabatic wall temperature is lower than the wall temperature (Fig. 2, a). Total temperature after launching is almost constant during the experiment. Wall temperature is lowering and in the limit it will reach its adiabatic meaning but during 90 seconds of this experiment it didn't happen.

Difference between total temperature and adiabatic wall temperature along the model length is presented on Fig. 2, b). For flow around plane model this meaning is about 17 degrees. For separated flow behind the rib the difference is higher: from 21 degress for 2 mm rib height to 27 degrees for 8 mm rib height.

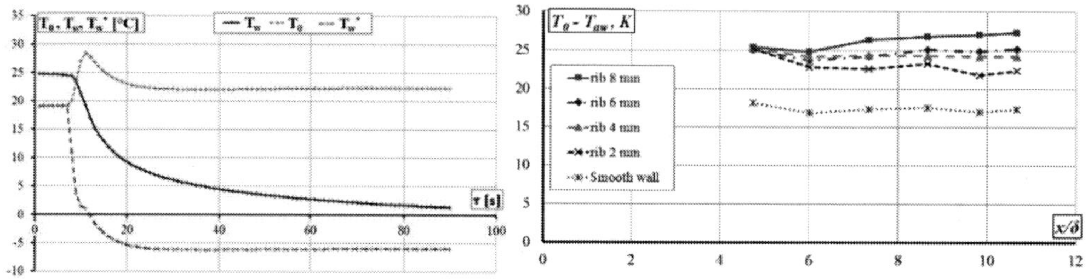

a) **b)**

Fig. 2. a) Total temperature, wall temperature and adiabatic wall temperature changing during launching of the wind tunnel; **b)** Difference between total temperature and adiabatic wall temperature for flow around smooth wall (M_0=2.25) and flow behind the rib of varied height (2, 4, 6 and 8 mm)

Temperature recovery factor is decreased in the region of boundary layer separation behind the rib (Fig. 3): from 3.6% for 2 mm rib height up to 7.7% for 8 mm rib height in comparison with the flow around plane model. While going down from the rib the recovery factor is decreasing in proportion to rib height.

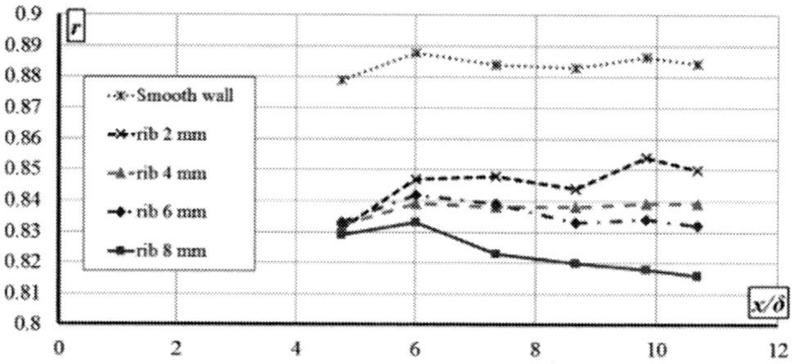

Fig. 3. Temperature recovery factor for flow around smooth wall (M_0=2.25) and flow behind the rib of varied height (2, 4, 6 and 8 mm)

Maximum increasing of heat transfer coefficient was about 34% for 8 mm rib height, while the highest 30% meaning for most thermocouples along the model length was registered for 6 mm rib height (Fig. 4). Comparison of the regimes with and without the rib was taken for the same input Mach number (M_0=2.25). Most of the thermocouples registered 20% increasing heat transfer rate for separated flow in comparison with the smooth model flow.

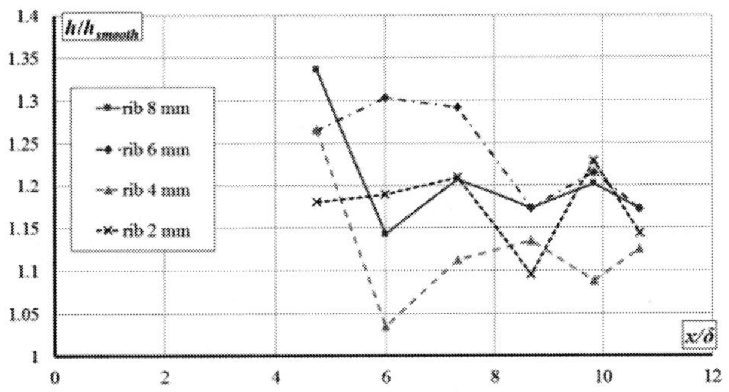

Fig. 4. Heat transfer coefficient for flow behind the rib (rib heights 2, 4, 6 and 8 mm) divided by the coefficient for flow around smooth wall

Conclusions

Experimental results are presented for boundary layer separation influence on adiabatic wall temperature and heat transfer coefficient in supersonic flow around plane surface. Separated flow was generated by means of a rib in front of the plane model. Temperature recovery factor is decreased in separation region by 3.6% for 2 mm rib height up to 7.7% for 8 mm rib height in comparison with the flow around smooth surface. Heat transfer coefficient is increased in the separation region by up to 34% compared to smooth model flow.

The research was supported by Russian Science Foundation (Project 14-19-00699).

References

1. S.S. Kutateladze, A.I. Leontiev, *Heat Transfer, Mass Transfer, and Friction in Turbulent Boundary Layers* (Taylor and Francis, New York, 1990)
2. H. Schlichting, *Boundary layer theory* (McGraw-Hill, New York, 1968)
3. D.L. Schultz, T.V. Jones, AGARDograph, 165 (1973)
4. J.R. Hayes, R.D. Neumann, AIAA, *Tactical Missile Aerodynamics*, **142** (1992)
5. H.A. Johnson, M.W. Rubesin, *Transactions of the ASME* (1948)
6. P.F. Brinich, NACA TN 4213 (1958)
7. B.J. Garland, J.R. Hall, NACA RM, L57L09 (1958)
8. G.E. Gadd, W.F. Cope, J.L. Attridge, A.R.C. Technical Report, 3148 (1960)
9. Yu.V. Baryshev, Yu.A.Vinogradov, A.I. Leont'ev, V.I. Rozhdestvenskii, Fluid Dynamics, **7**, 2, 303 (1972)
10. A.I. Leontiev, Yu.A. Vinogradov, S.M. Bednov, A.N. Golikov, I.K. Ermolaev, E.V. Dilevskaya, M.M. Strongin, Exp. Thermal Fluid Sci., **26**, 487-497 (2002)
11. A.I. Leont'ev, V.G. Lushchik, A.E. Yakubenko, Fluid Dynamics, **43**, 5 (2008)
12. R.J. Moffat, Experimental and Flud Science, 3-17 (1988)
13. S.S. Popovich, K.S. Egorov, Yu.A. Vinogradov, *Proceedings of 15th International Heat Transfer Conference*, IHTC-15-8962, 11 (2014)
14. S. Popovich, M. Strongin, Yu. Vinogradov, A. Zditovets, *Proceedings of the 8th International Symposium on Turbulence, Heat and Mass Transfer*, 865-868 (2015)
15. D.J. Mee, H.S. Chiu, P.T. Ireland, Int. J. of Heat and Mass Transfer, **45**, 3287-3297 (2002)

EPJ Web of Conferences **159**, 00031 (2017)
AVTFG2016
DOI: 10.1051/epjconf/201715900031

Condensation of HFE-7100 vapor in a loop heat pipe having a curvilinear fin

Yuriy Lyulin[1,2,*] *Munko* Gonchikzhapov[2], *Irina* Tueva[2] and *Igor* Marchuk[1,2,3]

[1]Kutateladze Institute of Thermophysics SB RAS, 630090, Novosibirsk, Russia
[2]Novosibirsk State University, 2, Pirogov Ave., 630090, Novosibirsk, Russia
[3]Novosibirsk State Agrarian University, 160, Dobrolyubov Ave., 630039, Novosibirsk, Russia

Abstract. Vapor condensation of the HFE-7100 in loop heat pipe was studied experimentally and theoretically. Numerical calculations of the vapor condensation on the curvilinear fin have been performed. Numerical, theoretical and experimental data are in a good agreement. Minimal condensate film thickness on the top of the fin has been determined and increases monotonously with the increase in the temperature drop.

1 Introduction

Condensation of a pure vapor with a small value of the mass flow is used in many industrial applications. Specifically, it plays an important role in thermal control systems for space applications. For instance, such systems as heat pipes, loop heat pipes, capillary pumped loops and mechanically pumped loops are widely used on satellites to ensure the heat transfer from the core module equipment's to the external radiators [1]. The main component of such devices is condenser. The latent heat of the fluid vaporization is used to transfer the heat efficiently at a nearly constant temperature. The miniaturization of the system requires new studies on the two-phase flows and a better understanding of the condensation phenomenon at this scale. The major goal of the studies is an increase of the performance of condensers. Using of internal curvilinear fins provide an essential heat transfer enhancement [2, 3]. Gregorig [4] first derived a theoretical model for condensation on curvilinear fin with account of surface tension. The fin was described by equation for a spiral with variable direction of rotation. Subsequent models for stationary vapour film condensation on parametric family of curvilinear fins with power dependence of the curvature have been presented in [5-8]. Some experimental results related to the film wise vapour condensation on curvilinear fins are presented in [9, 10]. Description of various theoretical models and experimental results on film wise condensation are presented in several books and reviews [11-14].

Correlations used for the design of condensers for space applications, have been developed and tested only under the ground conditions. It is known that length of the smooth in-tube condenser under microgravity conditions significantly exceeds that for the similar condenser under ground conditions, [15, 16]. Both, theoretical modelling and quantitative experimental results give a better understanding the physical phenomena and

* Corresponding author: lyulin@itp.nsc.ru

lead to correlations and design rules for the next generation of the space exploration of thermal control systems.

The main goals of this work are experimental testing of a new design condenser model with longitudinal fin for loop heat pipe and theoretical study of vapor condensation of the HFE-7100 on curvilinear fin surface.

2 Experimental setup

For the studies of vapor condensation in the condenser with curvilinear fins a new experimental rig was created, see Fig. 1. The experimental rig consists of the following components: test cell (1); high precision liquid pump (2); evaporator (3); cooling water liquid flow meter (4); working liquid flow meter (5); heating water liquid flow meter (6); personal computer (7); thermocouples (8); pressure sensor (9); vapor flow meter (10); pressure controller (11); vacuum pump (12); control and data acquisition system (13); CCD camera (14); heat insulated box (15); heater fan (16); power supply (17); thermostats (18,19); liquid container (20); valves (21-22). Before starting the experiment, air is pumped out from the experimental loop (EC) by a vacuum pump (Agilent IDP-15,12 on Fig.1). Further, the operating liquid from the container is injected into the EC using a liquid pump (ISMATEC REGLO-Z, 2 on Fig.1). The liquid fills the lower part of the condenser and half of the volume of the evaporator (3 on Fig.1). Plate heat exchanger Alfa Laval AXF10-10H-F is used as the vapor generator-evaporator. The heat exchanger has two independent lines, between which heat exchange takes place. The heat exchanger is installed in vertical position. The first line is used in the contour of loop heat pipe for generating vapor. The liquid from the test cell under the action of hydrostatic pressure fills the half of the first line. Hot water is pumped through the second line, thereby heating, an intense evaporation or boiling of the fluid occur in the first line. The temperature of water is controlled by thermostat № 2 (Huber, MPC-K6, 19 on Fig.1) and does not exceed 80 °C. The temperature is measured on input and output of the second line. The thermostat 2 has an internal fluid flow meter (6 on Fig.1). As a result of intense evaporation or boiling liquid vapor generates at the top of the second line and the pressure increases. Vapor from the evaporator moves to the test cell where the condensation occurs on the cooled curvilinear fin. The temperature of the condenser is controlled by pumping water from thermostat № 1 (Huber, MPC-K6, 18 on Fig.1). Vapor flow in the test cell is carried out by the pressure difference between the evaporator and the test cell. The pressure in the test cell is controlled by the condenser temperature and pressure controller (Bronkhorst El-Press, 10 on Fig.1) of vacuum pump. Vapor flow rate is measured by mass flow metter (Bronkhorst FLOW-DP F-201D, 11 on Fig.1) at the entrance to the test cell. Test cell pressure is measured by the pressure meter (WIKA p-30, 9 on Fig.1). The test cell with all monitoring sensors and devices is placed in the heat insulated casing (15 on Fig.1). In this casing a special thermal fan (17 on Fig.1) is placed. The temperature inside the casing is kept above the saturation temperature of 1-3 ° C to prevent uncontrolled condensation which may occur in the used monitoring devices and connecting tubes. As a result the uncontrolled condensation causes the pressure and the flow fluctuations, which can terminate the experiment. Condensation occurs only inside the test cell. Special valve before the vacuum pump is installed which cuts off the line after completion of the experiment, providing a contour sealing (22 on Fig.1). Container volume is 500 ml. Position of meniscus is fixed by CCD camera (14 on Fig.1). Temperature readings were recorded by measuring module NI TB 9214 (13 on Fig.1). All sensors of pressure and of flow are working in the analog mode, the control of which was produced using two DAQ NI 6001 modules (13 on Fig.1). Pressure in the system without liquid was increased by approximately 1 mbar for 24 hours.

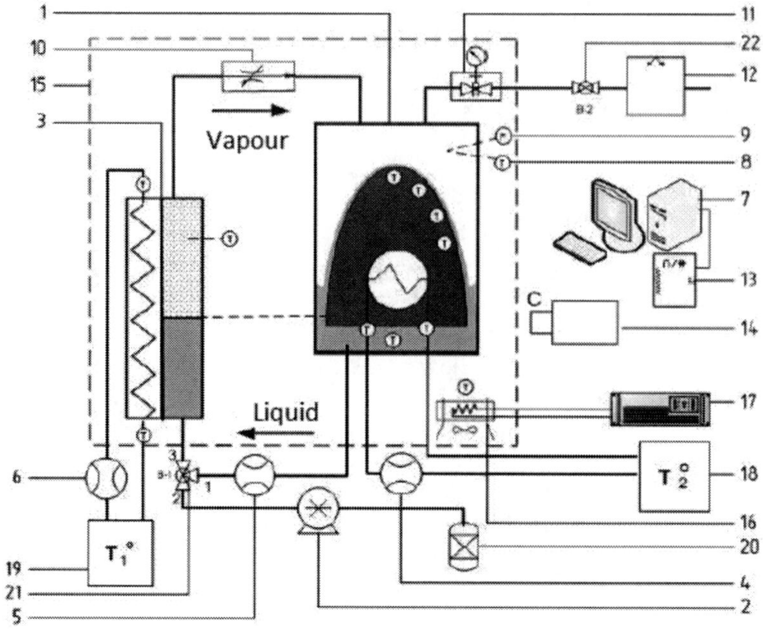

Fig. 1. Scheme of the experimental setup.

The most important part of the experimental rig is the test cell. The design and photo of the test cell with a longitudinal fin are shown in Fig. 2. The test cell consists of the condenser in shape of the longitudinal fin, vapor chamber composed of transparent walls and distribution cambers for cooling water. The condenser fin is made of highly thermally conductive material (Cu). The height of the fin is 54 mm, width 30 mm and length 200 mm. Cooling water from thermostat controls the temperature of the condenser. The cooling water is pumping through the distribution chambers and special channels in the fin body with a diameter of 5 mm. The distribution chambers are fitted at the housing face of test cell and have a water temperature measurement. Attachment of the distribution chambers with the condenser is made with the help of two bolts (M8 with 300 mm length) through housing face walls. Hermeticity of the test cell is achieved by using silicon sealant and rubber seal. The temperature of the condenser is measured with help of six thermocouples (OD = 0.5 mm) that installed by three pieces on each end of the fin. Thermocouples are installed in the fin body through the front sides and holes with a diameter of 1.5 mm. The implementation depth of the thermocouple tips is 100 mm and 30 mm for each side, correspondently. The holes are located on the top, in the middle and at the bottom of the fin at the distance of 5 mm from the surface of condensation. The housing of the test cell is made of transparent polycarbonate walls. The distance between the side walls and fin surface can be varied in the range of 0.5-1.5 mm. Vapor is supplied into the test cell through a stainless steel tube with inside diameter of 14 mm and length of 200 mm. The tube is placed at the distance of 20 mm over the top of the fin. Special holes with diameter of 2 mm with the step of 10 mm are located on the side of the tube providing uniform distribution of the vapor inside the test cell. The temperature of the vapor, condensed liquid and cooling water are measured by the thermocouples with diameter of 1.5 mm that embedded in the test cell using special fitting. Pressure sensor is installed in top cover. Fluid line for pumping out of the condensed liquid is connected from the bottom of test cell.

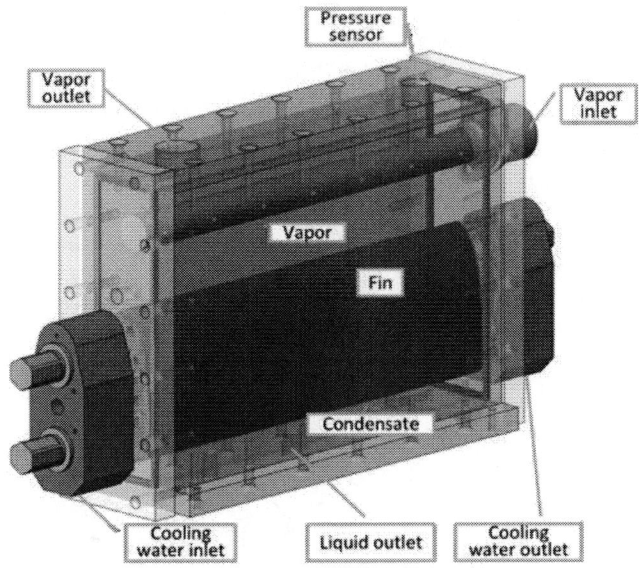

Fig. 2. Design of the test cell.

The average heat-transfer coefficient is determined by formula: $\alpha = q/A\cdot(T_s - T_w)$, where q is heat flux, A is the area of condenser surface, T_s is the vapor saturation temperature, T_w is the average temperature of condenser fin. The temperature of saturated vapor is determined using the dependence of saturated vapor temperature and pressure (P-T dependence). The pressure in the test cell is measured by high-precision sensor of pressure. Temperature of vapor is measured by carefully calibrated thermocouple. Difference between temperature measured by thermocouple and temperature determinated by P-T dependence should be no more than 0.1 K. Thus it is assumed that the vapor moving from the evaporator has no significant impurities affecting the condensation process. The temperature of the condenser fin is defined as the mean value of six thermocouples mounted in the body of the fin. The heat flux is determined by the formula: $q = r\cdot G_v$, where r – specific heat of vaporization, G_v–rate of condensed vapor. The method is based on measuring mass flow rate of vapor at the inlet of the test cell.

3 Results

Experiments are carried at saturation temperature 30,12 C° relevant to pressure 354 mbar. Pure HFE-7100 is used as the working liquid. The difference between the saturation and wall temperatures are changed during the experiments in range from 0.22 to 4.67 K. Shape of fin is calculated using the theoretical model and optimized to obtain the maximum condensate flow rate from the fin surface for given geometric parameters and microgravity conditions. The given generatrix curve of the fin surface is a solution of the corresponding variational problem presented in [4, 7]. The curvature of the generated curve of the fin is defined as a function of curve length s by formula: $k(s) = (as + b)^{-1/2} + c$. The formula constants of defined fin have the following values: a = 6.769519e + 000, b = 4.339554e-005, c = -2.762646e + 001, the curvature at the fin top - 1 mm^{-1}, the length of the generatrix curve S = 60 mm. Full rotation of the curve is equal to 90 °.

The dependence of the average heat transfer coefficient on the temperature drop is presented in Fig. 3. The temperature drop between the surface and vapor temperature varies from 0.1 K to 5 K. The diagram shows the experimental (curve 1) numerical (curve 2) and theoretical (curve 3) results. Numerical calculations of the vapor condensation of pure HFE-7100 liquid on the curvilinear fin have been performed using the theoretical model described in [7, 10]. The theoretical heat transfer coefficient is calculated by Nusselt formula for condensation on the vertical square plate: $\alpha \quad 4/3\sqrt[4]{\lambda^3\rho^2 rg/4\mu\Delta TL}$, where λ – heat conductivity of the liquid, W/(m·K); ρ - liquid density, kg/m^3; r – latent heat of vaporization, J/kg; g – gravitational acceleration, m/s^2; μ – dynamic viscosity Pa·s; $\Delta T \quad (T_s \quad T_w)$ –difference between the saturation and wall temperatures, K. L – linear size is chosen using the assumption that the surface area of the plate equals to the half surface of the longitudinal fin.

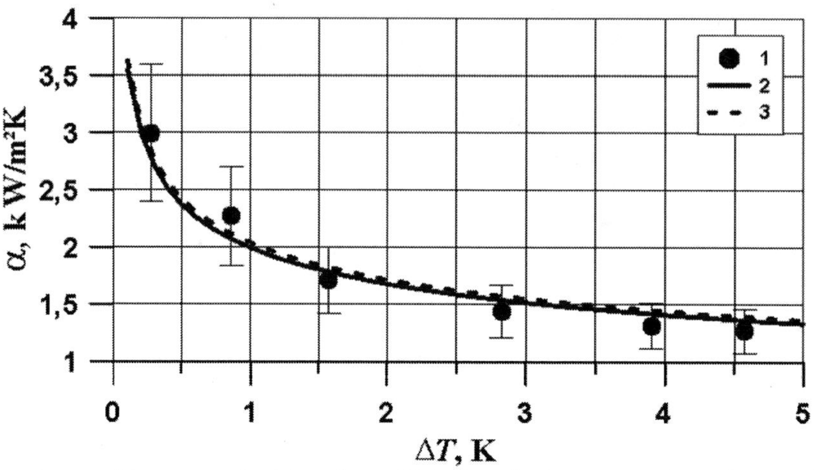

Fig. 3. Dependence of the average heat transfer coefficient on the temperature drop. 1 – experimental data, 1 – numerical calculation, 2- calculation by Nusselt formula.

The heat transfer coefficients display the usual trend that expected for the condensation of the pure vapor on vertical wall. The heat transfer coefficient is reduced monotonically with the increase of the temperature drop. The values of the average heat transfer coefficient are in the range of 1.27-2.9 kW/ m^{2}K for experimental data and 1.3-3.6 kW/ m^{2}K for the numerical and theoretical ones, Fig. 3. The distribution of the condensate film thickness for the temperature drop ΔT=5 K is presented in Fig. 4. Calculated minimal condensate film thickness is at the top of the fin and changes from 3 μm to 11.4 μm for the given temperature range, see Fig 4b. Minimal condensate film thickness increases monotonously with the increase in the temperature drop.

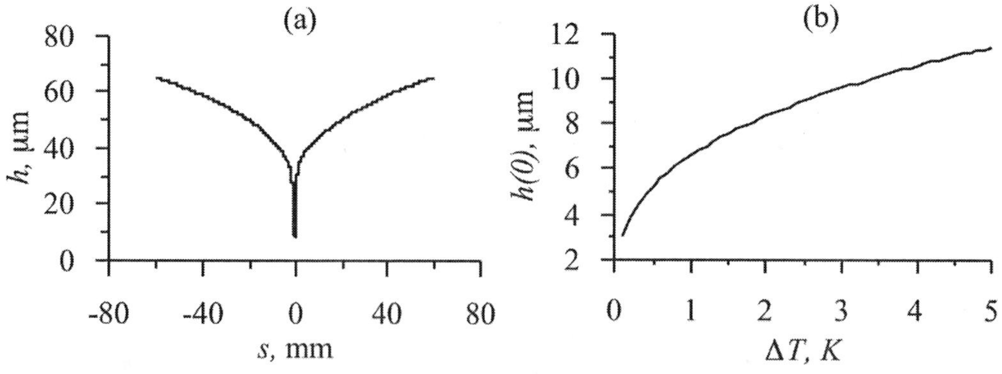

Fig.4. Condensate film thickness. (a)-along the fin surface, ΔT=5 K; (b) - on the fin top.

4 Conclusions

New experimental rig and numerical calculations for study of film vapor condensation HFE-7100 on the curvilinear fins has been performed. The results of experiment and calculations of condensation of HFE-7100 on the longitudinal fin have been obtained for different values of the temperature difference between saturation temperature and fin surface temperature. The calculated heat transfer coefficient is in the range of 1.3-3.6 kW/m^2K, the experimental data is in the range 1.27-2.9 kW/ m^2K. Numerical, theoretical and experimental data are in a good agreement. Minimal condensate film thickness on the top of the fin has been calculated and changes from 3 μm to 11.4 μm. It is shown the minimal condensate film thickness increases monotonously with the increase in the temperature drop.

The authors gratefully acknowledge the support of this work by the Ministry of Education and Science of the Russian Federation (Agreement 14.616.21.0016, project identifier RFMEFI61614X0016).

References

1. V. Serin, P. Lavieille, M. Miscevic, Microgravity Science and Technology, **21**, 1, 103–109 (2009)
2. O.A. Kabov, D.R. Kolyukhin, I.V. Marchuk, J.-C. Legros, Journal of Engineering Thermophysics **12**, 1, 199–210 (2003)
3. R.L. Webb, *Principles of Enhanced Heat Transfer*, p. 556 (1994)
4. R. Gregorig, Zeitschrift für angewandte Mathematik und Physik **Bd. 5**, 1, 36 – 49 (1954)
5. C. Zener and A. Lavi, Journal of Heat Transfer **96**, 205-209 (1974)
6. T. Adamek, Warme - und Stoffubertragung **15**, 255 – 270 (1981)
7. I.V. Marchuk and O.A. Kabov, *Encyclopedia of Two-Phase Heat Transfer and Flow II: Special Topics and Applications,* **Vol. 3, Chapter 5,** pp. 133-176 (2015)
8. I.V. Marchuk and O.A. Kabov, Doklady Physics, **61**, 1, 19-23 (2016)
9. A. Glushchuk, I.V. Marchuk and O.A. Kabov, Journal Microgravity Science and Technology **23**, 1, 65–74 (2011)
10. I.V. Marchuk, A.V. Gluschuk and O.A. Kabov, Technical Physics Letters **32**, 5, 388-391 (2006)
11. J.R. Thome, *Engineering Data Book III* (2008)
12. V.G. Rifert, H.F. Smirnov, *Condensation Heat Transfer Enhancement*, p. 371 (2004)
13. J. Rose, Enhanced condensation heat transfer, **49**, 3, 626-635 (2006)
14. A. Cavallini, G. Censi, D. Del Col, L. Doretti, G.A. Longo, L. Rossetto, C. Zilio, International Journal of Refrigeration **26**, pp. 373-392 (2003)
15. E. G. Keshok, M. S. Sadeghipour, Aste Astronautica **10**, 7, 505-511 (1983)
16. I. Da Riva, A. Sanz, Microgravity Science and Technology **4**, 3, 179-187 (1991)

Experimental study of air delivery into water-conveyance system of the radial-axial turbine

Alexandra Maslennikova[1], *Dmitry* Platonov[1,2 *], *Andrey* Minakov[1,2] , and *Dmitry* Dekterev[2]

[1]Siberean Federal University, 79 Svobodny pr., 660041 Krasnoyarsk, Russia
[2] Institute of Thermophysics, SB RAS, 1 Lavrentyev pr., 630090, Novosibirsk, Russia

Abstract. The paper presents an experimental study of oscillatory response in the Francis turbine of hydraulic unit. The experiment was performed on large-scale hydrodynamic test-bench with impeller diameter of 0.3 m. The effect of air injection on the intensity of pressure pulsations was studied at the maximum pressure pulsations in the hydraulic unit. It was revealed that air delivery into the water-conveyance system of the turbine results in almost two-fold reduction of pressure pulsations.

1 Introduction

Power control in the energy system is an important task of hydropower plant. When changing load of hydraulic unit, the latter repeatedly undergoes off-design modes of operation. In these modes, the flow remains essentially swirling after passing through the water turbine runner. The instability of swirling flow results in the appearance of intense low-frequency hydrodynamic pulsations, which threaten the safety and reliability of turbine design [1-4].

The main reason for the occurrence of low-frequency pressure pulsations is the precessing vortex core generated in the diffuser of the draft tube due to essential residual swirl of the flow.

Up to date, numerous methods are proposed to stabilize flow in the hydraulic units [5-8]. Each of them has its advantages and disadvantages, and requires further detailed study. In this study, the effect of air delivery on the oscillatory response of the flow was experimentally investigated on the model hydrodynamic Francis turbine test-bench.

2 Experimental techniques

The model hydroturbine with radial-axial impeller is part of the hydro-power test-bench at the Laboratory of Hydraulics and Hydraulic Machines of the Siberian Federal University (Fig. 1.). Maximum water head in the test-bench is 3.5 m, the diameter of the impeller is D=0.3 m.

* Corresponding author: platonov-08@yandex.ru

DOI: 10.1051/epjconf/201715900032

Hydraulic unit includes a pressure tank, performing the functions of the upper pool, power conduit with the diameter of 400 mm, and penstock valve, equipped with electric drive.

Water from the penstock enters the spiral chamber, where from flows through the guide vanes to the impeller and then passing the draft tube gets into the lower pool, which is a glass groove 1.52 m in width. Diffusion grid is designed to reduce fluctuations in water level before weir gauge. Weir gauge is used for measuring water flow rate and is fixed by water gauge.

After weir gauge the water is discharged into the collecting tank and then is diverted through pipe into the pool. From the pool, water is pumped into the pressure tank. The water intake in the collecting tank is performed through the suction cap with mesh. The water flow rate is controlled by gate valve with electric actuator.

To record the pressure pulses, the test-bench was equipped with a piezoelectric flow rate sensors (type 014MT) designed to convert rapidly varying and pulsed pressure into an electrical signal.

Fig. 1. Model hydro-turbine with radial-axial impeller.

3 Experimental results

In the course of experiments we carried out measurements within a wide range of the hydraulic unit operation. The pressure head and the degree of opening of the guide vanes were varied in the experiment. Then all the necessary parameters, namely, impeller rotation frequency, the flow rate through the hydraulic unit, and generating current and voltage were recorded.

The behavior of the frequency and the pressure pulsations intensity in the water-conveyance system of the turbine is of great interest for high-head power plants since the precession of a vortex core is a serious hazard to the hydroturbine equipment. Pressure pulsations were recorded at specified points during five minutes for each of the operation modes.

Figures 2 and 3 below present the results of the experiment carried out for one of the operating modes of the test-bench. The diagrams present time behavior of the pressure pulsation and the pressure pulsation spectrum in the draft tube diffuser. As is obvious, in this mode there is a pronounced frequency of pressure fluctuations associated with the precession of the vortex.

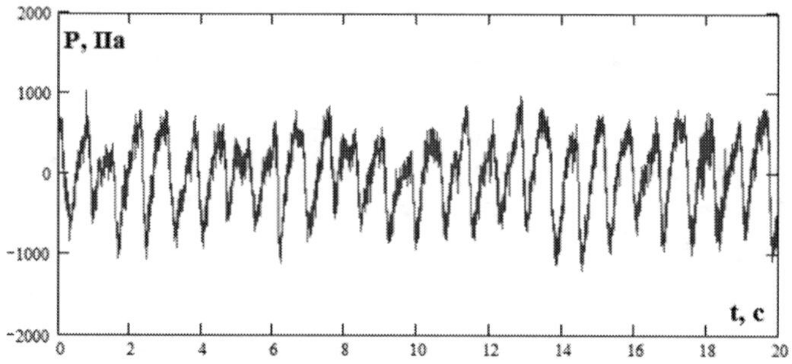

Fig. 2. Time dependence of the pressure pulsation in the draft tube diffuser *(opening **3/15**; head H=269 cm; Q=186.5 m³/h)*

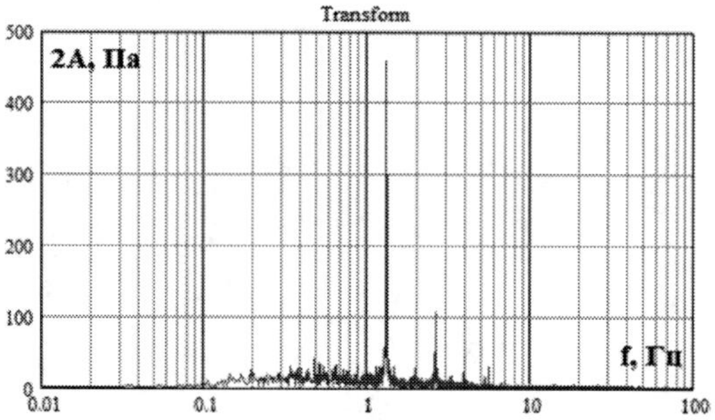

Fig. 3. Spectrum of the pressure pulsations in the draft tube diffuser *(opening **3/15**; head H=269 cm; Q=186.5 m³/h)*

Figure 4a shows the pressure pulsation in the draft tube diffuser behind the impeller depending on the opening of the guide vanes. Apparently, the pressure pulsations have a maximum in part-load operation mode at the opening of the guide vanes at 5 mm. Further increase in the opening of the guide vanes and transition of the turbine to the optimal operation mode leads to the decrease in pressure pulsations.

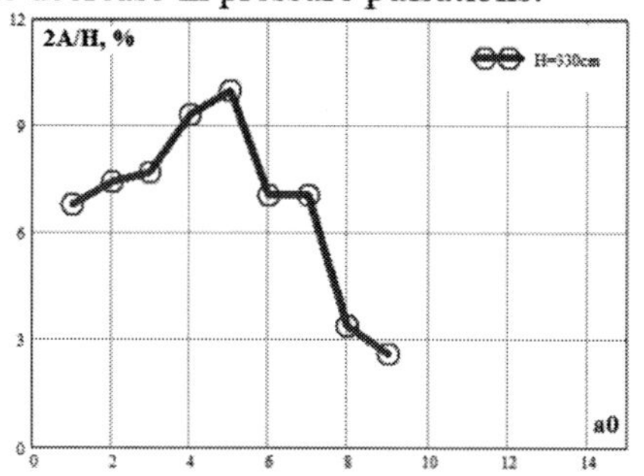

Fig. 4. Dependence of the pressure pulsation amplitude in the draft tube diffuser on the opening of the guide vanes.

Further, to suppress pressure pulsations, air was delivered into the water-conveyance system of the hydraulic turbine. Air was supplied into a scroll chamber of the turbine by two compressors. Air flow rate was monitored by ultrasonic flowmeter.

Figure 6 shows the pressure pulsations spectrum in the draft tube diffuser: (a) – corresponds to spectrum without air, while (b) – with air (air flow rate equals 16 m^3/h). These results are obtained for the mode of H=3m and the guide vanes opening at 5/15. This mode was accompanied by the maximum level of pressure pulsations (see Fig. 4).

The spectra presented in Figs. 6 show that the amplitude of pressure pulsations at the air injection is reduced by almost an order of magnitude. In addition, the fundamental frequency of the pressure pulsations changes upwards. Earlier similar behavior was noted in the works [9] for model test-benches with swirling flow.

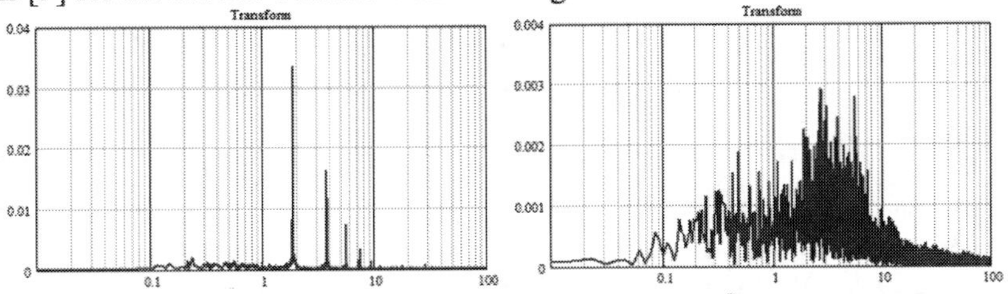

a **b**

Fig. 6. Pressure pulsations spectrum in the draft tube diffuser; (a) – without air; (b) – with air.

Figure 7 shows the mean-square pressure pulsations at two points located in the draft tube diffuser versus the volumetric flow rate of air. Here the pressure pulsations are reduced to those without air. It is clear that with increasing the flow rate of the intake air, pressure pulsations reduce quite significantly. Thus, at the flow rate of 16 m^3/h, that is approximately 2% of the water flow rate, pressure pulsations are reduced by about 40%.

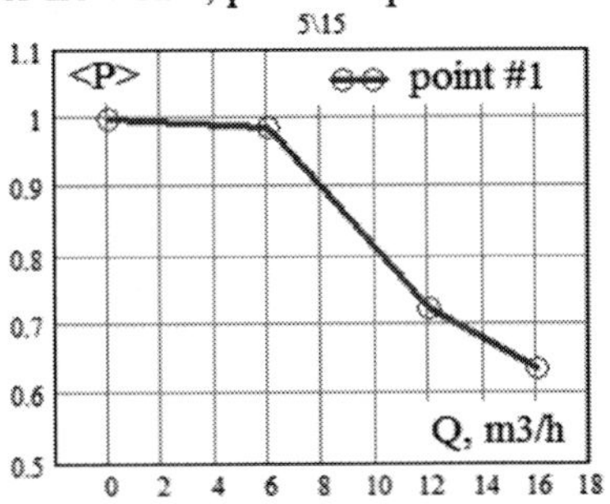

Fig. 7. The root mean square pressure pulsations in the draft tube diffuser depending on the opening of the guide vanes.

4 Conclusion

The work deals with experimental investigation of pressure pulsations for different flow regimes in the hydraulic unit of the medium-scale hydrodynamic test-bench. The authors obtained correlations between the pressure pulsations and the opening of the guide vanes for different pressure heads, as well as carried out a preliminary study on the effect of air delivery on pressure pulsations in the hydraulic turbine. The study was performed in the conditions of maximum pressure pulsations. Air was supplied into a spiral chamber of the turbine. This method is one of the most promising to reduce loads on the hydraulic unit, caused by the precession of the vortex core in the space behind the impeller. The results of the study have shown that it is possible to significantly reduce the pressure pulsations using air. However, this effect needs further investigation. In particular, it is necessary to investigate the influence of air on pressure pulsations in other operation modes of the turbine as well as to study the effect of air injection point.

The current work is performed at partial support of the projects funded by the Russian Foundation for Basic Research and Krasnoyarsk Regional Fund for Support of Scientific and Scientific-Technical Activities (16-48-243076).

References

1. A.A. Dekterev, A.V. Zakharov, A.V. Minakov, D.V. Platonov, I.M. Pylev, Fluid Dynamics, **50**, 5, 601-612
2. A.V. Minakov, A.V. Sentyabov, D.V. Platonov, A.A. Dekterev, A.V. Zakharov, Computer and fluids, 197-205
3. A.V. Minakov, A.V. Sentyabov, D.V. Platonov, A.A. Dekterev, A.V. Zakharov, International Journal of Heat and Fluid Flow, **53**, 183–194, (2015)
4. D. Platonov, A. Minakov, D. Dekterev, A. Sentyabov, A. Dekterev, Journal of Physics: Conference Series, **754**, (2016)
5. Falvey, H.T., REC-ERC-71-42, U.S. Bureau of Reclamation, 33, (1971)
6. Wahl, T.L., Colorado State University, 77, (1990)
7. Grein, H., Proceedings, 10th Symposium of the International Association for Hydraulic Research Section for Hydraulic Machinery Equipment and Cavitation, Tokyo, Japan, (1980)
8. Bhan, S Proceedings, 5th International Symposium on Hydro Power Fluid Machinery, Chicago, Illinois, 95-102, (1988)
9. Alekseenko S.V., Kuibin P.A., Okulov V.L., Shtork S.I., Heat transfer research, **41**, 4, 465-477, (2010)

Experimental study of moving liquid layer evaporation under gas flow into rectangular minichannel

Vyacheslav Maksimov[1,*], *Takhir* Sharifulin[1] , *Larisa* Zhidkova[1] and *Karapet* Eloyan[2]

[1]National Research Tomsk Polytechnic University, Tomsk, Russia
[2]Institute of Thermophysics SB RAS, Novosibirsk, Russia

Abstract. The paper presents the experimental results on the mass evaporation rate from the moving liquid layer surface streamlined by the gas flow for different layer thicknesses. The maximum value of the mass evaporation rate was observed for a 5 mm liquid layer. Further increase in the thickness of the liquid layer causes a decrease in the mass evaporation rate. The assumption about the change in the balance of forces was made. A self-organized structure of the square-shaped spatial cells is found to form in the moving liquid layer streamlined by the gas flow. This structure has the most visible form at the maximum possible velocity of the air.

1 Introduction

Convective motion in a liquid layer arising due to turning up a liquid with higher density is known as Rayleigh-Bénard convection. Rayleigh-Bénard convection has been studied mainly in a horizontal liquid layer between solid walls [1]. Convection caused by the surface tension (Marangoni convection) has been studied in [2]. At the same time, the convection in a liquid layer with free boundary (the phase boundary between the liquid and the gas or vapor located above) is still less studied.

There are two basic mechanisms affecting the instability in systems with a free boundary. The mechanism of the instability arising due to the buoyancy force is more important for relatively thick liquid layers. While the instability caused by thermocapillary effect is dominant in thin liquid layer. Many works are devoted to the determination of the boundary of so-called "pure convection" caused by turning up a liquid with higher density (Rayleigh–Bénard convection). In systems with a free boundary there are two kinds of instability: stationary [3] and periodic [4]. In the case of Marangoni convection in systems with free boundary it was found that there are many kinds of instabilities. They are cells [5], free surface deformation leading to the appearance of dry spots [6], longitudinal [7], and transverse [8, 9] periodic instabilities. But the most common cases are in which there are two mechanisms of instability formation simultaneously.

Convection in the liquid layers which are continuously heated from below with a constant temperature gradient and excluding evaporation is most often studied [10]. There

* Corresponding author: author@email.org

is a lack of studies devoted to evaporating liquid layers since evaporation greatly complicates the problem. Experimental studies devoted to research of evaporating single-component liquid layer and observation of the convective flows formation in small amount of liquid have been conducted in [11]. Four basic modes of convection for the single-component liquids have been analyzed in the work. For all liquids with 2 mm or less thickness, the dominant mode of convection is a mode with formation of cells. For moderately volatile liquids, such as alcohols, there is a pseudo stable flow pattern, temporary cells. There is either growth or reduction of cells or accidental movement of cells due to interaction with neighboring cells. For intensively evaporating liquids (for example, different types of ether), unstable state of liquid can be observed.

The influence of non-linear temperature profiles before convection on the order of the critical Marangoni number in the case of the flow caused by the surface tension in the evaporating liquid layer was analyzed using the linear stability theory [12]. It is found that in case of a strong non-linearity of the temperature profile, the critical Marangoni number may be much higher than for the corresponding linear temperature profile.

The evaporation in the liquid layer is mainly considered only as a means of cooling the surface layer. The role of evaporation is ignored in the problem of convective stability. There are very small numbers of experimental studies on the comparative analysis of the instability with a short-wave nature that may arise in volatile liquid layers. There are experimental results on the effect of the evaporation on the thermocapillary convection in thin liquid layer [13]. The evaporation was found to affect not only the cooling of the surface layer but also to be the driving force influencing the development of convective instability.

The dynamics of instabilities caused by the action of surface tension and interfacial phenomena of convection in two-layer system was studied in [14]. The analysis of the interaction between the natural convection, thermo-capillary convection and convection caused by the shear stress [15] was conducted. The analytical and numerical solution of a liquid flow in an open horizontal space under the influence of the gas flow was obtained. The topology of convective flows with non-deformable interphase boundary due to the impact of the inert gas flow is presented. The effect of the stress tangential component caused by the action of the gas flow was studied. The exact solutions of the stationary problem of convective flow with thermo capillary interfacial boundary in a horizontal liquid layer under the influence of the inert gas flow were obtained. It demonstrated the presence of a reverse flow with respect to the direction of gas flow near the interface.

It was shown theoretically and experimentally that the structural changes in the flow resulting to backward movement of the liquid film can occur under the action of thermo-capillary forces arising due to the temperature gradient at the interface between liquid and gas [16].

Thermocapillary convection has a significant effect on the motion of the liquid in two-phase systems [17]. Significant temperature differences at the interface cause waves, vortex and reversal flows, and can lead to rupture of the film [18].

It can be concluded based on the analysis of scientific publications [19-25] that the evaporation of the liquid film under the influence of the gas flow has not been studied at a level that would allow for predictive modeling of high-technology equipment with mini- and micro-channels.

The dynamics of evaporation depending on the temperature of the two-phase system, gas flow velocity, liquid layer thickness, and, in addition, the influence of normal and shear stresses, thermocapillary effect, instabilities forming in the liquid on evaporation were studied in [26]. It is noted that the viscosity and surface tension depend on the temperature and lead to a change in the evaporation rate; by increasing the thickness of the liquid film, thermocapillary effect becomes the main factor of evaporation. In this case, causes of changes in the evaporation rate at different parameters cannot be explained from a physical point of view.

In view of the above, the study of the heat exchange processes in the interaction of the moving liquid film with a parallel gas flow is relevant.

The purpose of this study was to determine the dependence of the mass evaporation rate on the liquid layer thickness moving in minichannel under the action of the following gas flow.

2 Experimental apparatus

Experimental studies were conducted on the setup [27, 28] for the study of the dynamics of evaporation and convective flows in minichannel.

The conception of the experiment includes the application of previously tested methods [27] for studying heat and mass transfer across the interface between liquid and gas.

The experiments have been conducted under the atmospheric pressure in the test cell. The thickness of moving liquid layer was varied from 1 to 8 mm. As a working liquid we used ethanol (95% ppb mass). The evaporation surface blown by a gas flow (air) was 100mm^2 with 10x10 mm cut in the plate. The gas flow rate was ranged from 100 to 1000 ml/min with a 100 ml/min step, which corresponds to an average gas velocity from 0.014 m/sec to 0.125 m/sec. The main influencing factors are summarized in Table 1.

Table 1. The main influencing factors.

Factors	Values
Gas velocity, m/sec	0.014 – 0.125
Interface area, mm^2	100
Velocity of the moving liquid layer, m/sec	$8.3 \cdot 10^{-5}$
Pressure in the circuit	atmospheric
Liquid in the system	ethanol (95% ppb mass)
Gas in the system	air

3 Results and discussion

When conducting the experiment, the temperatures of the liquid and gas were kept constant. Figure 1 presents the dependence of the mass evaporation rate on the liquid layer thickness at the ethanol/air system temperature of T=20°C in the gas velocity range from 0.014 to 0.125 m/sec. The value of the mass flow rate is found to have a local maximum. An increase in the evaporation liquid layer thickness from 1 to 5 mm leads to an increase in the intensity of evaporation, which reaches a maximum at a 5 mm layer thickness. A further increase in the liquid layer thickness causes a decrease in the mass evaporation rate.

DOI: 10.1051/epjconf/201715900033

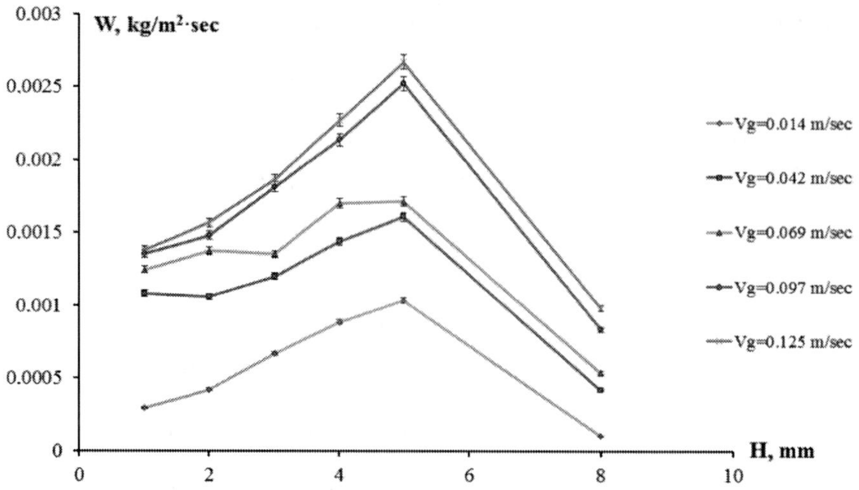

Fig. 1. Mass evaporation rate versus liquid layer thickness at the ethanol/air system temperature of T=20°C.

Experimental dependences of the mass evaporation rate on the liquid layer thickness in the gas flow velocity range from 0.028 to 0.139 m/sec at the ethanol/air system temperatures of T=30°C, T=40°C, T=50°C are presented in Fig. 2-4.

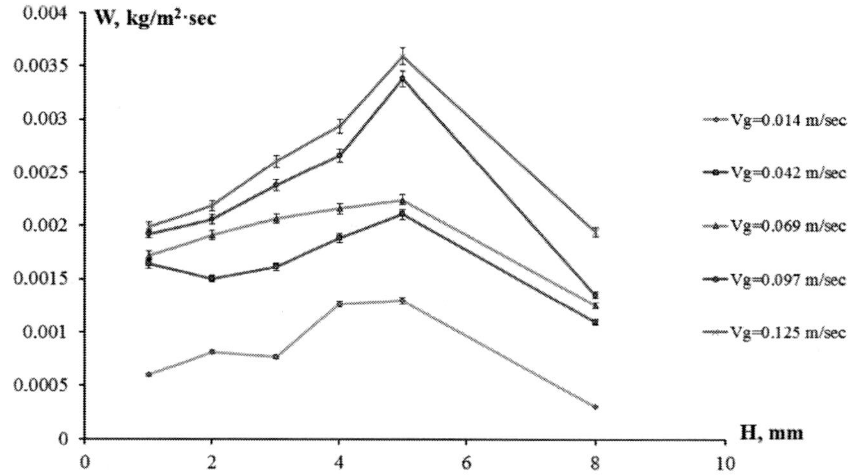

Fig. 2. Mass evaporation rate versus liquid layer thickness at the ethanol/air system temperature of T=30°C.

According to analyses of presented dependences (Fig. 1-4) W is found to increase reaching a maximum value at a 5 mm liquid layer thickness. Further increase in the thickness of the liquid layer causes a decrease in the mass evaporation rate.

Most likely, it is due to the change in the balance of forces caused by thermocapillary effect, shear stresses, which appear because of the gas flow and natural convection. Natural convection is dependent on the liquid layer thickness. Possibly, there is some convection mode with the maximum evaporation rate for each liquid layer thickness and temperature of the liquid and gas. It should be noted that the geometrical dimensions of the rectangular channel through which the gas and liquid is transported can significantly affect the convection mode.

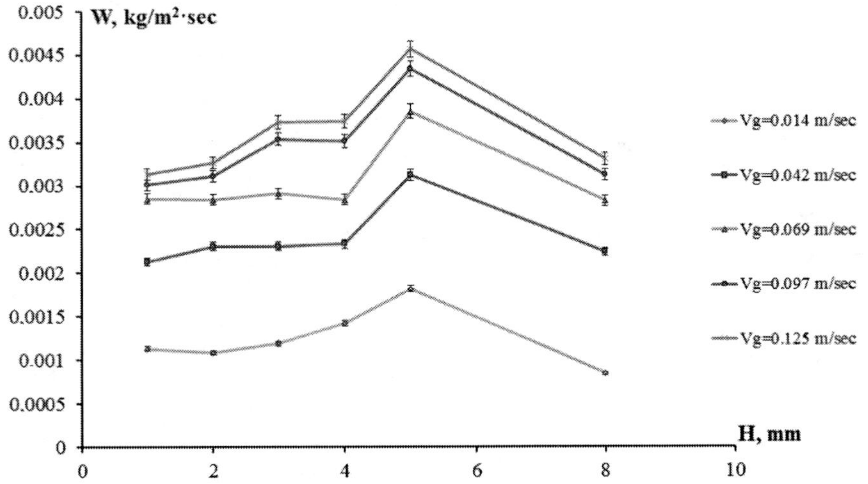

Fig. 3. Mass evaporation rate versus liquid layer thickness at the ethanol/air system temperature of T=40°C.

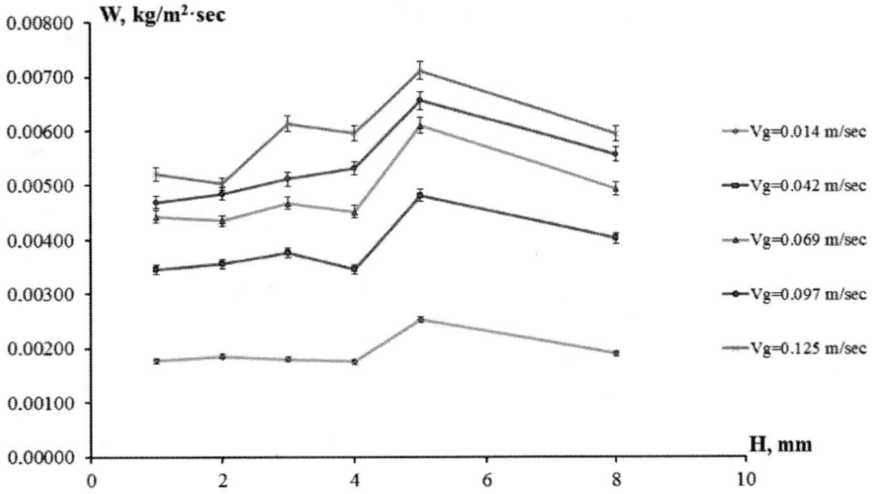

Fig. 4. Mass evaporation rate versus liquid layer thickness at the ethanol/air system temperature of T=50°C.

The most common method for visualizing convective flows is the shadow method based on the temperature dependence of the refractive index. Cold downflows focus the light and look bright projected on the screen. And warm upward flows act dissipating and appear dark.

Photographic images of experimental studies obtained using the shadow method are shown in Table 2-3.

According to analyses of the photographic images a self-organized structure of the square-shaped spatial cells is found to form in the moving liquid layer streamlined by the gas flow. Formation of cells occurs in proportion to the average velocity of the air flow; at the maximum possible velocity of the air cells have the most visible form.

 DOI: 10.1051/epjconf/201715900033

Table 2. Photographic images of the interface at T=50°C (H=1÷3 mm).

	V_g=0.014 m/sec	V_g=0.042m/sec	V_g=0.069m/sec	V_g=0.097m/sec	V_g=0.125m/sec
1mm	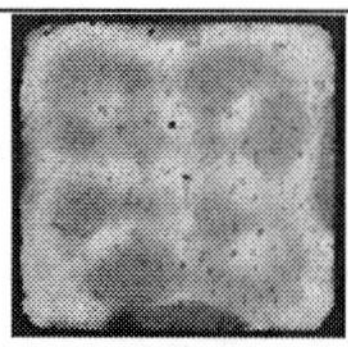	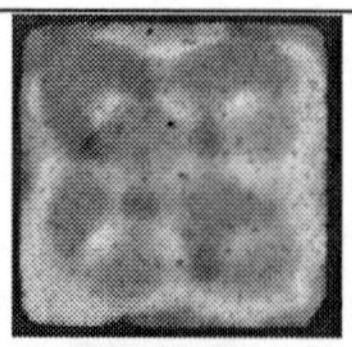	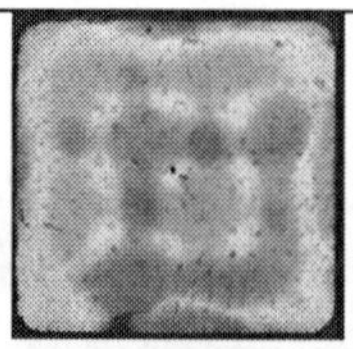	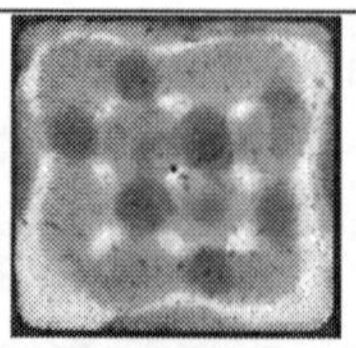	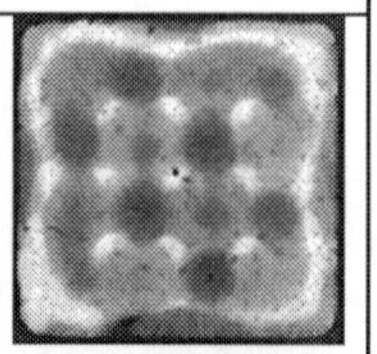
2mm			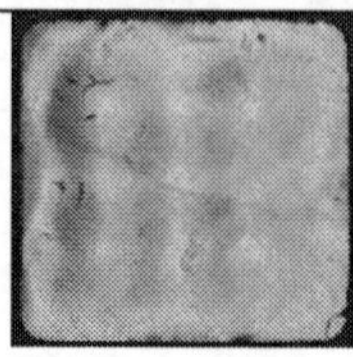	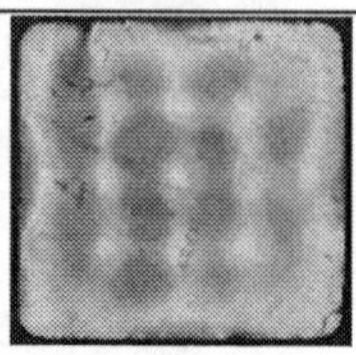	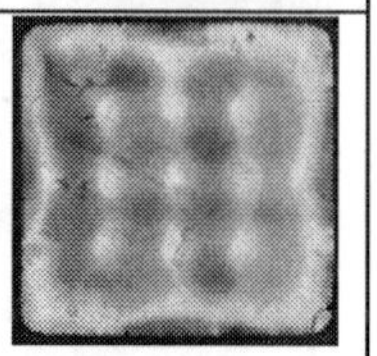
3mm	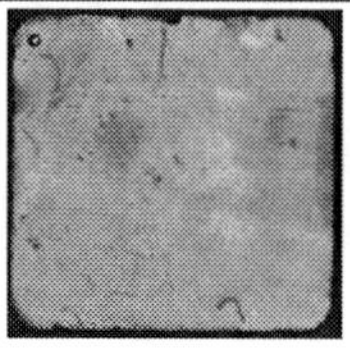	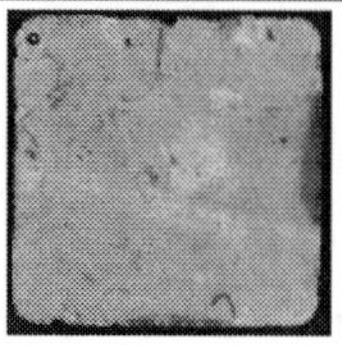	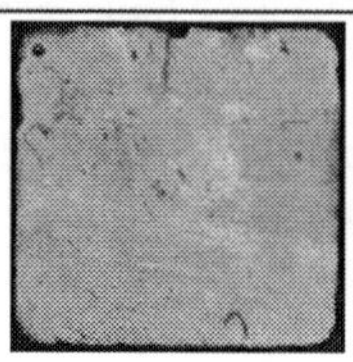	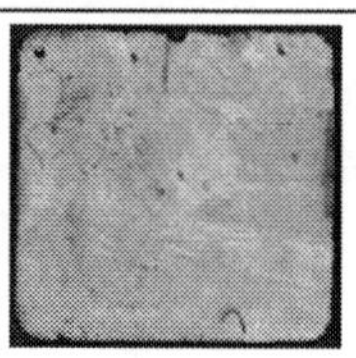	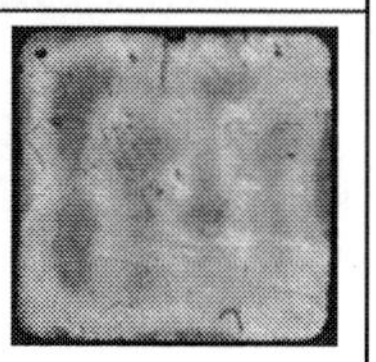

Table 3. Photographic images of the interface at T=50°C (H=4÷8 mm).

	V_g=0.014 m/sec	V_g=0.042m/sec	V_g=0.069m/sec	V_g=0.097m/sec	V_g=0.125m/sec
4mm	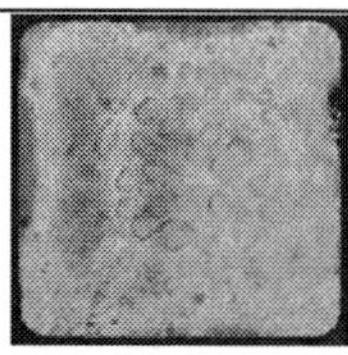	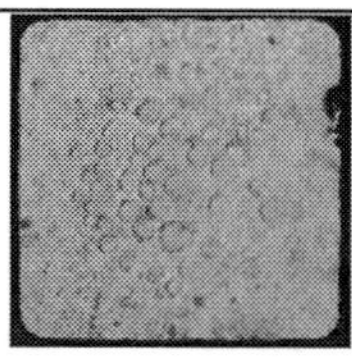	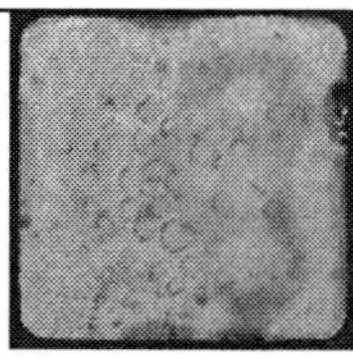	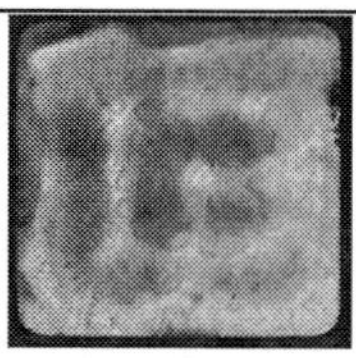	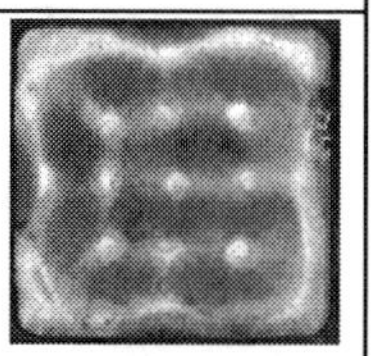
5mm	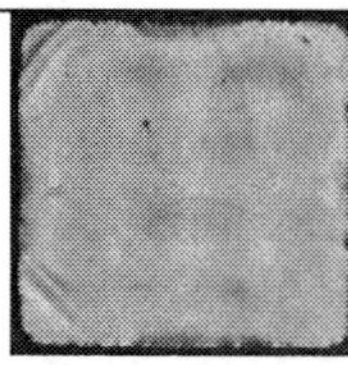	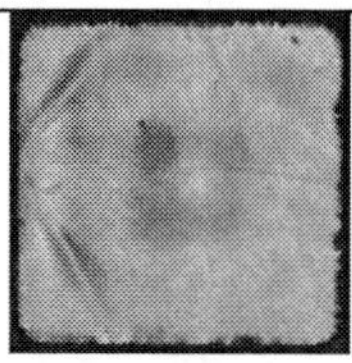	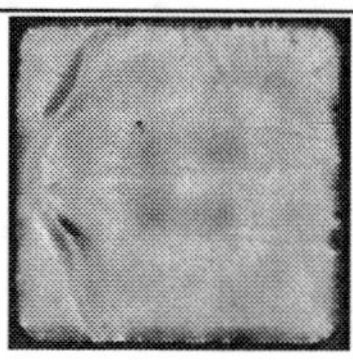	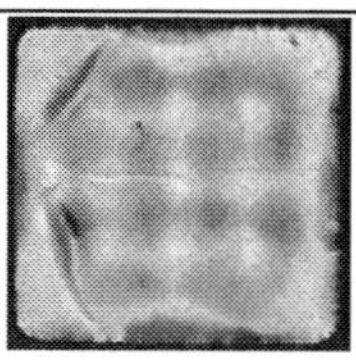	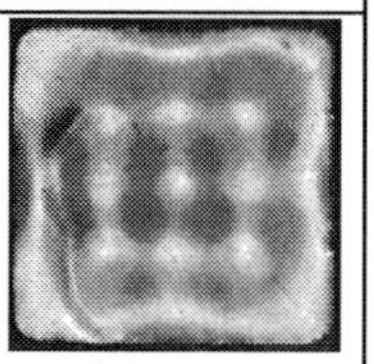
8mm	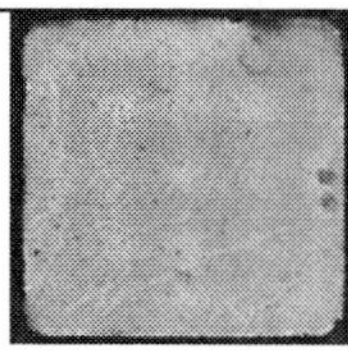	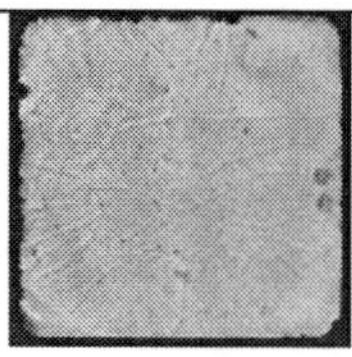	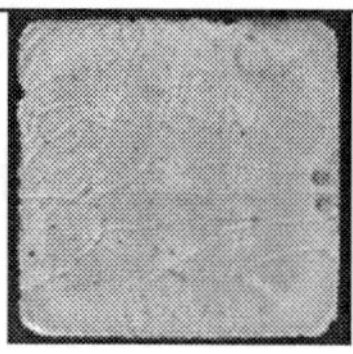	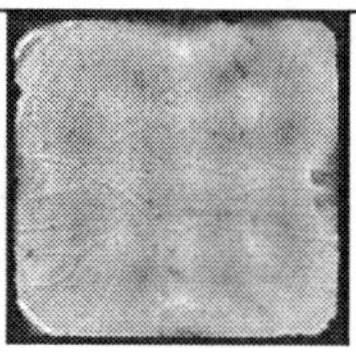	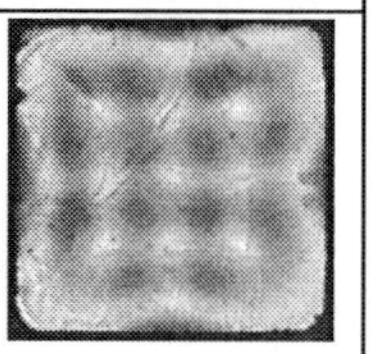

Also it is worth noting that at the 5 mm liquid layer thickness a regular structure in the form of liquid roller occurs from the side of the gas inlet. This structure becomes more visible with increasing gas flow velocity. Further increase in the liquid layer thickness up to 8 mm results in the disappearance of the structure, while the convective cells form at the interface.

In earlier work [29] the effect of the film structure on the evaporation intensity was studied. It was noted that various types of instabilities can increase the mass evaporation rate. Thus, the occurrence of the maximum evaporation rate of liquid layer with a 5 mm thickness, and reduce in the evaporation rate when further increase in the liquid layer thickness can be caused by a regular structure, which is formed at a 5 mm layer thickness.

The research was supported by the Ministry of Education and Science of Russia (Agreement No 14.613.21.0011, project identifier RFMEFI61314X0011).

References

1. J. R. A. Pearson, J. Fluid Mech. **4**, 489 (1958)
2. I. Simanovskii, Fluid Dyn. **14**, 637 (1979)
3. P. Colinet, J.C. Legros. Phys. Fluids **6**, 2631 (1994)
4. K. Eckert, M. Bestehorn, A. Thess. J. Fluid Mech. **356**, 155 (1998)
5. S. J. Van-Hook, M. F. Schatz, W. D. McCormick, J. B. Swift, H. Swinney. J. Fluid Mech. **345**, 45 (1997)
6. C. V. Sterling, L. E. Scriven. AIChE J. **5**, 514 (1959)
7. P. L. Garcia–Ybarra, M. G. Velarde. Phys. Fluids **30**, 1649 (1987)
8. J. C. Berg, M. Boudart and A. Acrivos. J. Fluid Mech. **24**, 721 (1966)
9. J. C. Berg, A. Acrivos, M. Boudart, *Evaporative convection*, Advanced Chemical Engineering, Vol. 6, Ed. By T. B. Drew et al. (Academic Press, New York, 1966)
10. O.A. Kabov, D.V. Zaitsev, Multiphase Science and Technology **21**, 249 (2009)
11. A. Vidal, A. Acrivos. I & EC Fundam. **7**, 53 (1968)
12. N. Zhang, A.-T. Chai. Exp. Heat Transfer **11**, 187 (1998)
13. A. Nepomnyaschy, I. Simanovskii, L. Braverman. J. Fluid Mech. **442**, 141 (2001)
14. C. S. Iorio, O. N Goncharova, O. A. Kabov. Microgravity Sci. Technol. **23**, 373 (2011)
15. O. A. Kabov, B. Scheid, I. A. Sharina, J. Legros. Int. J. Therm. Sci. **41**, 664 (2002)
16. M. A. Frank, O. A. Kabov. Phys. Fluid **18**, 032107, 2006.
17. P. Colinet, J.-C. Legros, M. G. Velarde, *Nonlinear Dynamics of Surface - Tension Driven Instabilities* (Wiley-VCH, Berlin, 2001)
18. O. A. Kabov, E. A. Chinnov. High Temp. **39**, 703 (2001)
19. V. E. Nakoryakov, S. L. Elistratov, S. Y. Misyura. J. Eng. Thermophys-Rus. **20**, 338 (2011)
20. V. E. Nakoryakov, S. Y. Misyura, S. L. Elistratov, A. Y. Manakov, A. A. Sizikov. J. Eng. Thermophys-Rus. **22**, 169 (2013)
21. V.E. Nakoryakov, S.Ya. Misyura, S.L. Elistratov, R.A. Dekhtyar. J. Eng. Thermophys-Rus. **23**, 257 (2014)
22. G. V. Kuznetsov, N. I. Kurilenko, V. I. Maksimov, G. Ya. Mamontov, T. A. Nagornova. J. Eng. Phys. Thermophys. **86**, 519 (2013)
23. G. V. Kuznetsov, P. A. Strizhak. J. Eng. Phys. Thermophys. **87**, 103 (2014)
24. G.V. Kuznetsov, D.V. Feoktistov, E.G. Orlova, K.A. Batishcheva, Colloid J. **78**, 335 (2016)
25. G. Kuznetsov, D. Feoktistov, E. Orlova, Thermophys. Aeromech, **23**, 17 (2016)
26. D. V. Feoktistov, E. G. Orlova, V. D. Velicanov. MATEC Web of Conferences **23**, 01056 (2015)
27. Y. V. Lyulin, D. V. Feoktistov, I. A. Afanas'ev, E. S. Chachilo, O. A. Kabov, G. V. Kuznetsov. Tech. Phys. Lett+ **41**, 665 (2015)
28. I. Afanasyev, E. Orlova, D. Feoktistov. EPJ Web of Conferences **82**, 01054 (2015)
29. D. V. Zaitsev, D. A. Rodionov, O. A. Kabov. Technical Physics Letters **35**, 680 (2009)

Vortical structures in a shallow wake of the flow around a confined cylinder at *Re*=3750

Egor Palkin[1,2] , and *Rustam* Mullyadzhanov[1,2]

[1]Institute of Thermophysics SB RAS, 630090, Novosibirsk, Russia

[2]Novosibirsk State University, 630090, Novosibirsk, Russia

Abstract. We study the flow over a cylinder placed between two parallel rigid walls using Large-eddy simulations. The Reynolds number based on the inflow velocity and diameter of the cylinder is 3750 corresponding to the subcritical regime with laminar separation. The comparison of the time-averaged flow patterns shows very good agreement with experiments. The unsteadiness of the recirculating bubble features the low-frequency modulation observed for the first time.

1 Introduction

Flows over obstacles in a duct are common in many engineering applications such as cooling systems, bridge piers, heat exchangers, building sections, junctions in wing-body and turbine blade-rotor systems, among others. In such configurations a horseshoe vortex system appears prior to a bluff body increasing the local shear stress and heat transfer. Figure 1 demonstrates the changes introduced by the flow confinement comparing the flow over an infinite cylinder (left) and with side walls (right) at close Reynolds numbers. Aside from the horseshoe vortex, the wake becomes dominated by streaky structures. In the present work we focus our attention on the basic time-averaged characteristics of the flow and unsteady behaviour of the recirculating bubble.

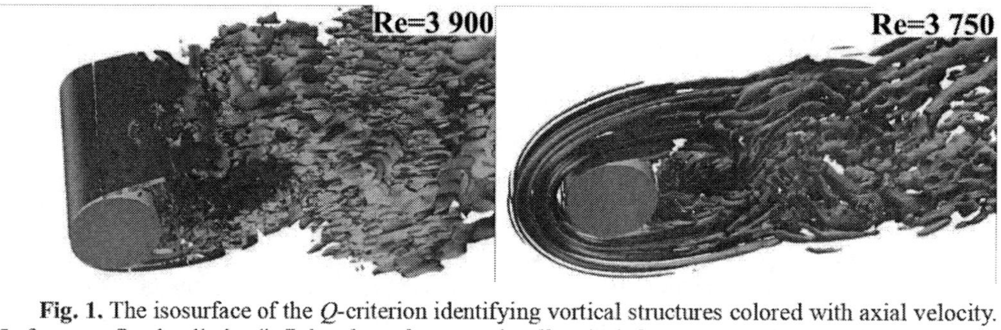

Fig. 1. The isosurface of the *Q*-criterion identifying vortical structures colored with axial velocity. Left: unconfined cylinder (infinite along the spanwise direction) from our previous results [1] at Re = 3900; right: quasi-2D present configuration at Re = 3750 with H/D = 0.4, where H is the distance between parallel walls and D is the diameter of the cylinder.

2 Computational details

The Large-eddy simulation (LES) was carried out with the unstructured finite-volume computational code T-FlowS. The filtered Navier-Stokes and continuity equations for incompressible fluid are closed by the dynamic Smagorinsky subgrid-scale model. The diffusion and convection terms in the momentum equations are discretized by the second-order central-difference scheme, whereas the time-marching is performed using a fully-implicit three-level time scheme. The velocity and pressure are coupled with the SIMPLE algorithm.

The computational domain is shown in Fig. 2. We study the flow over a circular cylinder fixed between a pair of side walls at the Reynolds number of $Re = 3750$ based on the bulk inflow velocity U_b and cylinder diameter D. The domain represents a rectangle of the size $x \times y \times z = 29D \times 20D \times H$ leading to 8.7 mln cells grid, where $H = 0.4D$ is the distance between parallel walls corresponding to z=0 and z=0.4D planes. The inflow is provided by a steady parabolic laminar profile, while no-slip conditions are set at all walls. The total computational time was around $tU_\infty/D=10^3$ with a timestep $\Delta tU_\infty/D=0.0025$. The statistics was gathered for 500 non-dimensional time units.

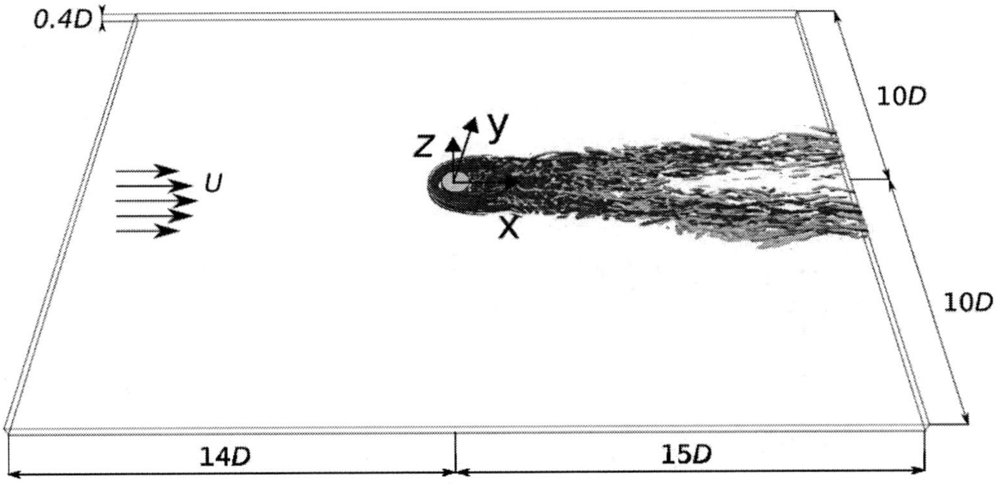

Fig. 2. Computational domain. Surface on the left corresponds to the inflow boundary condition, the outflow is on the right. All other surfaces are rigid walls, yellow surface denotes the cylinder.

3 Results

At this low Reynolds numbers the laminar inflow turbulizes only in the wake with a slow transition (growing of disturbances) in the shear layer. The typical vortical structure in the near wake represents a meandering elongated vortex, Fig. 1. Further downstream one can still recognize elongated coherent structures. The presence of walls strongly affects dynamics of the flow resulting in the change of time-averaged characteristics. Figure 3 (left) shows the comparison of the time-averaged axial velocity and streamlines for the confined (present) case in the mid-plane (z=0.5H=0.2D) and without walls from previous simulations [1]. While the pattern is the same, the recirculating bubble shrinks and becomes narrower in the presence of walls. Turbulent kinetic energy is shown in Fig. 3 very close to the wall denoting the footprint of a slightly fluctuating horseshoe vortex appearing in front of the cylinder [2]. The streamlines near the wall differ significantly from those at the mid-plane pointing at strong three-dimensionality of the flow.

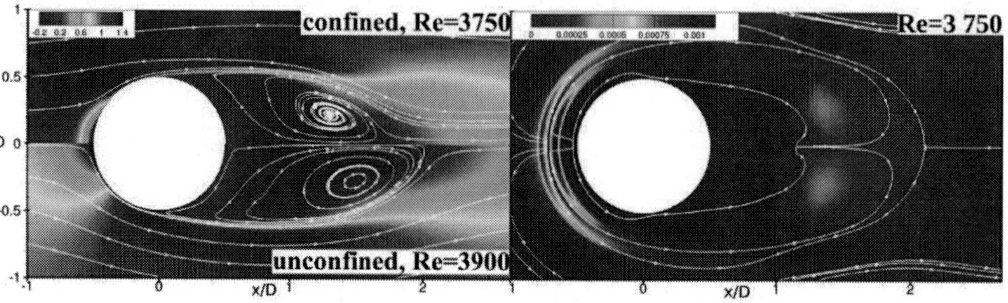

Fig. 3. Left: Time-averaged axial velocity for the present confined case of the flow over a cylinder at z=0.5H=0.2D (top) and the one without walls from previous simulations [1] (bottom). Right: turbulent kinetic energy and streamlines near the wall.

Figure 4 (top) presents the streamlines from particle image velocimetry (PIV) measurements at Re=4000 in x-z plane at y=0 before and after the cylinder [2] showing time-averaged vortical structure of the flow. In the bottom of Fig. 4 we compare it with LES results, namely the time-averaged vertical velocity fluctuations field and streamlines indicating very good comparison with experiments. The position of main vortices is well reproduced.

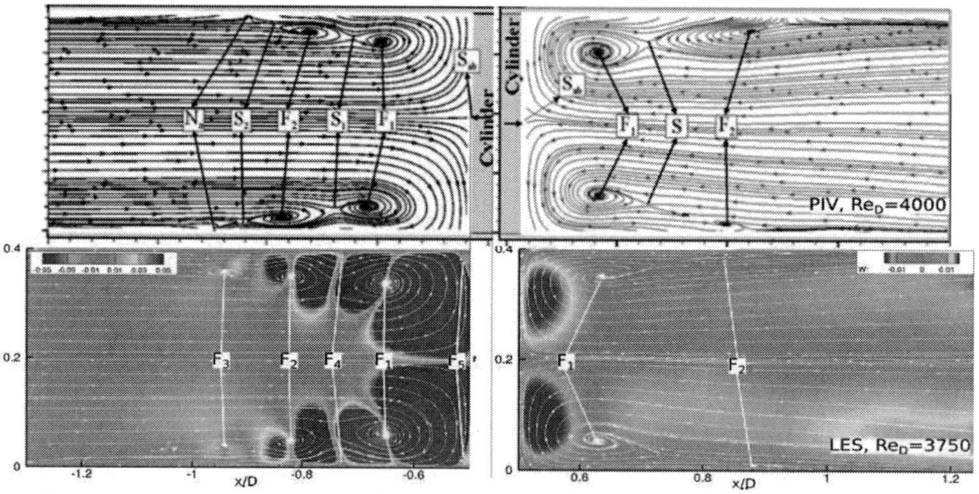

Fig. 4. Top: time-averaged streamlines in x-z plane and y=0 from PIV measurements [2]. Bottom: vertical velocity field and corresponding streamlines from LES.

Figure 5 shows the comparison of the instantaneous velocity signal in the shear layer at z=0.5H. The filtered signal reveals the existence of low-frequency oscillations of the recirculation zone. The non-dimensional period is around 28.6D/U_b corresponding to the frequency f=0.035U_b/D. These results are in agreement with numerical simulations [1, 3] of unconfined cylinder where such modulations have also been detected.

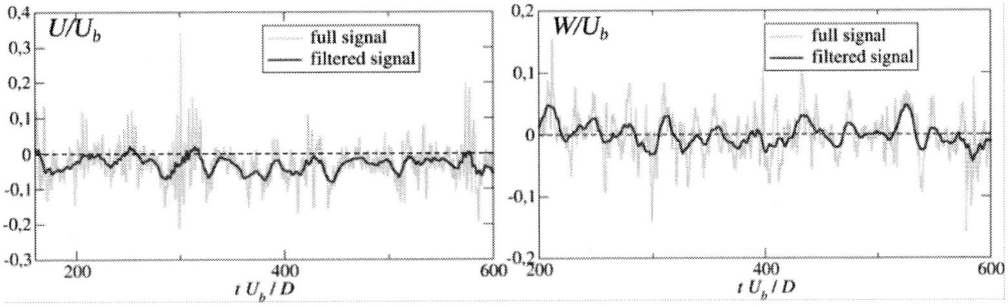

Fig. 5. Left: Instantaneous signal of axial velocity in the point (x/D, y/D, z/D) = (0.5, 0.5, 0.2). Right: The same for the vertical velocity component.

We have presented the results of Large-eddy simulations of the confined flow over a cylinder at Re=3750. The direct comparison with PIV experiments showed very good agreement of the position of main time-averaged vortices. We find that the recirculating bubble manifests low-frequency oscillations similar to the unconfined case [1, 3]. Further work will be devoted to studying the elongated vortical structures found also in confined turbulent jets [4-7] and their role in heat transfer across the channel.

This work is funded by the Russian Science Foundation grant No. 14-19-01685. The computational resources are provided by Siberian Supercomputer Center SB RAS and Supercomputing Center of the Novosibirsk State University.

References

1. E. Palkin, R. Mullyadzhanov, M. Hadziabdic, K. Hanjalic, Flow Turbul. Combust., appeared online (2016), doi:10.1007/s10494-016-9772-z
2. N.A. Ozturk, A. Akkoca, B. Sahin, Exp. Fluids **44** (6), 1001-1014 (2008)
3. O. Lehmkuhl, I. Rodriguez, R. Borrell, A. Oliva, Phys. Fluids, **25** (8), 085109 (2013)
4. R. Mullyadzhanov, B. Ilyushin, K. Hanjalic, Int. J. Heat Fluid Flow, **56**, 284-289 (2015)
5. K. Hanjalic, R. Mullyadzhanov, Phys. Fluids, **27** (3), 031703 (2015)
6. R. Mullyadzhanov, B. Ilyushin, M. Hadziabdic, K. Hanjalic, In: *Direct and Large-Eddy Simulation IX*, Springer, Netherlands, (2015).
7. M.V. Shestakov, R.I. Mullyadzhanov, M.P. Tokarev, D.M. Markovich, J. Eng. Thermophys. **25**, 159 (2016)
8. SThermophys., **25** (2), 159-165 (2016)

EPJ Web of Conferences **159**, 00035 (2017) DOI: 10.1051/epjconf/201715900035
AVTFG2016

The flows structure in unsteady gas flow in pipes with different cross-sections

Leonid Plotnikov[1,*], *Alexandr* Nevolin[1], and *Dmitrij* Nikolaev[1]

[1]Ural Federal University named after the first President of Russia B.N. Yeltsin, Turbines and engines Department, 620002 Ekaterinburg, ul. Mira, 19, Russian Federation

Abstract. The results of numerical simulation and experimental study of the structure of unsteady flows in pipes with different cross sections are presented in the article. It is shown that the unsteady gas flow in a circular pipe is axisymmetric without secondary currents. Steady vortex structures (secondary flows) are observed in pipes with cross sections in the form of a square and an equilateral triangle. It was found that these secondary flows have a significant impact on gas flows in pipes of complex configuration. On the basis of experimental researches it is established that the strong oscillatory phenomena exist in the inlet pipe of the piston engine arising after the closing of the intake valve. The placement of the profiled plots (with a cross section of a square or an equilateral triangle) in the intake pipe leads to the damping of the oscillatory phenomena and a more rapid stabilization of pulsating flow. This is due to the stabilizing effect of the vortex structures formed in the corners of this configuration.

1 Introduction

It is known that the stagnation zone and reverse flow can be formed in the flowing parts of energy machines and plants (e.g., gas-turbine units or internal combustion engines). It has a negative influence on hydraulic resistance and consequently on the flow characteristics of the system. One of the ways of improving the gas-dynamic characteristics of gas flows is to use pipes with cross sections in the form of a square or an equilateral triangle. Steady vortex structures occur in such pipes, which can stabilize the flow and prevent the formation of stagnant zones and reverse pressure waves.

The structure of gas flows in pipes with different cross sections is investigated in a large number of scientific papers. In Russia, S. S. Kutateladze and I. E. Idelchik [1-3] have actively performed these studies. In their work, it is shown that the gas flow in a long circular pipe is axisymmetric. In such flows, sustainable large-scale turbulent structures (secondary flows) do not occur. Another picture is observed for gas flows in pipes with square or triangular cross plots (Fig. 1).

Secondary flows are formed in the corners of such pipes, which lead to the formation of a steady longitudinal vortex structures. It is determined that isotachs in appearance similar to the shape of the cross section are observed in cross-section of the pipe [1-3].

* Corresponding author: plotnikovlv@mail.ru

DOI: 10.1051/epjconf/201715900035

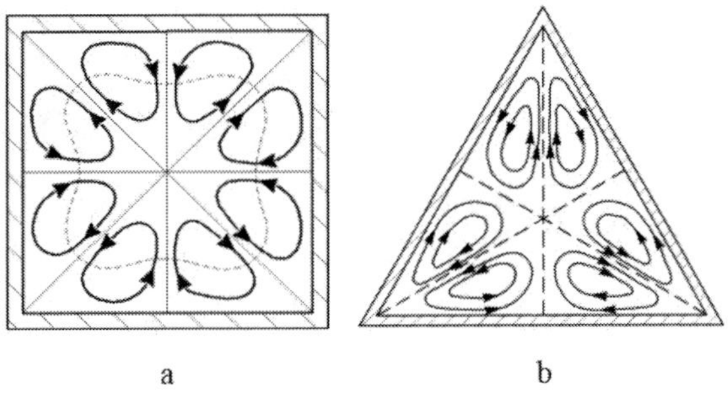

Fig. 1. Scheme of the secondary flows in a square pipe (a) [1] and in an equilateral triangular pipe (b) [2]

2 Experimental setups and software for numerical simulation

A number of numerical models based on STAR-CCM+ software was designed to study the structure of gas flows in pipes with different configurations. The analyzed pipes were straight channels with different cross sections (see figures below). All pipes had a length of 500 mm and the equivalent hydraulic diameter of 32 mm. Dry air was used as the working medium. The air in the pipes was studied in a turbulent flow regime with an average speed of from 10 to 100 m/s. The mass flow was specified at the inlet of the mathematical model. The output of the mathematical model was defined as the outlet pressure (outlet pressure of the pipes was equal to barometric pressure). k-e turbulence model was used for mathematical modeling. The flow in the pipes was modeled in the unsteady formulation.

Experimental setup for experimental investigations of gas-dynamics of the inlet and exhaust processes were designed and manufactured. They were full-scale model of single-cylinder engine of 8.2/7.1. The valve control mechanism for the experimental setups is taken from the engine of the car VAZ-OKA. Valve timing and valve lift of the experimental setups consistent with those for this engine. The drive of a crankshaft was carried out by using an asynchronous motor, the rotational speed of which was regulated by the frequency Converter in the range from 600 to 3000 rpm. A detailed description of the experimental facilities is presented in [4, 5].

To make the necessary measurements on the basis of the analog-to-digital Converter was created an automated data collection system, which passed the experimental data in the personal computer. To determine the speed of the air flow (w) was used constant temperature anemometer [6]. Sensor probes was the nichrome filament diameter of 5 μm and a length of 5 mm. Measuring speed and position of the crankshaft of the engine produced by the tachometer.

3 The results of numerical simulation

Some results of mathematical simulation of flow in pipes with cross sections in the form of a circle, square and triangle are shown in Fig. 2.

It was established (Fig. 2), that mathematical modeling confirms the data from the classical literature. Flow in a circular pipe (Fig. 2, a) is axisymmetric without any secondary flows and turbulent structures. The secondary flows (vortical structures) were detected in pipes with cross sections in the form of a square and an equilateral triangle (Fig. 2, b and c). Secondary flows (vortical structures) were formed in the corners of the channels.

DOI: 10.1051/epjconf/201715900035

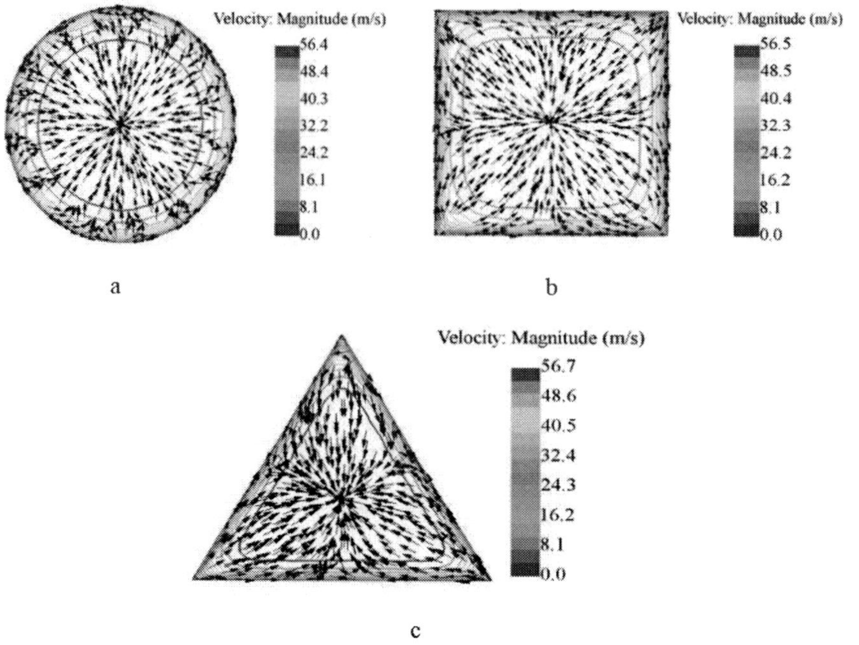

Fig. 2. Visualization of stationary flows in pipes of different cross sections in the form of isotach and tangential velocity vectors projected from different cross-sections at an initial flow rate of 50 m/s. Cross sections: a) round; b) square; c) triangle

The hypothesis that these turbulent structures (secondary flows) are sustainable and have a significant impact on the flow structure in pipes of complex configuration was tested in the second phase of the research. The complex configuration of the pipe was the changing shape of cross section downstream. The scheme of this configuration with the indication of the control cross-sections is shown in Fig. 3.

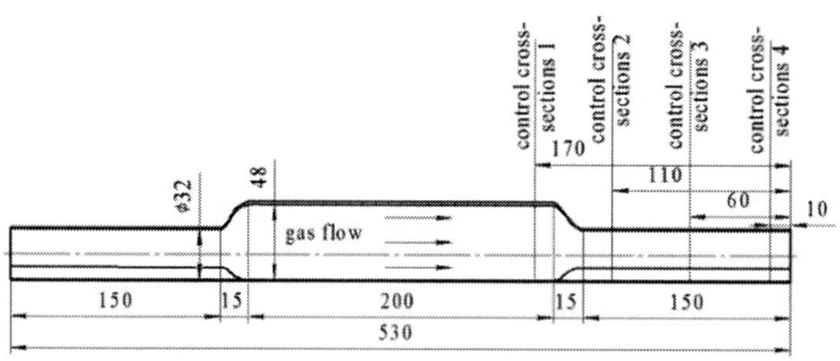

Fig. 3. The configuration of the investigated pipe with a triangular plot

The results of numerical simulation of unsteady flow in the pipe with a plot with a cross section of a triangle are shown in Fig. 4.

The plot with triangular cross section leads to a significant deformation of the flow structure along the whole length of the pipe (Fig. 4). First, it can be noted that the shape of isotach in cross section of a circular plot has a triangular shape, it is maintained until the output from the pipe. Secondly, secondary flows (turbulent structure) in the cross section of a circular plot are observed. It should be noted that these secondary flows were generated in the triangular plot.

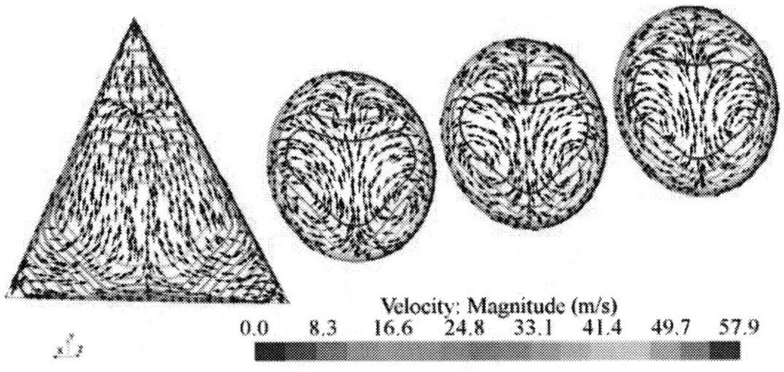

Fig. 4. Visualization of the unsteady flow in the pipe with a triangular section in the form of isotach and tangential velocity vectors projected from different cross-sections at an initial flow rate of 50 m/s

Thus, we can state that the stable vortex structures (they are formed in the triangular plot) have a significant impact on gas flow and they are stored throughout the pipe. Therefore, we can assume that this effect can be used in intake and exhaust systems of energy machines and plants to stabilize the gas flows.

4 The results of experimental studies

This assumption was tested in the experimental investigation on piston internal combustion engines. The gas flows in the intake and exhaust systems have been investigated on single-cylinder engine.

The geometric configuration of the intake pipe is fully consistent with the dimensions specified in Fig. 3.

It was found that significant oscillatory phenomena exist in the intake pipe. These oscillatory phenomena occur after closing of the intake valve (Fig. 5). In this case, more rapid damping of oscillatory phenomena is observed when using triangular and square plots in the intake system. According to the authors, this is due to the stabilizing effect of the longitudinal vortex structures generated at the corners of triangular and square cross-sections [4].

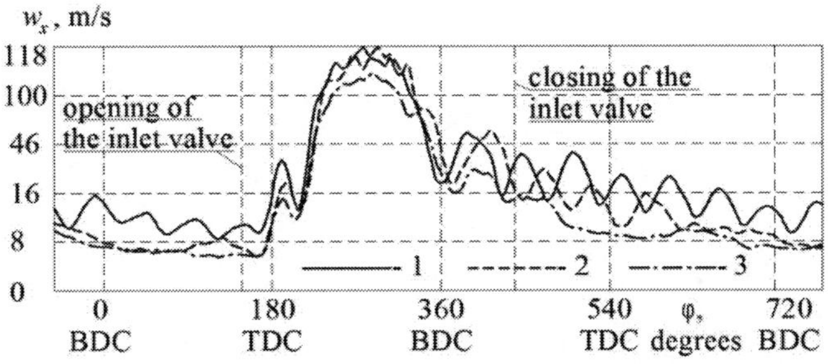

Fig. 5. The dependences of the local (l_x = 110 mm, d = 32 mm) velocity of the fresh charge w_x in the inlet pipes of different cross-section from the crankshaft rotation angle φ for the rotational speed n = 3000 rpm. Cross section shape: 1 – circle; 2 – square; 3 – triangle

It was found that we can have a number of advantages due to the placement of the triangular and square plots in the intake pipe in comparison with a pipe of circular cross-section. In particular, there are an increase of the volumetric flow of fresh charge through the intake system (in average 22 %) and the increase of the slope of the dependence of the flow of fresh charge from the crankshaft rotation speed *n* in the operating range *n* when using the triangular plot in the intake pipe [5]. This will lead to increase in power and throttle response of the piston engine.

Summary

Thus, in this study it was found that the cross-sectional shape of the pipe has a significant impact on the structure of the gas flow. The secondary flows (steady vortex structures) are formed in pipes with cross sections in the form of a square and an equilateral triangle. Secondary flows (steady vortex structures) in the pipe can stabilize the pulsing flows. This phenomenon can be used in the design of the intake and exhaust systems of energy machines and plants with a view to their gas-dynamic improvement.

The work has been supported by the Russian Science Foundation (grant No. 17-69-00002).

References

1. S.S. Kutateladze, *Heat transfer and flow resistance: A Reference Guide* (Jenergoatomizdat, Moscow, 1990).
2. I.E. Idel'chik, *Manual of hydraulic resistance* (Mashinostroenie, Moscow, 1975).
3. I.E. Idel'chik, *Aerohydrodynamics of technological devices. Supply, drainage and flow distribution over the cross section of devices* (Mashinostroenie, Moscow, 1983).
4. L.V. Plotnikov, B.P. Zhilkin, Polzunovskij vestnik **3**, 178-183 (2012).
5. B.P. Zhilkin, Ju.M. Brodova, *Process Improvement in gas path of reciprocating internal combustion engines* (Izd-vo Ural, Ekaterinburg, 2015).
6. S.N. Plohov, L.V. Plotnikov, B.P. Zhilkin, RU Patent 81338 (2009).

DOI: 10.1051/epjconf/201715900036

Experimental investigation of heat transfer in a falling down rivulet of liquid FC-72 on the vertical heated foil

Tatyana Ponomarenko*

Novosibirsk State Technical University, 630073, K. Marx Ave., 20, Novosibirsk, Russia,
Kutateladze Institute of Thermophysics SB RAS, 630090, Lavrentiev ave. 1, Novosibirsk, Russia

Abstract. Heat transfer at rivulet FC-72 flow over the constantan vertical heated foil with the length of 80 mm, width of 35 mm, and thickness of 25 mm was studied experimentally. Distributions of temperature on the bottom side of foil from the rivulet flow using the infrared thermographic data were obtained. In the future heat flux from the foil will be calculated with help of Koshi problem solution. It was shown that the rivulet width as the liquid flow rate increases, and it narrows as heat power increases due to the high evaporation rate. In the case with water, the rivulet narrows not so intensively as compared with the rivulet of FC-72 liquid. Good wettability and low heat of vaporization of FC-72 liquid contribute to using it to develop effective cooling systems.

1 Introduction

The well-known regularity, which was observed by Gordon Moore in 1965, that the number of transistors in microprocessors almost doubles each two years, as their new models are introduced is still relevant. Due to the growth of transistor counts of modern computer microprocessors, many parameters, including such a major one as throughput, improve, but the heat output that should be dissipated from the processing unit also increases. Moreover, an increase in the temperature by 10°C shortens the lifespan of the processor by half. Therefore, creating an effective cooling system is one of the major objectives for today. The main cause for cooling is an unstable operation, and eventually the breakdown of the process as the result of exceeding a certain critical temperature threshold for a certain period of time.

The liquid cooling system with the rivulet flow, which is a particular case of the liquid film flow limited by the interface from both sides - the contact line of three phases, has prospects. The high heat transfer coefficient is observed in the small field of three-phase contact line – the micro-region [1]. Therefore, the use of rivulet flows in the cooling systems is a promising practice.

Numerous studies that deal with the rivulet flows and its specific characteristics have been conducted. Many studies focused on the rivulet flow over the inclined plane and the

* Corresponding author: t.evans2010@yandex.kz

impact of surface tension and viscosity on the emergency of varieties of rivulet flows [2]. When liquid is flowing down over the inclined flat surface, a number of various flow regimes, depending on the flow rate, the physical properties of liquid and the material of the inclined surface as well as the incline itself, can be observed. The results of such studies were used to plot the maps of varieties of rivulet flows of different liquids (water, water solution of ethyl alcohol, water solution of glycerine) over the inclined plate. As the liquid flow rate increased, the drop flow, straight flow, meandering flow, which converted into the film flow [2-4], were observed.

The LIF method (Laser Induced Fluorescence method) was used in [5] to measure the local thickness of the rivulet flows of water solutions of glycerine and ethanol that have the wave structure and flow down the inclined plate, and the maps of varieties of wave formation at the rivulet flows were plotted.

Heat transfer in an evaporating droplet on the horizontal heated foil and in a water rivulet flowing down over the vertical heated foil was studied in [6,7]. The fluctuating temperature measurements made by the IR-scanner were used to solve the Cauchy problem and to measure heat flux density on the surface with the droplet and on the surface with the rivulet. This showed that maximal heat flux density was observed near the contact line of an evaporating droplet, which 5-7 times higher than the average heat flux density from the entire foil surface. The experiment with a water rivulet confirmed that the heat transfer intensity in the micro-region near the contact line of wetting is higher than that in average in the rivulet in the micro-region and leads to a local temperature drop in this region.

The determination of heat flux density using the infrared thermographic data is an effective method of experimental studies of three-phase contact line. This method is used in the current study.

Therefore, the current study aims to determine heat flux density in the region of three-phase contact line of the liquid rivulet flowing down over the heated surface, using the IR-scanner and the method for solving the ill-conditioned problems.

Unlike the previous similar experiments [7] when water was used as the liquid, FC-72 liquid developed for cooling supercomputers was used in the current study.

Fluorinert Electronic Liquid (FC-72) is a transparent, colourless, thermally and chemically stable liquid cooling agent. It is compatible with sensitive materials, non-flammable and almost nontoxic and leaves no residue after evaporation. The heat of vaporization of the liquid is substantially lower than that of water. The unique combination of the properties makes FC-72 ideal for use in many areas of electronics, for example immersion cooling systems and film cooling systems that are under active development at present.

The value of the contact angle on the substrate of the constantan foil with black paint (graphite 33) for FC-72 - 10 ± 0,4° that was determined with the help of the gas (air) bubble method testifies to better wettability as compared with that of water – 63° (Fig.1).

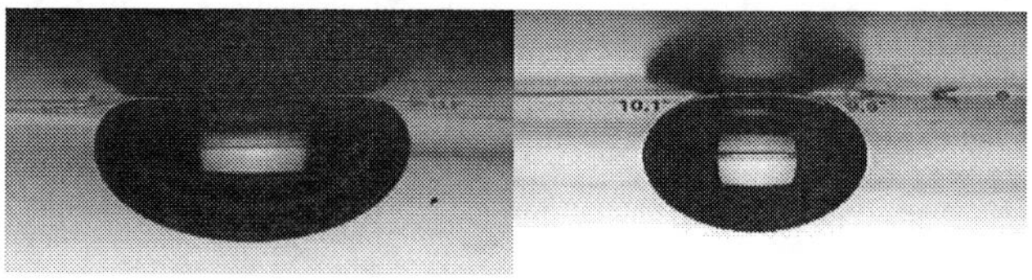

Fig.1. The determination of the value of the contact angle of the constantan foil with black paint for FC-72, using the gas (air) bubble method.

This illustrates once again that good wettability and low heat of vaporization of FC-72 liquid contribute to using it to develop effective cooling systems

2 Experimental setup

The scheme of the experimental setup and the photograph of the setup in operation are presented on Fig. 2 a and b.

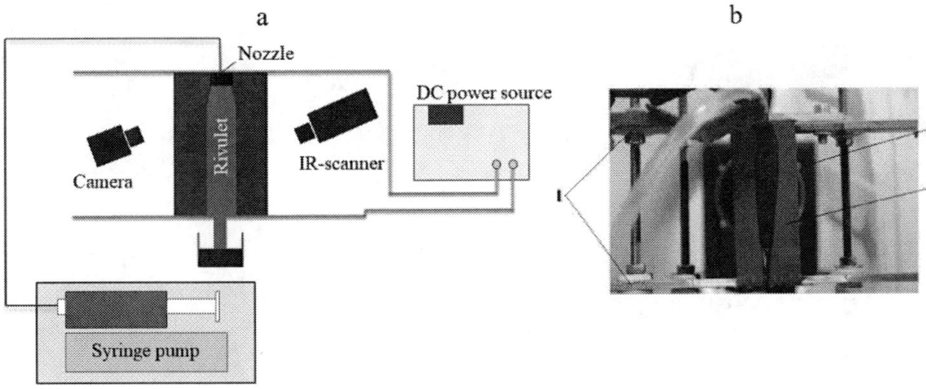

Fig.2. a – the scheme of the experimental setup; b – the photograph of the FC-72 liquid rivulet flowing down over the heated foil at Q = 1 ml/min, P = 0,09W: 1- brass electrodes, 2 - IR-camera, 3 – foil with rivulet.

The vertical foil with the length (l) of 80 mm, the width (w) of 35 mm and the thickness (h) of 25 microns was used to conduct the experiments for the current study. The material of the foil is constantan, the density is 8.9 g/cm3, heat capacity is 0.41 J/(g·K) and heat conductivity is 23 W/(K·m). Constantan is a thermostable copper (Cu) containing alloy (about 59%) with the addition of nickel (Ni) (39-41%) and manganese (Mn) (1-2%). The liquid was injected with the flow rate of 0.1-5 mL/min to the foil through the tube with a special nozzle using the Cole-Parmer EW-74905-54 syringe pump. The foil was connected to the DC power source TTi QPX 1200L through the brass electrode holders. The power source was used to heat the foil and regulate with heat power that was generated on the foil in the range of 0.09 to 1.47W. The Nikon D 7000 camera and the Titanium 570 M infrared (IR) scanner were used to make the observations. The foil was coated with black graphite paint with the emissivity factor of about 1. The ambient air temperature was 30,2-30,5°C, humidity ranged from 35 to 40%. The Testo 646 multimeter was used to measure the ambient air temperature and humidity.

3 Experimental findings

An IR-scanner was used to measure distributions of temperature on the bottom side of foil from the rivulet flow (Fig. 3). The experimental findings were compared with the findings of similar experiment when super-purified water obtained by the Milli-Q system was used as a liquid. It was found out experimentally that the flow regimes of two liquids differ from each other due to substantial difference in the wetting properties.

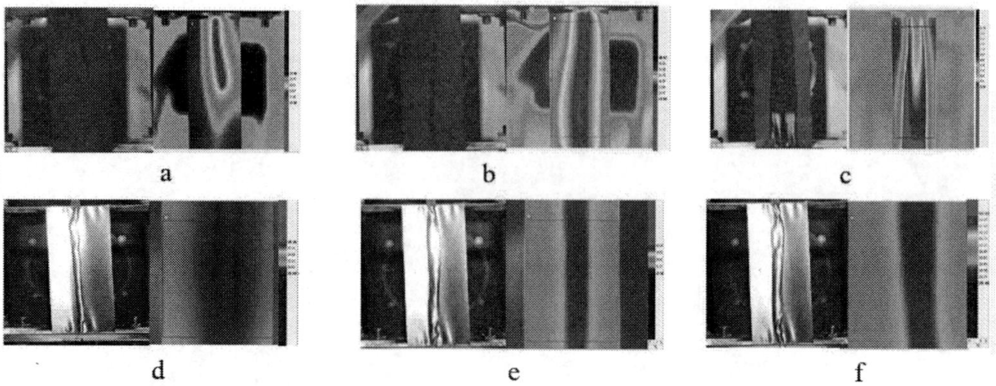

Fig.3. The photographs of the rivulet flowing down over the heated foil and the thermograms from the side opposite to the rivulet. FC-72: a - Q = 0,1 ml/min, q_{av} = 15,049 W/m^2; b - Q = 1 ml/min, q_{av} = 97,45 W/m^2; c - Q = 5 ml/min, q_{av} = 237,637 W/m^2. Water: d - Q = 10 ml/min, q_{av} = 16,7 W/m^2; e - Q = 10 ml/min, q_{av} = 98,0 W/m^2; f - Q = 100 ml/min, q_{av} = 226,0 W/m^2.

The photographs were used to measure the width of the rivulet. The experimental findings were used to draw the curves illustrating the dependence of the width of the rivulet on heat power at the given rates and the dependence of the width of the rivulet on the liquid flow rate at the given heat power rates (Fig. 4 and 5).

FC-72:

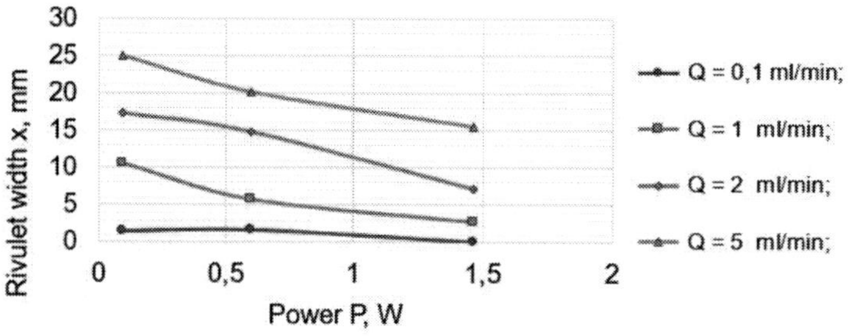

Water:

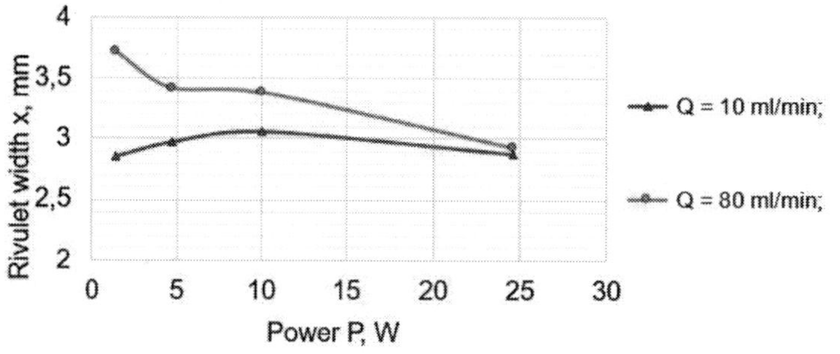

Fig.4. The curves illustrating the dependence of the width of the FC-72 rivulet and water rivulet on the power.

FC-72:

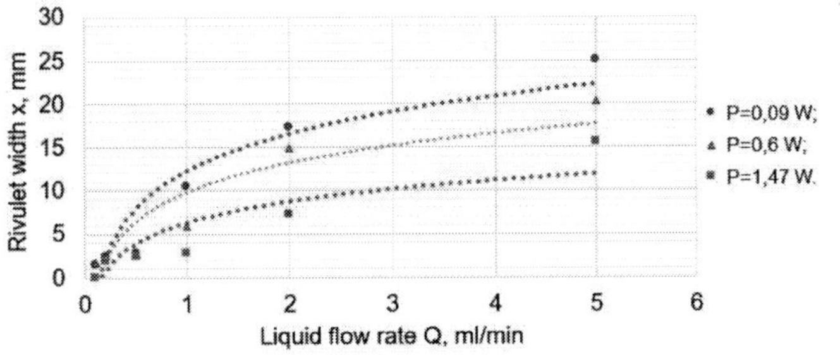

Water:

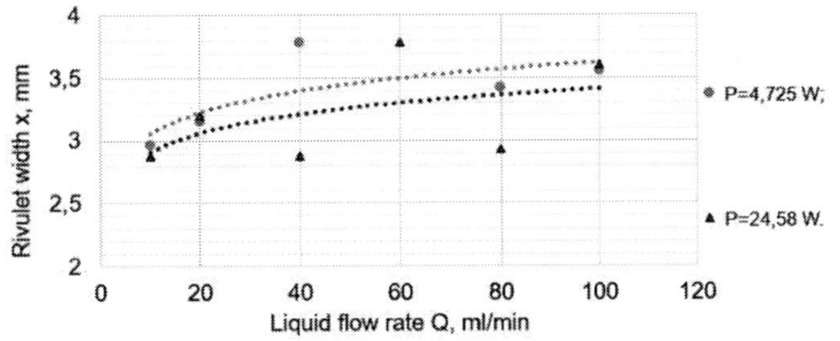

Fig.5. The curves illustrating the dependence of the width of the FC-72 rivulet and water rivulet on the liquid flow rate.

The curves show that the rivulet widens as the liquid flow rate increases, and it narrows as heat power increases due to the high evaporation rate. In the case with water, the rivulet narrows not so intensively as compared with the rivulet of FC-72 liquid.

This study was performed in the framework of the Russian Science Foundation grant (Agreement No. 14-19-01755).

References

1. Stephan P., Brandt C, Heat Transfer Engng. **25**, 78–85 (2004)
2. Schmuki P., Laso M., J. Fluid Mechanics. **215**, 125–143 (1990)
3. Hirt C.W, Los Alamos: Flow Sci., 7 (1998)
4. Su G.W., Geller J.T., Pruess K., Wen F., Water Resources Research. **35**, 1019–1037 (1999)
5. Alekseenko S.V., Bobylev A.V., Kharlamov S.M., Markovich D.M., Proc. of 14th Int. Symp. on Applications of Laser Techniques to Fluid Mechanics, Lisbon, Portugal, 7-10 July, 8 (2008)
6. Marchuk I., Karchevsky A., Surtaev A., Kabov O., Intern. J. of Aerospace Engng. Art. ID 391036. 5 (2015)
7. Cheverda V.V., Marchuk I.V., Karchevsky A.L., Orlik E.V., Kabov O.A., Thermophysics and Aerodynamics, **23**, 3 (2016)

Forced wetting of stainless steel by distilled water drop

Konstantin Ponomarev[1,*], *Yuri* Popov[1], *Evgeniya* Orlova[1] and *Dmitry* Kirichenko[2, 3]

[1]National Research Tomsk Polytechnic University, Tomsk, Russia
[2]Kutateladze Institute of Thermophysics SB RAS, Novosibirsk, Russia
[3]National Research Novosibirsk State University, Novosibirsk, Russia

Abstract. Experimental results analyses of macroscopic contact angle change from the drop growth rate of distilled water on stainless steel substrates is presented. The value of contact angle is found to depend on the drop growth rate (m): contact angle decreases at m up to 0.02ml/s and increases at m more than 0.02ml/s. At the final stage of spreading the contact line pinning occurs, depending on the drop growth rate and surface roughness.

1 Introduction

Theoretical, experimental research and modelling of the fundamental laws of heat and mass transfer processes [1-9] in multiphase systems are extremely important for the optimization of modern energy technologies. Nowadays irrigation technologies are paid full attention in the chemical, food, agricultural and other fields. It is connected with the reason of possible increase in the efficiency of production processes through the use of drop systems.

The spreading of a drop on a surface is known to depend on surface microstructure and properties of the liquid [10-16]. There is a lack in the experimental research studying the effect of the drop growth rate on the spreading.

The purpose of the work is to find experimentally the dependence of the macroscopic contact angle on the drop growth rate on microstructured surface of stainless steel substrate.

2 Experimental technique

Experimental studies were carried out by using the shadow and Schlieren optical methods. The scheme of the experimental setup is presented in Fig. 1 [11-13]. To implement the optical shadow method, the light source 1, ground glass 2, shield with an opening 3, and the collimating lens 4 were used to produce a beam of plane-parallel light illuminating a drop on the substrate 5.

The condensing lens 6 and camera lens 8 were used to project the image on the camera sensor. To implement the Schlieren method, source of incoherent light 17, ground glass 16, and condensing filter 15 were used to obtain the light flux with a stepwise decrease in

* Corresponding author: kop.tpu@gmail.com

intensity over the space. The light beam from the source 17 passed through the collimating lens 14, which transformed it into a plane-parallel. Then it reflected on the beam splitter 13, passed to the substrate 5 and then to the lens 12, focused on the filter 11 and projected onto the camera sensor 10 by the lens.

A drop was squeezed by a syringe electronic pump 9 from the bottom side of the substrate 5 through an 2 mm opening. The flow rate of distilled water was controlled in the experiment from 0.005 mL/sec to 0.16 mL/sec.

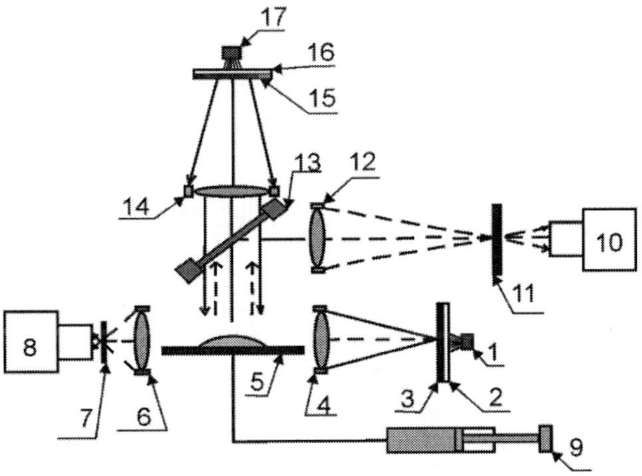

Fig. 1. Schematic diagram of an experimental setup: 1, 17 – light sources; 2, 16 – ground glasses; 3,7,11 - transparent shields with an opening; 4, 14 - collimating lenses; 5 – substrate; 6, 12 - condensing lenses; 8, 10 - high-speed video cameras; 9 – syringe pump; 13 - beam splitter; 15 - coding filter.

Three substrates made of stainless steel were used in the experiment. Their surfaces have different roughness: 1.5; 1,554; 4.59 μm. The photographic images of the substrates are shown in Fig.2.

(a) (b) (c)

Fig. 2. Photographic images of the substrates: (a) not bombarded by aluminum oxide particles; (b) bombarded by 10 μm aluminum oxide particles; (c) bombarded by 100 μm aluminum oxide particles

It is known the condition of wetting the material by liquid 90 °> θ> of 0 °; the increase in the roughness causes a decrease in the macroscopic contact angle. In the absence of wetting θ> 90 ° an increase in roughness increases the macroscopic contact angle.

Wenzel-Derjaguin equation does not consider the location of the cavities and asperities forming the roughness and the drop growth rate. The calculation the macroscopic contact angles with using thermodynamic equations is only possible if the surface roughness is formed by spaced parallel grooves. When the grooves are located chaotically, the values of angles may not be matched with the conclusions obtained from thermodynamic analysis. To date, there are no experimental data on the analysis of the influence of the microstructure (arrangement of the grooves on the surface) at different drop growth rate on the macroscopic contact angle.

3 Results and discussion

According to the experimental results (Table 1) the drop growth rate is found to affect the contact angle. It decreases with increasing flow rate (drop growth rate) reaching a minimum value at m=0.02 mL/sec. When values of flow rates are more than 0.02 mL/sec, the contact angle was observed to increase. It is worth noting that such dependence was obtained for surfaces with roughness formed both longitudinally and chaotically arranged grooves.

Table 1. Macroscopic contact angles.

Substrate	Flow rate of distilled water (drop growth rate), mL/sec					
	0.005	0.010	0.020	0.040	0.080	0.160
Sample No 1 (Ra 1.500 μm)	78°	77°	76°	80°	81°	82°
Sample No 2 (Ra 1.554 μm)	89°	84°	83°	91°	93°	94°
Sample No 3 (Ra 4.590 μm)	97°	95°	83°	84°	85°	89°

According to analyses of obtained angles (Table 1), the macroscopic angle is found to increase with an increase in the roughness when the drop growth rate is up to 0.02 mL/sec. The lower flow rates correspond to the larger values of the contact angle, and also its increase with an increase in the roughness.

The angle is found to be more than 90° (θ> 90 °) on the surface with the roughness 4.590 μm and the liquid flow rate up to 0.01 mL/sec. It means dewetting condition. With increasing liquid flow rate up to 0.02 mL/sec θ varies between 90 °> θ> 0 ° (liquid wets the surface).

The macroscopic contact angle increases with an increase in the roughness and at the drop growth rate more than 0.02 mL/sec. It is worth noting that the greatest values of θ are recorded on the substrate with Ra=1.554 μm (the roughness of substrate was formed by chaotically arranged cavities and asperities). Moreover, θ> 90 °, which is characterized by the absence of wetting. The value of angle varies in the range of 90 °> θ> 0 ° on the sample №3 at Ra = 4.590 μm (liquid wets the surface).

According to the analysis of experimental data (Table 1) obtained for samples No 2 and 3 (with the arithmetic average roughness 1.554 μm and 4.590 μm, respectively) it was found that when the cavities and asperities are arranged chaotically on the surface, the conclusions obtained on the basis of thermodynamic analysis of the Wenzel-Derjaguin equation do not agree with the experimental data.

Most likely this is due to the fact that the hydrophobic and hydrophilic properties are affected significantly, in addition to the roughness, the drop growth rate.

The study was financially supported by the grant of Russian Science Foundation (agreement No. 14-19-01755).

References

1. D. Glushkov, J. Legros, P. Strizhak, A. Zakharevich, Fuel, **175**, 105 (2016)
2. D. Glushkov, G. Kuznetsov, P. Strizhak, R. Volkov, Therm. Sci., **20**, 131 (2016)
3. D. Zaitsev, D. Rodionov, O. Kabov, Tech. Phys. Lett., **35**, 680 (2009)
4. D. Zaitsev, D. Kirichenko, O. Kabov, Tech. Phys. Lett., **41**, 551 (2015)

5. A. Sivkov, A. Ivashutenko, Y. Shanenkova, I. Shanenkov, Adv. Powder Technol., **27**, 4 (2016)
6. V. Maksimov, T. Nagornova, I. Shestakov, EPJ Web of Conferences, **82,** 01048 (2015)
7. A. Sivkov, I. Shanenkov, A. Pak, D. Gerasimov, Y. Shanenkova, Surf. Coat. Technol., **291,** 1 (2016)
8. G. Kuznetsov, P. Kuybin, P. Strizhak, High Temp., **53**, 254 (2015)
9. G. Kuznetsov, A. Zakharevich, N. Bel'kov, Chem. Pet. Eng., **50**, 1 (2014)
10. G. Kuznetsov, M. Sheremet, Fluid Dyn., **41**, 881 (2006)
11. G. Kuznetsov, D. Feoktistov, E. Orlova, J. Eng. Phys. Thermophys., **89**, 317 (2016)
12. D. Feoktistov, E. Orlova, A. Islamova, MATEC Web of Conferences, **23,** 01054 (2015)
13. G. Kuznetsov, D. Feoktistov, E. Orlova, Thermophys. Aeromech., **23**, 17 (2016)
14. V. Nakoryakov, S. Misyura, Dokl. Phys., **59**, 441 (2014)
15. V. Nakoryakov, S. Misyura, S. Elistratov, R. Dekhtyar, J. Eng. Thermophys., **23**, 257 (2014)
16. K. Batischeva, E. Orlova, D. Feoktistov, MATEC Web of Conferences, **19,** 01001 (2014)

Numerical study of magnetic nanofluids flow in the round channel located in the constant magnetic field

Maxim Pryazhnikov[1,2,*], *Dmitriy* Guzei[1], *Andrey* Minakov[1,2] and *Tatyana* Rodionova[1]

[1]Siberian Federal University, 79 Svobodny pr., Krasnoyarsk, 660041, Russia
[2]Institute of Thermophysics SB RAS, 1 Acad. Lavrentiev pr., Novosibirsk, 630090, Russia

Abstract. In this paper, the study of ferromagnetic nanoparticles behaviour in the constant magnetic field is carried out. For numerical simulation we have used Euler-Lagrange two-component approach. Using numerical simulation we have studied the growth of deposition of nanoparticles on the channel walls depending on the Reynolds number and the position of the magnet. The flow pattern, the concentration field and the trajectory of nanoparticles as a function of the Reynolds number were obtained. The good qualitative and quantitative agreement between numerical simulation and experiments was shown.

1 Introduction

One of the promising directions in the treatment of cancer may be the use of magnetic nanoparticles [1-3]. The essence of this approach is that the magnetic nanoparticles of sizes of 5-100 nm are injected into the bloodstream and concentrated in the tumor by an external magnetic field. Further, nanoparticles are heated in the tumor either by laser radiation or electromagnetic alternating fields. This leads to the death of cancer cells. The minimum harm is caused to healthy cells. At present, the effectiveness of this approach has been confirmed by many experiments on animals and humans during the treatment of cancer. There are lots of works study of nanoparticles transport in the magnetic field in the channels [1-4]. There are so much unresolved problems and issues related to transfer mechanisms, concentration and heating of nanoparticles in fluids and tissues, and their control under the influence of external electromagnetic fields.

In this paper, we conducted the numerical simulation of ferromagnetic nanoparticles behavior flowing in the liquid in the round pipe in the constant magnetic field.

Detailed description of the experimental study is presented in the paper [5]. The round channel is a glass tube with the inner diameter of 6 mm and the length of 50 cm. The tube is connected with a pump and a container with liquid. We used gear pump WT3000IJA to pump the liquid. The permanent neodymium magnet 50x30x10 mm in size is fixed under the glass tube. Magnetic induction of the constant magnetic field was measured by

* Corresponding author: arrivent@yandex.ru

milliteslameter TPU. Distilled water with the nanoparticles was pumping through the tube. The particles interacted with the magnetic field and deposited on the tube wall.

Investigation of particles deposition process was conducted in the ferromagnetic fluid based on distilled water contained the nanoparticles Fe_3O_4. Volumetric concentration of particles was very low (10^{-4} %). The size of the nanoparticles was 50 nm.

2 Mathematical model

The numerical simulation of nanoparticles transport in the constant magnetic field has been carried out. We used two-component model of nanofluids to describe the motion of nanoparticles in the magnetic field. This model considers the nanofluids as a binary mixture of base fluid and nanoparticles. The equations of this model are as follows:

$$\frac{\partial \rho}{\partial t} + \nabla \cdot (\rho \mathbf{v}) = 0 \tag{1}$$

$$\frac{\partial \rho \mathbf{v}}{\partial t} + \nabla \cdot (\rho \mathbf{v}\mathbf{v}) = -\nabla p + \nabla \cdot \mathbf{T} + \mathbf{F}_{\mathrm{m}} \tag{2}$$

$$\frac{\partial \rho C_{\mathrm{m}}}{\partial t} + \nabla \cdot (\rho \mathbf{v}_{\mathrm{p}} C_{\mathrm{m}}) = \nabla \cdot (\rho D C_{\mathrm{m}}) \tag{3}$$

$$\mathbf{T} = \mu \cdot \left(\nabla \mathbf{v} + \nabla \mathbf{v}^T - \frac{2}{3} \nabla \cdot \mathbf{v} \mathbf{E} \right) \tag{4}$$

where ρ – density of nanofluid, $\mathbf{v}$ – velocity vector, p – pressure, $\mathbf{T}$ – viscous stress tensor, $\mathbf{F}_{\mathrm{m}}$ – electromagnetic force, C_{m} – mass flow rate of nanoparticles, $\mathbf{v}_{\mathrm{p}}$ – particle velocity, D – diffusion coefficient, μ – dynamic viscosity, $\mathbf{E}$ – single tensor.

Volumetric electromagnetic force is found using the following formula:

$$\mathbf{F}_m = \varphi \cdot \mu_0 \frac{3\chi}{\chi + 3} (\mathbf{H} \cdot \nabla) \mathbf{H} \tag{5}$$

where φ – volume concentration, μ_0 – magnetic constant, χ – magnetic susceptibility of the material, $\mathbf{H}$ – magnetic field.

The particle velocity is calculated using the following formula:

$$\mathbf{V}_p = \mathbf{v} + \frac{\mathbf{F}_m \cdot V_{\mathrm{d}}}{\varphi \cdot 6\pi\mu r_{\mathrm{p}}} \tag{6}$$

where V_{d} – volume of the particle, r_{p} – radius of the particle.

Density and heat capacity are computed as follows:

$$\rho = \varphi \rho_{\mathrm{p}} + (1 - \varphi) \rho_{\mathrm{w}} \tag{7}$$

$$C_{\mathrm{p}} = \frac{\varphi \rho_{\mathrm{p}} C_{\mathrm{p,p}} + (1 - \varphi) \rho_{\mathrm{w}} C_{\mathrm{p,w}}}{\rho} \tag{8}$$

We have to add equations of magnetostatics to obtain a closed system of equations.

Component of the magnetic field around the magnet can be found by the magnetic field vector potential:

$$\mathbf{B} = \nabla \times \mathbf{A} \tag{9}$$

The equation of the magnetic field vector potential in the absence of vortex is as follows:

$$\nabla \times \left(\frac{1}{\mu_0} (\nabla \times \mathbf{A} - \mathbf{B}_\mathrm{R}) \right) = 0 \tag{10}$$

where $\mathbf{B}_\mathrm{R}$ – residual magnetic induction which is equal to zero everywhere except the region occupied by the magnet.

The magnetic field intensity can be found using the following formula:

$$\mathbf{H} = \mathbf{B} / \mu_0 \tag{11}$$

3 Numerical results

The problem statement was the same as experimental study [5]. The computational domain was represented as the round channel with the inner diameter of 6 mm and the length of 50 cm. Computational mesh consisted of half a million nodes.

Numerical simulation of nanoparticles transport in the constant magnetic field are carried out for various Reynolds numbers. The flow patterns in channel without a magnetic field are shown in figures 1-2. As you can see nanoparticles move strictly along the streamlines. They do not deposit on the wall of the channel. Velocity vector field in the channel and trajectory of nanoparticles are shown in figures 3-4. Large deposits of particles are formed on the channel wall. The numerical results correlate with the experimental data. This can be seen in figure 5. You can also see that the interaction of the magnetic field with the particles flow leads to formation of the flow recirculation zones. The deposits are formed in the flow recirculation zones. In this case nanoparticles move like it is shown in figure 4. It shows that at low flow rate all the nanoparticles deposit on the channel walls.

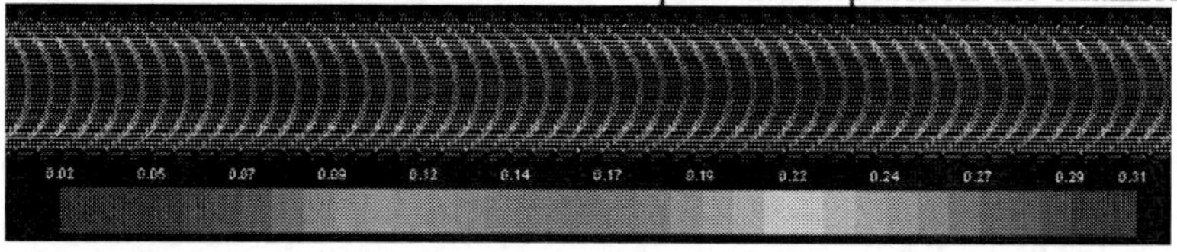

Fig. 1. Velocity vector field in the channel at Re = 1200. Without the magnetic field.

Fig. 2. Velocity vector field in the channel at Re = 1200. Without the magnetic field.

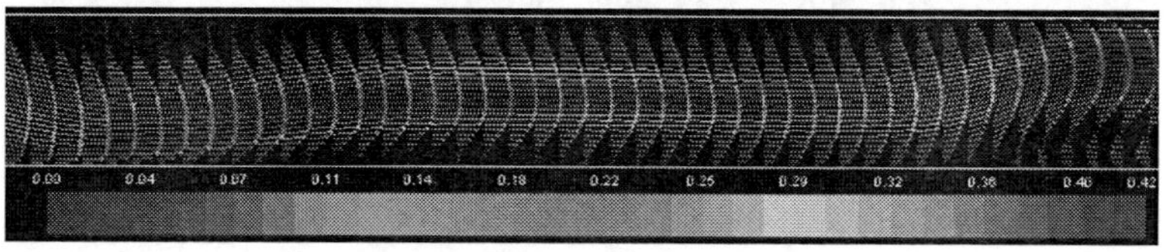

Fig. 3. Velocity vector field in the channel at Re = 1200.

Fig. 4. Trajectory of nanoparticles at Re = 1200.

Fig. 5. Comparison of particle deposition shape in the channel at Re = 1200 (top – experimental photo, bottom - simulation)

The higher the Reynolds number, the more complicated the flow. The shape of the vortex region has changed near the channel wall (see fig. 6). The deposits form has also changed. However, the numerical simulation reproduces perfectly the shape of deposition obtained during the experiment.

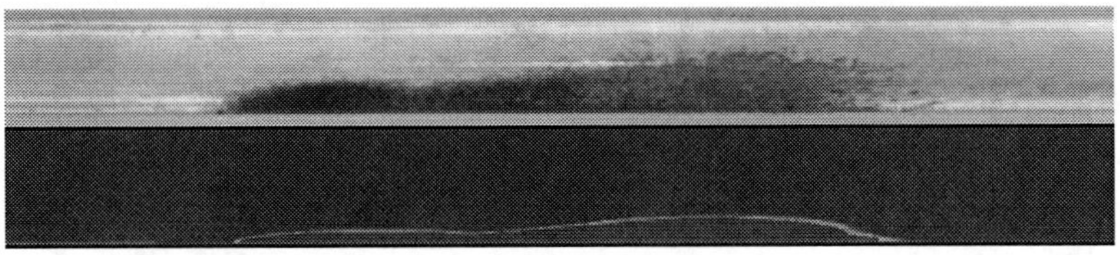

Fig. 6. Comparison of particle deposition shape in the channel at Re = 2810 (top – experimental photo, bottom - simulation)

4 Conclusion

The study of ferromagnetic nanoparticles behaviour flowing in the round channel with the liquid in the constant magnetic field was carried out. To simulate the transport of nanoparticles in the magnetic field a novel Euler-Lagrange combined two-component approach was used. The verification and adaptation of mathematical models of magnetic nanoparticles transport under the influence of external electromagnetic fields based on the experimental data was performed. The numerical results were compared with experimental data. Good qualitative and quantitative agreement is obtained between numerical simulation and experiment. In future it is planned to measure the velocity profiles for these regimes using particle image velocimetry (PIV) method.

The reported study was funded by Russian Foundation for Basic Research, Government of Krasnoyarsk Territory, Krasnoyarsk Region Science and Technology Support Fund to the research project №16-48-243061.

References

1. J. Dobson, Drug Dev. Res. **67** (2006)
2. P.A. Voltairas, D.I. Fotiadis, L.K. Michalis, J. Biomech. **35** (2002)
3. H.D. Liu, W. Xu, S.G. Wang, Z.J. Ke, Appl. Math. Mech.-Engl. Ed. **29** (2008)
4. C. M. Oldenburg, S. E. Borglin, G. J. Moridis, Transport in Porous Media **38** (2000)
5. M. Pryazhnikov, D. Guzei, A. Minakov, V. Zhigarev, A. Shebeleva, MATEC Web of Conferences **72** (2016)

Modeling of liquid and gas flows in the horizontal layer with evaporation

Yuri Lyulin[1], *Ekaterina* Rezanova[1,2,*]

[1]Kutateladze Institute of Thermophysics SB RAS, 1, Lavrentyieva ave., 630090, Novosibirsk, Russia
[2]Altai State University, Faculty of Mathematics and Information Technologies, 61, Lenina, 656049 Barnaul, Russia

Abstract. Mathematical modeling of two-layer flows in the "ethanol-nitrogen" system on the basis of the exact solutions of a special type is carried out. The influence of the gas flow, temperature and Soret effect on the flow patterns and evaporating processes at the interface is investigated. The results of comparison of the experimental and theoretical data are presented; the dependence of the evaporation intensity at interface of the gas flow rate and temperature is studied.

1 Introduction

The convective fluid flows under action of a gas flow with respect to the evaporation are actively studied experimentally and analytically nowadays. The problem of construction of the exact solutions to describe the two-layer fluid flows in an infinite channel with an interface generates a considerable interest [1]. The solutions, which are studied in this work, can be considered as a generalization of the solution of the convection problem in a horizontal layer with a free boundary [2]. One of the first examples of the exact solutions, that takes into account the mass transfer at the interface of two media, has been obtained in [3]. The solutions of the equations of convection in the two-layer systems subject to the Dufour and Soret effects in the gas layer are studied in [4, 5, 6].

In the present paper the analytical study of the liquid and gas flows with evaporation at the interface and a comparison of the theoretical and experimental results are carried out. Mathematical modeling of the problem is based on the exact solutions of the Navier-Stokes equations in the Oberbeck-Boussinesq approximation [7]. Some characteristics of the liquid and gas flows and the intensity of evaporation are studied with the help of the system like ethanol-air. The influence of gas flow rate and temperature on the velocity and temperature profiles and evaporation intensity at the interface is investigated.

2 Mathematical model of the process

* Corresponding author: katerezanova@mail.ru

A viscous incompressible liquid and gas-vapor mixture fill the infinite horizontal layers of thickness l and h, respectively. The coordinate system is chosen so that the vector of the gravity acceleration **g** is directed opposite to the Oy axis. The effects of thermodiffusion and diffusive heat conductivity (the Soret and Dufour effects) are taken into account in the upper layer [7, 8]. The mathematical model in the two-dimensional case represents a system of differential equations of the following form [7]:

$$(\mathrm{v}\cdot\nabla)\mathrm{v} = -\frac{1}{\rho}p' + \nu\Delta\mathrm{v} + \mathbf{g}\left(\beta T + \underline{\gamma C}\right), \quad (1)$$

$$div\,\mathrm{v} = 0, \quad (2)$$

$$\mathrm{v}\cdot\nabla T = \chi\left(\Delta T + \underline{\delta\Delta C}\right), \quad (3)$$

$$\mathrm{v}\cdot\nabla C = D\left(\Delta C + \alpha\Delta T\right), \quad (4)$$

The underlined terms and equation (5) are used for simulation of the flow in the upper layer. Here $\mathrm{v} = (u, \upsilon)$ is the velocity vector, p' is the modifed pressure or deviation from the hydrostatic pressure, T is the temperature, C is the vapor concentration, ρ is the density, ν is the kinemtic viscosity, χ is the coefficient of heat diffusivity, D is the coefficient of vapor diffusion, β is the coefficient of thermal expansion, γ is the concentration density coefficient, the coefficients δ and α defines the Dufour and Soret effects in the gas-vapor layer.

Let us seek the solution of the equations (1)-(5) as follows (see [4, 5, 6]):

$$u_i = u_i(y),\ \ \upsilon_i = 0,\ \ p_i' = p_i'(x, y),\ \ T_i = (a_i^1 + a_i^2 y)x + \vartheta_i(y),\ \ C = (b_1 + b_2 y)x + \varphi(y). \quad (5)$$

Here the index i (i = 1,2, subscript or superscript) is responsible for belonging to the liquid layer if $i = 1$ or to the gas layer if $i = 2$. In the course of solving the problem (1)-(4) with account (5), the functions u_i are defined as fourth-degree polynomials, ϑ_i and φ is the polynomials of the seventh degree, p' is the eighth degree polynomial on the transverse coordinate y.

The no-slip conditions for the velocity should be imposed on the upper $y = h$ and lower $y = -l$ fixed boundaries and the conditions for temperature are as follows $T_1|_{y=-l} = A_1 x + \vartheta^-$, $T_2|_{y=h} = A_2 x + \vartheta^+$ (here A_1, A_2, ϑ^+, ϑ^- = const). The vapor concentration on the upper boundary is equal to zero.

On the thermocapillary interface $y = 0$ the kinematic condition is fulfilled automatically; the dynamic conditions can be written as follows: $\rho_1\nu_1 u_{1y} = \rho_2\nu_2 u_{2y} + \sigma_T T_{1x}$, $p_1 = p_2$ (here σ_T is the temperature coefficient of the surface tension σ, $\sigma = \sigma_0 + \sigma_T(T - T_0)$). The conditions of the velocity and temperature continuity are set at the interface. The heat transfer condition at interface should be written with respect to the Dufour effect: $\kappa_1 T_{1y} - \kappa_2 T_{2y} - \delta\kappa_2 C_y = -\lambda M$ (here λ is the latent heat of evaporation, κ_i are the thermal conductivity coefficients, M is the evaporation mass flow rate supposed here being constant). A mass balance equation at $y = 0$ has the following form with respect to the Soret effect: $M = -D\rho_2(C_y - \alpha T_{2y})$. The saturated vapor concentration is determined with the help of the relation $C = C_* [1 + \varepsilon(T_{2y} - T_0)]$ that is a result of the Clapeyron-Clausius equation and Mendeleev-Clapeyron equation for the ideal gas. Here C_* is the saturated vapor concentration at $T_2 = T_0$, T_0 is a some initial temperature, $\varepsilon = \lambda\mu / RT_0^2$, μ is the molar mass of the evaporating liquid, R is the universal gas constant.

The gas flow rate in the upper layer is given and defined by the relation $Q = \int_0^h \rho_2 u_2(y) dy$.

These boundary conditions allow us to determine the integration constants arising from the finding of exact solutions of the system (1)-(4) and unknown parameters of the problem.

3 Results of analytical calculations. Comparison with physical experiments

We consider the two-layer system, where the lower and upper layers are filled by ethanol and air, respectively. The experimental studies were conducted for different temperature regimes of the system. This fact is reflected in analytical studies, and some values of the physico-chemical parameters of the liquid and gas are selected depending on the initial temperature. The values of ethanol and air are presented in Tab. 1 and [9]-[11]. The Dufour and Soret parameters are $\delta = 10^{-3}$(K) and $\alpha = -5 \cdot 10^{-3}$ (1/K), respectively.

Table 1. Values of some physico-chemical parameters of the liquid and gas.

Parameter	ethanol		air	
	20 °C	30 °C	20 °C	30 °C
ρ, kg/m^3	789	781	1.20	1.16
ν, m^2/s	$1.5 \cdot 10^{-7}$	$1.285 \cdot 10^{-7}$	$1.6 \cdot 10^{-5}$	$1.5 \cdot 10^{-5}$
C_*	0.1	0.15	-	-
ε, 1/K	0.059	0.0549	-	-

In Fig. 1 shows the velocity profiles in the case of selecting physical parameters of the system for temperature 20 °C (fig. 1a) and the temperature profiles in the section $x = 0$ for the physical parameters of the system, corresponding to 20 °C (fig. 1b) and 30 °C (fig. 1c) for different values of gas flow rate Q. Increasing the gas flow rate values in the upper layer entails the increase of the flow intensity (see fig. 1a). More intensive cooling of the interface for different temperature regimes is also noticeably (see fig 1b, 1c). Note, that in the case when the physical parameters of the system selected for temperature 30 °C the temperature difference between solid walls and interface much more. However, the qualitative picture of temperature distribution in the system is maintained.

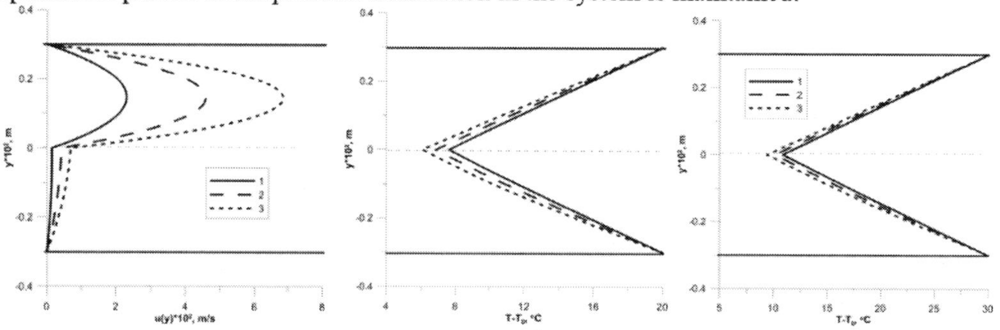

a b c

Fig. 1. a) Velocity and b) temperature profiles at different values of the gas flow rate, $T = 20$ °C; c) temperature profiles at different values of the gas flow rate, $T = 30$ °C. $A_1 = 10$ K/m, $h = 0.3 \cdot 10^{-2}$ m, $l = 0.3 \cdot 10^{-2}$ m (1 – Q = $5.625 \cdot 10^{-5}$ kg/(m·s), 2 – Q = $1.125 \cdot 10^{-4}$ kg/(m·s), 3 – Q = $1.6875 \cdot 10^{-4}$ kg/(m·s)).

Figures 2 illustrate the results of investigation of the mass flow rate M dependence from the gas flow rate Q obtained experimentally (Fig. 2, lines 1 and 2) and with help of the

DOI: 10.1051/epjconf/201715900039

analytical formulas (Fig. 2, lines 3 and 4). The intensity of the liquid evaporation increases with increasing of the gas flow rate. This effect occurs by extension of the vapor concentration gradient in the gas phase. The mass flow rate of evaporation increases with the increase of the temperature in the channel, which agrees well with the experiments. Analytical calculations are performed for two-dimensional case and did not take into account all the factors that are present in the experiment. However, for smaller values of temperature (20 ° C) or small values of the gas flow are observed fairly good quantitative agreement between theoretical calculations and experiments. Note, that the analytical results are given for values of the longitudinal temperature gradient at the lower solid boundary A_1 = -100 K/m, whereas the temperature gradients in the experiment are not measured.

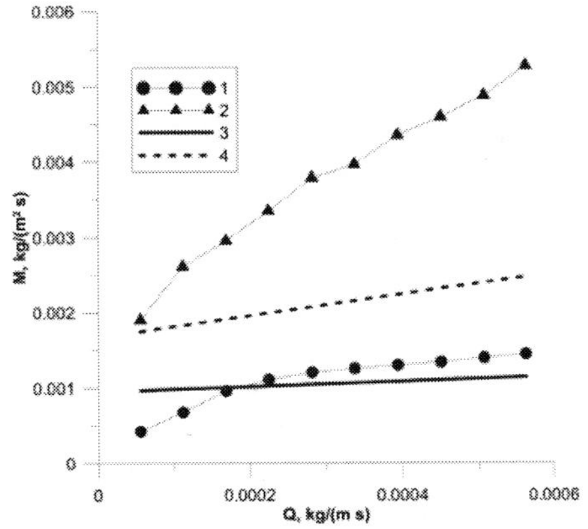

Fig. 2. Dependence of the evaporation intensity on the gas flow rate. $A_1 = -100$ K/m, $h = 0.3\cdot10^{-2}$ m, $l = 0.3\cdot10^{-2}$ m (1 experimental data at $T = 20$ °C; 2 experimental data at $T = 30$ °C; 3 calculations at $T = 20$ °C; 4 3 calculations at $T = 30$ °C).

The research has been supported by the Russian Foundation for Basic Research (Project No. 14-08-00163).

References

1. O.N. Goncharova, O.A. Kabov, Int. J. of Heat and Mass Transfer, **53**, 2795-2807 (2010)
2. R.V. Birikh, J. Appl. Mech. Tech. Phys, **3**, 69-72 (1969)
3. M.I. Shliomis, V.I. Yakushin, Uchenye zapiski Permskogo Gosuniversiteta, **4**, 129-140 (1972)
4. O.N. Goncharova, M. Hennenberg, E.V. Rezanova, O.A. Kabov, IPHT, **1**, 3, 317-338 (2013)
5. O.N. Goncharova, E.V. Rezanova, J. of Appl. Mech. and Tech. Phys. **55**, 2, 247-257 (2014).
6. O.N.Goncharova, E.V. Rezanova, Yu.V. Lyulin, O.A. Kabov, Thermophys. and Aeromech., **5**, 631-637 (2015)
7. V.K. Andreev, Yu.A. Gaponenko, O.N. Goncharova, V.V. Pukhnachov, *Mathematical models of convection* (de Gruyter Studies in Mathematical Physics, Berlin/Boston: De Gruyter, 2012)
8. B. Gebhart, Y. Jaluria, R.L. Mahajan, B. Sammakia, *Bouyancy-induced flows and transport* (Springer Verlag, 1988)

9. I. Prigogine, R. Dufay, *Chemical thermodynamics* (Longmans, Green, London, New York, 1954)
10. *Quick Reference physico-chemical quantities* (SPb, Spetc. lit., 1998)
11. R.C. Weast *CRC Handbook of Chemistry and Physics* (CRC Press, 1978-1979)

EPJ Web of Conferences **159**, 00040 (2017) DOI: 10.1051/epjconf/201715900040
AVTFG2016

Features of two-phase flow in a microchannel of 0.05×20 mm

Fedor Ronshin[1,2*]

[1]Kutateladze Institute of Thermophysics, 630090, 1 Acad. Lavrentiev ave., Novosibirsk, Russia
[2]Novosibirsk State University, 630090, 2 Pirogov st., Novosibirsk, Russia

Abstract. We have studied the two-phase flow in a microchannel with cross-section of 0.05×20 mm^2. The following two-phase flow regimes have been registered: jet, bubble, stratified, annular, and churn ones. The main features of flow regimes in this channel such as formation of liquid droplets in all two-phase flows have been distinguished.

1 Introduction

Miniaturization of devices in various fields of technology, such as aerospace, electronics, transport, power engineering and medicine significantly increases the interest in hydrodynamics of gas-liquid flow and heat transfer in microsystems and microchannels. The studies show that heat exchange systems of mini- and microsizes are more energy efficient than macrosystem with free-flow of liquid [1-3] or a typical channel size greater than 1 mm [4].

The review on the regimes of two-phase flows in the channels of different geometries under different conditions is presented in [5, 6]. It is shown that the boundaries between the regimes differ significantly depending on the experimental conditions. According to analysis, the structure of two-phase flow is mainly influenced by the following parameters: channel geometry and dimensions, parameters of the inlet section and liquid properties such as viscosity and surface tension. In [7-9], the influence of the inlet section on the two-phase flow structure is analyzed, and it is shown that the conditions of gas and liquid input affect substantially the borders of two-phase flows in microchannels. The flows in the rectangular channels of various heights and widths are compared in [10-12]. It has been determined that the height and width of the horizontal microchannel affect substantially the boundaries between the flow regimes in a microchannel. The mechanisms that influence the formation of two-phase flow are studied in [11, 12]. Two new types of instability (frontal and lateral), responsible for the flow regime formation in microchannels depending on the channel sizes and liquid and gas flow rates, are distinguished.

The work deals with the studies of the features of two-phase flow in a short (the length from the liquid inlet to the microchannel is 90 mm) rectangular horizontal microchannel with the cross-section of 0.05×20 mm^2.

* Corresponding author: f.ronshin@gmail.com

DOI: 10.1051/epjconf/201715900040

2 Experimental setup and results

The test section consists of two parallel plates with the length of 160 mm and width of 55 mm (the upper plate is of glass, the lower plate is of stainless steel); the distance between them is set by two constantan strips with the height of 50 μm. A nozzle is made in the bottom plate at the angle of 11° made, and liquid is fed in-between the plates through this nozzle via a precision syringe pump. Microchannel dimensions are as follows: length is 160 mm, width is 20 mm, and height is 50 μm. The gas mixture is fed into the central part of the microchannel. The gas flow is regulated and maintained constant by the El-Flow flow controller of Bronkhorst company. After the assembly, the microchannel height has been measured at several points using the confocal technique. The average microchannel height in the observation area is 48.7 ± 5.3 μm.

The liquid flow rates have been ranged from 0.5 to 50 ml/min, gas flows have been varied from 100 to 5000 ml/min. Superpure distilled deionized nano-filtered water, pre-cleared by Direct-Q® 3 UV installation, have been used as liquid. High-purity nitrogen have been used as gas. The distance between the gas and liquid nozzle has been about 70 mm. The pressure in the gas chamber is measured by the pressure sensor WIKA Type P–30. The readings of this pressure sensor and current flow rate of gas are written in a file on a PC. Liquid and gas interaction in the microchannel have been visualized by the digital video and photocameras in the mode of Schlieren photography. The Schlieren method is used to register and visualize deformation of a thin liquid film surface. The detailed description of the Schlieren technique is presented in [13].

The following regimes of two-phase flow have been registered during the experiment: jet, bubble, stratified, annular, and churn ones. The regime map has been plotted, Fig. 1. Superficial velocities of liquid and gas, determined as the ratio of volumetric flow rate to cross-sectional area of the channel, have been used as the coordinates.

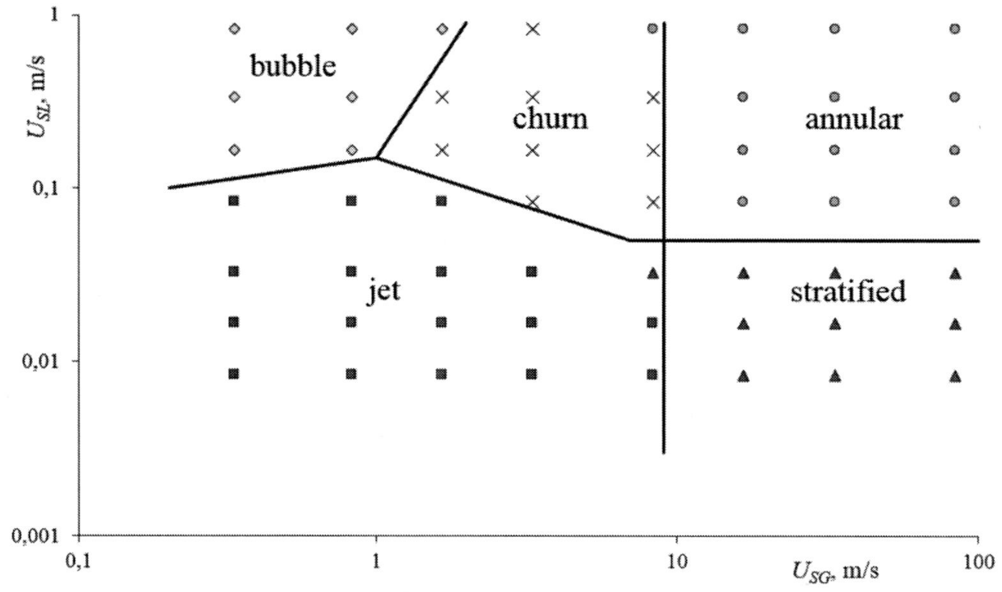

Fig. 1. Regime map in microchannel with cross-section of 0.05×20 mm^2.

At very low superficial liquid velocities, gas moves in the central part of the microchannel, and the bulk of liquid moves at the periphery along the sidewalls and as the jets in the center of microchannel. In microchannels higher than 100 microns, liquid flows only along the sidewalls of microchannel [11]. With increasing superficial liquid velocity, it starts occupying a larger part of microchannel, and a gas flow moves in the center. Disturbances on the surface of the liquid are not observed. The stationary jet flow is observed at low superficial liquid and gas velocities, when the gas flow occupies less than a half microchannel width. The jet regime is specific for the flat mini- and microchannels. An increase in the superficial velocity of liquid leads to an increase in pulsation frequency and amplitude, and loss of stability of the jet regime of the two-phase flow. At low superficial velocities of gas, the amplitude of liquid perturbations in the lateral parts of microchannel reaches its half, forming the stable liquid bridges, and the slug or bubble flow starts.

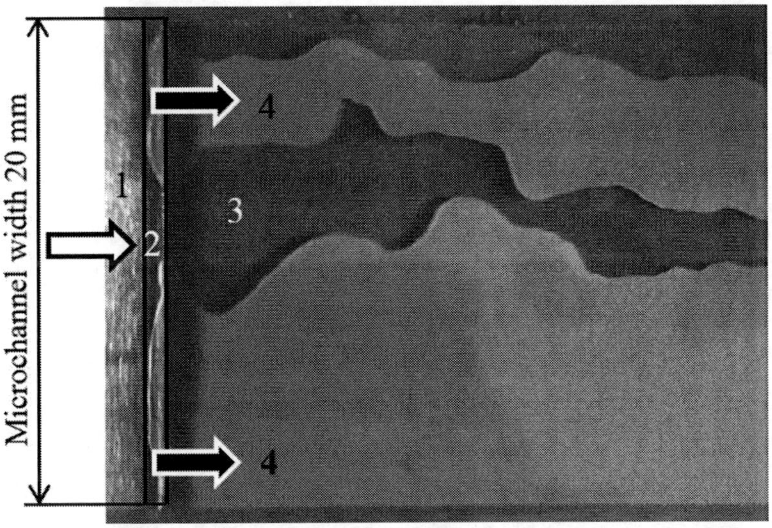

Fig. 2. Picture of the jet flow at U_{SG} = 0.333 m/s, U_{SL}= 0.017 m/s, top view.
1 – gas inlet to microchannel, 2 – slot in the bottom wall of microchannel for liquid input, 3 – unwetted zone on the upper wall of microchannel, 4 – liquid.
white arrow – direction of gas flow, black arrows – direction of liquid motion.

Figure 2 shows the jet flow at U_{SG} = 0.333 m/s, U_{SL}= 0.017 m/s. Liquid occupies a greater volume than gas. Under this regime, the motion of a gas jet in the microchannel center and along one wall is observed. Such asymmetry can be caused by inhomogeneity in the inlet region. In the channels of a greater height, liquid spreads along the sidewalls due to the effect of capillary forces. In the studied microchannel, liquid does not always spread along the walls; this can be caused by a large height to width ratio. By increasing the superficial gas velocity to U_{SG} = 1.667 m/s, the volume occupied by liquid, decreased significantly. Several liquid jets move in the microchannel. The motion of liquid droplets is also observed in the microchannel. With an increase in the superficial gas and liquid velocities, intensity of droplet formation increases, and the flow pattern becomes more complicated. After the passage of liquid droplets, a thin liquid film stays on the microchannel walls.

With increasing superficial velocity of liquid, the stable liquid bridges start being formed, and transition from the bubble to the jet regime occurs. Under this mode, liquid with many small gas bubbles moves along the channel. The size and number of bubbles change depending on the liquid and gas flow rates, but the bubble sizes are always much smaller than the channel width. With increasing superficial velocities of liquid and gas, bubble frequency increases. At U_{SG} = 0.333 m/s, U_{SL}= 0.167 m/s, both small and large gas bubbles move in the channel. There are also many small liquid droplets inside the bubbles. Formation of such droplets is caused by destruction of bridges or liquid films. With increasing gas and liquid velocities to U_{SG} = 0.833 m/s, U_{SL} = 0.333 m/s, the frequency of bubbles formation increases, they move along the left and right walls of the microchannel, and stationary small bubbles are observed in the middle.

At low superficial liquid velocities and high superficial velocities of gas, the stratified regime is observed. Under this regime, a part of liquid moves along the bottom wall of the microchannel in the form of a film entrained by a gas flow. The upper wall of microchannel is dry. Gas under this regime occupies more than a half microchannel cross-section. The stratified regime is characteristic only of the non-circular microchannels because in the circular microchannels the film closes, forming the annular flow. At U_{SG} = 16.667 m/s, U_{SL} = 0.017 m/s, gas moves in the microchannel center, occupying the most part of its cross-section. The liquid droplets move along one of the microchannel sidewalls, leaving the traces on the liquid film. On the other side of microchannel, the film is not formed and droplets do not move. We can distinguish two kinds of droplets: large ones with the diameter of 1 to 3 mm, which represent the liquid bridges between the top and bottom walls of microchannel, and small droplets with the diameter of several micrometers on the microchannel wall or on the liquid film. With increasing superficial gas and liquid velocities to U_{SG} = 83.333 m/s, U_{SL}= 0.0833 m/s, the number of formed droplets is considerably smaller. The liquid film (sometimes with moving liquid droplets, which leave the traces) moves along the bottom wall of microchannel. Under this regime, the stationary droplets are not observed.

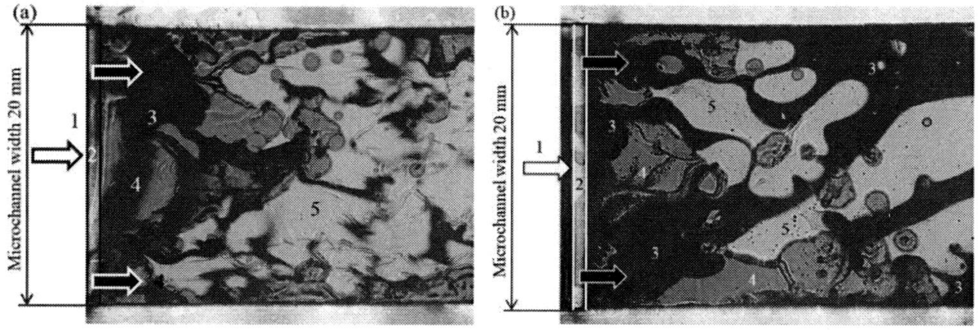

Fig. 3. Shlieren images (a) of the annular flow at U_{SG} = 83.333 m/s, U_{SL}= 0.667 m/s, (b) churn flow at U_{SG} = 6.667 m/s, U_{SL}= 0.5 m/s, top view.
1 – gas inlet to microchannel, 2 – slot in a lower wall of microchannel for liquid input,
3 – channel zones filled with liquid, 4 – film on the lower wall of microchannel, 5 – film on the upper wall of microchannel.
White arrow – direction of gas flow, black wall – direction of liquid flow.

With increasing superficial liquid velocity, the film is formed on the upper wall of microchannel, and transition to the annular regime occurs. Under this regime, we can see the mobile liquid droplets that slide over the liquid films on the top and bottom walls of microchannel. The transition from the stratified to the annular flow regime is determined using the Schlieren method. In the annular regime, liquid moves along the microchannel walls in the form of a film; in the central part, gas with droplets forms the flow core. Gas takes up much more volume than liquid. The liquid film is formed on the top wall of microchannel at a distance of several millimeters from the zone of liquid input to the microchannel. In previous experiments, in the microchannel with the height of less than 200 μm, the film has been formed on the top wall of microchannel directly near the liquid input into the microchannel due to frontal instability [11]. Figure 3a represents the annular regime at U_{SG} = 83.333 m/s, U_{SL}= 0.667 m/s. It can be seen that under this regime, the main part is occupied by the gas core with liquid droplets of a complex shape, the liquid film on the upper wall is firstly formed in the side parts of microchannel, and then in the center. It has been possible to register the film on the bottom (4) and upper (5) walls of the microchannel. It can be seen that the film on the top wall of microchannel is formed at a distance of several millimeters from the liquid nozzle. Liquid droplets (3) move over the films in the microchannel, leaving the characteristic traces on the liquid film.

For the superficial velocities of liquid from 0.1 to 1 m/s and superficial gas velocities from 1 to 10 m/s, the churn flow is observed. This regime is characterized by the features of the jet and bubble flows. This mode is characteristic of the vertical channels, and it occurs in the wide horizontal microchannels. The broken bridges are typical of this regime. The churn flow is caused by development of jet flow instability and increasing pulsation frequency of liquid, moving near the sidewalls of microchannel under the influence of the gas flow. Both gas flow (with discontinuity) and gas bubbles can be seen in Fig. 3b. Formation of a liquid film on the bottom wall of microchannel (4) and formation of regions, where liquid fills the entire height of the channel (3), can be observed near the microchannel inlet. Film formation on the top wall of the microchannel (5) can be seen at a distance of a few millimeters from the liquid nozzle, gas bubbles are formed there. The churn flow occupies a significant area on the map (Fig. 1). The transition from the bubble (solid liquid-filled bridges) to the churn flow is accompanied by breakdowns of bridges. Vice versa, the transition from the jet to the churn flow is accompanied by formation of filled bridges, which are stable and their number exceeds one in the microchannel. At transition to the annular flow, the continuous liquid-filled bridges disappear.

In conclusion, it should be noted that in the channel of rectangular cross-section with the height of 50 μm and width of 20 mm, the liquid film has been registered on in the lower and upper walls of the microchannel. Such features of the two-phase flow as droplet formation under all regimes have been determined. Several kinds of droplets have been distinguished: moving and sessile droplets, the large droplets with the diameter from 1 to 3 mm in the form of the liquid bridge between the upper and lower walls of microchannel, and small droplets with the diameter of several micrometers on the microchannel walls or on the liquid film.

The study was supported by the grant of Russian Science Foundation (Agreement no. 14-19-01755).

References

1. E. A. Chinnov, Journal of Heat Transfer **138**, 9 (2016)
2. E. A. Chinnov, E. N. Shatskii, High Temperature **52**, 461 (2014)
3. E. A. Chinnov, E. N. Shatskiy, Tech. Phys. Lett. **40**, 7 (2014)
4. V. V. Zamashchikov, A. A. Korzhavin, E. A. Chinnov, IJHMT **102**, 470 (2016)
5. E. V. Rebrov, Theor. Found. Chem. Eng. **44**, 355 (2010)
6. N. Shao, A. Gavriilidis, P. Angeli, Chemical Engineering Science **64**, 2749 (2009)
7. V. Haverkamp, V. Hessel, H. Löwe et al., Chem. Eng. Technol. **29**, 1015 (2006)
8. V. V. Cheverda, O. A. Kabov, Tech. Phys. Lett. **43**, 293 (2017)
9. V. V. Cheverda, I. V. Marchuk, A.L. Karchevsky, E.V. Orlik, O.A. Kabov, Thermophysics and Aeromechanics **23**, 415 (2016)
10. E. A. Chinnov, F. V. Ron'shin, O. A. Kabov, Tech. Phys. Lett. **41**, 817 (2015)
11. E. A. Chinnov, F. V. Ron'shin, O. A. Kabov, Int. J. Multiphase Flow **80**, 57 (2016)
12. E. A. Chinnov, F. V. Ron'shin, O. A. Kabov, Interfacial Phenomena and Heat Transfer **3**, 243 (2015)
13. V. Cheverda, A. Glushchuk, P. Queeckers, S. B. Chikov, O.A. Kabov, Microgravity sci. technol. **25**, 73 (2013)

Spectral-element simulations of variable-density turbulent flow in a plane channel

Vladimir Ryzhenkov[1,2], *Vladislav* Ivashchenko[1,2], *Ricardo* Vinuesa[3] and *Rustam* Mullyadzhanov[1,2]

[1]Institute of Thermophysics SB RAS, 630090, Novosibirsk, Russia

[2]Novosibirsk State University, 630090, Novosibirsk, Russia

[3]Linne FLOW Centre, KTH Mechanics and Swedish e-Science Research Centre, SE-100 44, Stockholm, Sweden

Abstract. We perform Large-eddy simulations (LES) of the turbulent flow in a channel with isothermal heated walls with the temperature ratio equal to 2. The variable properties of the fluid are accounted for by using the low Mach number approximation. The Reynolds number based on the bulk velocity, half-width of the channel, density and dynamic viscosity near the cold wall is 6800. We study the effect of spatial resolution on the accuracy of Large-eddy simulations with dynamic Smagorinsky model. The very good agreement of LES results is shown even for the coarse meshes which is attributed to the high accuracy of the spectral method.

1 Introduction

The accuracy of numerical simulations hardly relies on the chosen discretization method. In the present work we employ spectral-element approach [1] which combines the accuracy of spectral methods and flexibility of finite-element approach allowing unstructured grids. We further assess the performance of the nek5000 code in the low Mach approximation targeting the turbulent channel flow [2, 3].

2 The governing equations and computational details

The fluid motion at low speed can be described within the low Mach number approximation with the non-dimensional mass, momentum and energy transport equations [2, 4]:

$$\frac{\partial \rho}{\partial t} + \frac{\partial}{\partial x_j}\left(\rho u_j\right) = 0 \qquad (1)$$

$$\frac{\partial(\rho u_i)}{\partial t}+\frac{\partial}{\partial x_j}(\rho u_j u_i)=-\frac{\partial p}{\partial x_j}+\frac{1}{Re}\frac{\partial \sigma_{ij}}{\partial x_j} \quad (2)$$

$$C_p\left(\frac{\partial(\rho T)}{\partial t}+\frac{\partial(\rho u_j T)}{\partial x_j}\right)=\frac{1}{Re\cdot Pr}\frac{\partial}{\partial x_j}\left(k\frac{\partial T}{\partial x_j}\right)+\frac{\gamma-1}{\gamma}\frac{dp_0}{dt}, \quad (3)$$

where $\rho, T, \boldsymbol{u}, p$ are the non-dimensional density, temperature, velocity and hydrodynamic pressure fields respectively, $\sigma_{ij} = \mu(\partial u_i/\partial x_j + \partial u_j/\partial x_i - (2/3)\delta_{ij}\partial u_k/\partial x_k)$ is the viscous stress tensor, γ is the ratio of specific heats and p_0 is the thermodynamic pressure, which does not depend on space and is determined from a proper equation of state. All the variables are normalized using the reference state $u^{ref}, T^{ref} = p_0^{ref}/\rho^{ref}, C_p^{ref} = C_p^*(T^{ref}), \mu^{ref} = \mu^*(T^{ref})$ and $k^{ref} = k^*(T^{ref})$, where the superscript * represent dimensional quantities [2]. The Reynolds and Prandtl numbers are defined as follows $Re = u^{ref}L^{ref}/\nu^{ref}$, $Pr = \rho^{ref}C_p^{ref}\nu^{ref}/k^{ref}$. Note that $C_p = 1$ in Eq. (3). We consider the ideal gas equation of state:

$$p_0 = \rho T \quad (4)$$

The volume integration of (4) gives:

$$p_0 = \frac{M}{\int_V (1/T)dV}, \quad (5)$$

where $M = \int_V \rho dV$ is the total mass of fluid in the computational domain.

We perform Large-eddy simulations (LES) of the Navier-Stokes equations where the spatial low-pass filtering denoted by an overline is applied to the equations of motion eq. (1)-(3):

$$\bar{\boldsymbol{u}}(\boldsymbol{x},t) = \int G(\boldsymbol{x}-\boldsymbol{r})\boldsymbol{u}(\boldsymbol{r},t)d\boldsymbol{r}, \quad (6)$$

where G is the filter function. It is common to use the Favre-averaging in order to preserve the form of equations:

$$\tilde{f} = \overline{\rho f}/\bar{\rho} \quad (7)$$

Filtered equations contain unclosed terms $-\partial\tau_{kl}/\partial x_k$ and $-\partial q_k/\partial x_k$ in the right-hand side of momentum and energy equations, respectively, corresponding to the sub-grid scale stresses and heat flux. These terms have the following form:

$$\tau_{kl} = \bar{\rho}(\widetilde{u_k u_l} - \tilde{u}_k\tilde{u}_l), \quad q_k = \bar{\rho}(\widetilde{u_k T} - \tilde{u}_k\tilde{T}) \quad (9)$$

The Boussinesq hypothesis and Smagorinsky model are used to express sub-grid terms:

$$\tau_{kl} - \tau_{mm}\delta_{kl}/3 = -2\bar{\rho}\nu_t(\tilde{S}_{kl} - \tilde{S}_{mm}\delta_{kl}/3) = -2C\bar{\rho}\Delta^2|\tilde{S}|(\tilde{S}_{kl} - \tilde{S}_{mm}\delta_{kl}/3), \quad (10)$$

where $\tilde{S}_{kl} = (\partial\tilde{u}_k/\partial x_l + \partial\tilde{u}_l/\partial x_k)/2$, $|\tilde{S}| = \sqrt{2\tilde{S}_{kl}\tilde{S}_{kl}}$, $\nu_t = C\Delta^2|\tilde{S}|$ is the eddy viscosity, Δ the local filter size. The term τ_{mm} is lumped into the pressure field. The Smagorinsky constant C is defined according to the dynamic procedure [5]:

$$C = \frac{\langle [\widehat{\bar{\rho}\tilde{u}_k\tilde{u}_l} - (\widehat{\bar{\rho}\tilde{u}_k}\widehat{\bar{\rho}\tilde{u}_l})/\hat{\bar{\rho}}]\tilde{S}_{kl} - \tilde{S}_{mm}(\mathcal{T}_{jj} - \hat{\tau}_{jj})/3\rangle}{\langle -2\hat{\bar{\rho}}\hat{\Delta}^2|\hat{\tilde{S}}|\left(\hat{\tilde{S}}_{kl}\tilde{S}_{kl} - \hat{\tilde{S}}_{mm}\tilde{S}_{jj}/3\right) + 2\Delta^2\left(\widehat{\bar{\rho}|\tilde{S}|\tilde{S}_{kl}}\tilde{S}_{kl} - \widehat{\bar{\rho}|\tilde{S}|\tilde{S}_{mm}}\tilde{S}_{jj}/3\right)\rangle}, \quad (11)$$

where the brackets $\langle ... \rangle$ denote the averaging along homogeneous directions. Dynamic procedure implies the introduction of the test filter $\hat{\Delta} = 2\Delta$. The test-filtered stresses used in Eq. (11) are defined as:

$$\mathcal{T}_{kl} = \widehat{\bar{\rho}\tilde{u}_k\tilde{u}_l} - (\widehat{\bar{\rho}\tilde{u}_k}\,\widehat{\bar{\rho}\tilde{u}_l}/\hat{\bar{\rho}}) \quad (8)$$

Similarly, the heat flux vector is modeled as follows:

$$q_k = -\frac{\bar{\rho}\nu_t}{Pr_t}\frac{\partial\tilde{T}}{\partial x_k} \quad (12)$$

The turbulent Prandtl number is defined in a similar manner:

$$Pr_t = C\,\langle \hat{\Delta}^2\hat{\bar{\rho}}\left|\hat{\tilde{S}}\right|\frac{\partial\hat{\tilde{T}}}{\partial x_k}\frac{\partial\tilde{T}}{\partial x_k} - (\widehat{\Delta^2\bar{\rho}|\tilde{S}|\frac{\partial\tilde{T}}{\partial x_k}})\frac{\partial\tilde{T}}{\partial x_k}\rangle \times \langle(\frac{\widehat{\bar{\rho}\tilde{u}_k}\widehat{\bar{\rho}\tilde{T}}}{\hat{\bar{\rho}}} - \widehat{\bar{\rho}\tilde{u}_k\tilde{T}})\frac{\partial\tilde{T}}{\partial x_k}\rangle^{-1} \quad (13)$$

The above described governing equations are implemented in the nek5000 code [6] with a spatial discretization based on the spectral-element method using Lagrange polynomials. The Navier–Stokes equations are cast in a weak form and discretized in space by means of Galerkin approximation using N^{th}-order Lagrange polynomial interpolants on the Gauss-Lobatto-Legendre points for the velocity field and pressure ($\mathbb{P}_N$ - $\mathbb{P}_N$ formulation). In the present work the polynomial order N is set to 7. The semi-implicit time-stepping scheme is third-order accurate.

3 Problem formulation and results of simulations

We study a turbulent flow between two infinite parallel isothermal heated walls at a Reynolds number of $Re_c = \rho_c U_b H/\mu_c$, where U_b is the bulk velocity, H is the half-width of the channel, ρ_c, μ_c the density and dynamic viscosity near the cold wall providing the reference values for the non-dimensionalization of Navier-Stokes equations described above. A rectangular computational domain of the size $L_x \times L_y \times L_z = 2\pi H \times 2H \times \pi H$ is used where $x \in [0, 2\pi H]$ is the streamwise, $y \in [-H, H]$ the wall-normal and $z \in [0, \pi H]$ the spanwise coordinates. The planes $y = \pm H$ correspond to the position of parallel walls. The infinite flow domain is modelled by periodic boundary conditions in x and z directions. No-slip condition and constant temperature for hot upper ($T = T_h$) and cold lower ($T - T_c$) wall are imposed. A dynamically adjusted body force is added in equations (1) and (3) to keep a constant *volume* flux through the plane $y - z$. The dynamic viscosity and thermal conductivity depend on the temperature according to the Sutherland's law:

$$\mu(T) = T^{3/2}[1 + S_\mu]/[T + S_\mu], \qquad k(T) = T^{3/2}[1 + S_k]/[T + S_k], \quad (14)$$

where $S_\mu = 0.368$, $S_k = 0.648$. We consider a mild temperature difference, i.e. $T_h / T_c = 2$. Note that using these relations the Reynolds number calculated with hot wall characteristics correspond to $Re_h = \rho_h U_b H/\mu_h = 2081$. The difference between instantaneous temperature fields for constant- [7] and variable-density flows is shown in Fig. 1.

We consider different meshes for variable-density LES simulations with $N_{SE} = 20 \times 16 \times 16$ (fine) and $14 \times 10 \times 10$ (coarse), where N_{SE} is the number of spectral elements along each direction (x, y, z), corresponding to $N_{total} = 2.6 \times 10^6$ and 0.72×10^6, with N_{total} being total number of points. Both meshes satisfy the near-wall LES criteria.

a b

Fig. 1. Instantaneous temperature field for (a) the constant-density flow, (b) variable-density flow.

Previously we have compared constant-density simulations with the data from the literature finding excellent agreement for velocity and temperature fields. Figure 2 shows the comparison of $\boldsymbol{u}, T$ and their fluctuations for coarse and fine LES of variable-density flow. Note the inevitable asymmetry in profiles with respect to $y \to -y$ due to the effect of density/temperature on the velocity field.

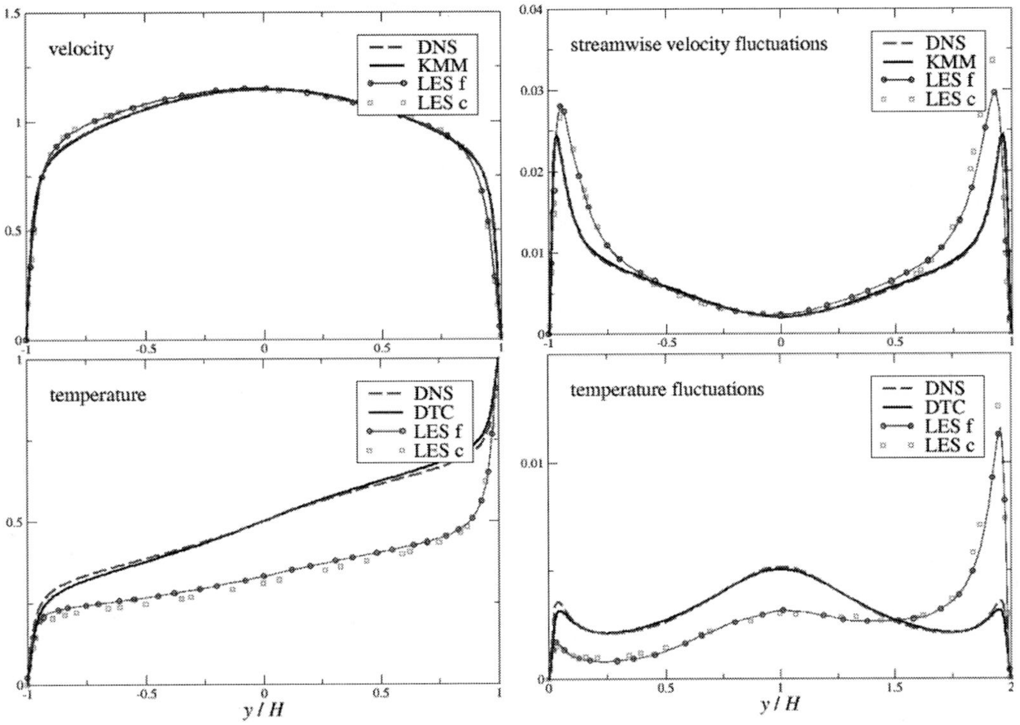

Fig. 2. Black and red dashed line shows the constant-density results [7-9]. Symbols "f" and "c" denote the coarse and fine LES profiles for the variable-density flow respectively.

4 Conclusions

We have performed a series of Large-eddy simulations of a variable-density turbulent channel flow at the Reynolds number of 6800 using the spectral-element method. The difference between constant- and variable density flow has been outlined. The very good agreement of LES results is shown even for the coarse mesh which is attributed to the high accuracy of the spectral method.

This work is funded by RFBR grants No. 15-08-06166 and No. 14-08-01233a. The computational resources are provided by Siberian Supercomputer Center SB RAS (Novosibirsk) and Supercomputing Center of the Novosibirsk State University (Novosibirsk).

References

1. M. O. Deville, P. F. Fischer, E. H. Mund, *High-order Methods for Incompressible Fluid Flow* (Cambridge University Press, New York, 2004)
2. F. C. Nicoud, *Numerical study of a channel flow with variable properties* (CTR, Annu. Res. Briefs, 1998)
3. A. Tyliszczak, Comp. Fluids **127,** 131-145, 2016
4. S. Paolucci, *On the filtering of sound from the Navier-Stokes equations* (SAND82-8253, Sandia National Laboratories, Livermore, 1982)
5. P. Moin, K. Squires, W. Cabot, S. Lee, Phys. Fluids A **3,** 2746, 1991
6. P. Fischer, J. Lottes, S. Kerkemeier, Nek5000 web page (http://nek5000.mcs.anl.gov)
7. V. Ryzhenkov, V. Ivashchenko, R. Vinuesa, R. Mullyadzhanov, J. Phys.: Conf. Ser. **754 (6)**, 062009, 2016
8. R. D. Moser, J. Kim, N. N. Mansour, Phys. Fluids **11 (4)**, 943-945, 1999
9. S. Dharmarathne, M. Tutkun, G. Araya, L. Castillo, Eur. J. Mech.-B/Fluids **55**, 259-271, 2016

The influence of precursor gas pressure on structure and properties of fluoropolymer coatings by hot wire CVD

Alexey Safonov[1,*], *Veronica* Sulyaeva[2], *Konstantin* Kubrak[1], *Sergey* Starinsky[1], and *Nikolay* Timoshenko[1]

[1]Kutateladze Institute of Thermophysics SB RAS, 630090, Lavrentyev Str. 1, Novosibirsk, Russia
[2]Nikolaev Institute of Inorganic Chemistry SB RAS, 630090, Lavrentyev Str. 3, Novosibirsk, Russia

Abstract. In this paper we investigated the deposition process of the fluoropolymer coatings by the method of Hot Wire Chemical Vapor Deposition. The dependence of precursor gas pressure on deposition rate, structure and properties of the formed film was investigated. The study has shown that the increasing of precursor gas pressure leads to change of the film structure from porous to continuous with globules and further to solid that have different wettability. Depending on the mode of deposition was obtained the fluoropolymer coating with different structure: the wetting angle changed from 60 to 170°. A mechanism of the formation of fluoropolymer coating was discussed.

1 Introduction

The surface of thin fluoropolymer films with high hydrophobic properties and chemical inertness can protect from corrosive environments, to intensify heat transfer at phase transitions, improve the surface antifrictional properties, etc. The structure of the film surface determines its properties. Thus, there is a need to develop methods to control of surface structure of the deposited coating.

The existing methods of deposition fluoropolymer films from the gas phase, such as plasma polymerization perfluoromonomer [1], magnetron sputtering [2], laser sputtering of the target, activation a gas precursor on the catalytic surface followed by deposition of a fluoropolymer on a substrate [3-5], require the improvement and the study of deposition processes.

The present work aims to study the possibilities of controlling the properties of fluoropolymer coatings deposited by Hot Wire Chemical Vapor Deposition (HWCVD) or CatCVD [4-6]. The method allows to control manages the structure of the resulting film by changing process parameters of deposition: the activating filament temperature [7], the substrate temperature, the precursor gas pressure, the mesh-substrate distance, etc.

* Corresponding author: safonov@itp.nsc.ru

2 Experimental setup and research techniques

The scheme of experimental setup is shown in Fig. 1. The vacuum chamber for deposition of fluoropolymer coatings was pumped out by oil-free vacuum pump to the base pressure of 10^{-3} Torr. As the precursor hexafluoropropylene oxide C_3F_6O was used. The precursor gas was fed through the flow controllers MKS 1179BX. The flow rate was constant and equal 20 sccm. In the center of the chamber the substrate was fixed. Above the substrate at a distance of 50 mm the catalytic activator from heated grid of nichrome wire was located. The temperature of the mesh 680 °C was monitored by electrical resistance and two contact chromel-alumel thermocouples. The substrate temperature was equal 30°C and was measured by the thermocouples. The pressure in the chamber was measured by a capacitive MKS Baratron sensor 623B and was regulated from 0.05 to 5 Torr with changing the pumping speed of the camera. As the substrates the plates of silicon (100) or copper with size 15×15 mm^2 were used.

The surface morphology and the thickness of the obtained coatings were determined by scanning electron microscopy (SEM) using the JEOL JSM-6700F microscope.

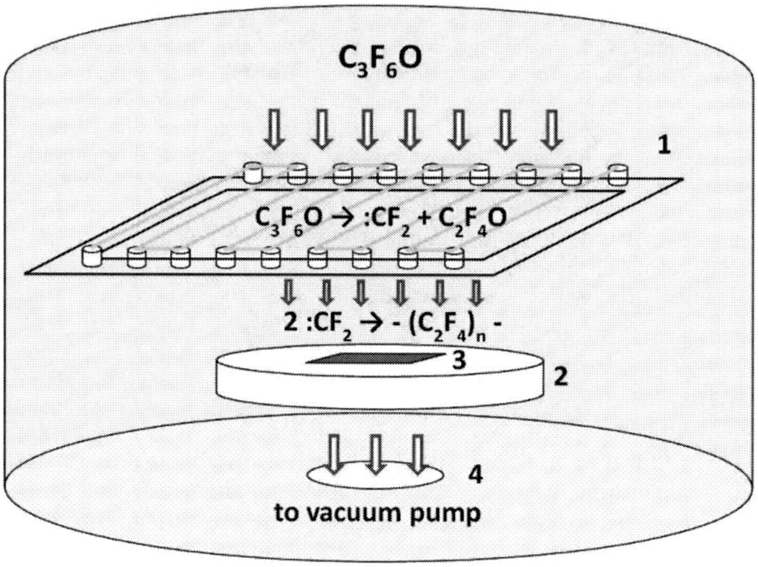

Fig. 1. The scheme of Hot Wire CVD method: 1 – wire activator of precursor gas; 2 – substrate holder with water cooling; 3 – silicon (100) and copper substrates; 4 – vacuum-pumping system.

3 Results and discussion

Fig. 2 shows the surface morphology of the obtained fluoropolymer coatings. In the SEM images are shown that at the precursor gas pressure of 0.1 Torr the films have a friable and porous structure. The cracks with length of several micrometers and a width of 5-10 nm were observed. With increasing pressure to 0.5 Torr the structure of the films varied, it become more dense and non-porous. The globule with size of 20 - 100 nm on the surface began to appear. A further increase of pressure led to an increasing of the number of globules on the surface and its agglomeration. At a pressure of about 5 Torr the globules finally coalesced into continuous film.

The measurement of the thickness of the obtained coatings showed that at different precursor gas pressures during the coating deposition the formation rate varied substantially. The formation rate increased from 0.1 to 11.2 nm/s at changing precursor gas pressure from 0.03 to 0.2 Torr, correspondingly. Further the formation rate reduced to 0.4 nm/s at pressure of 1 Torr and stabilized at 0.1 nm/s at 10 Torr.

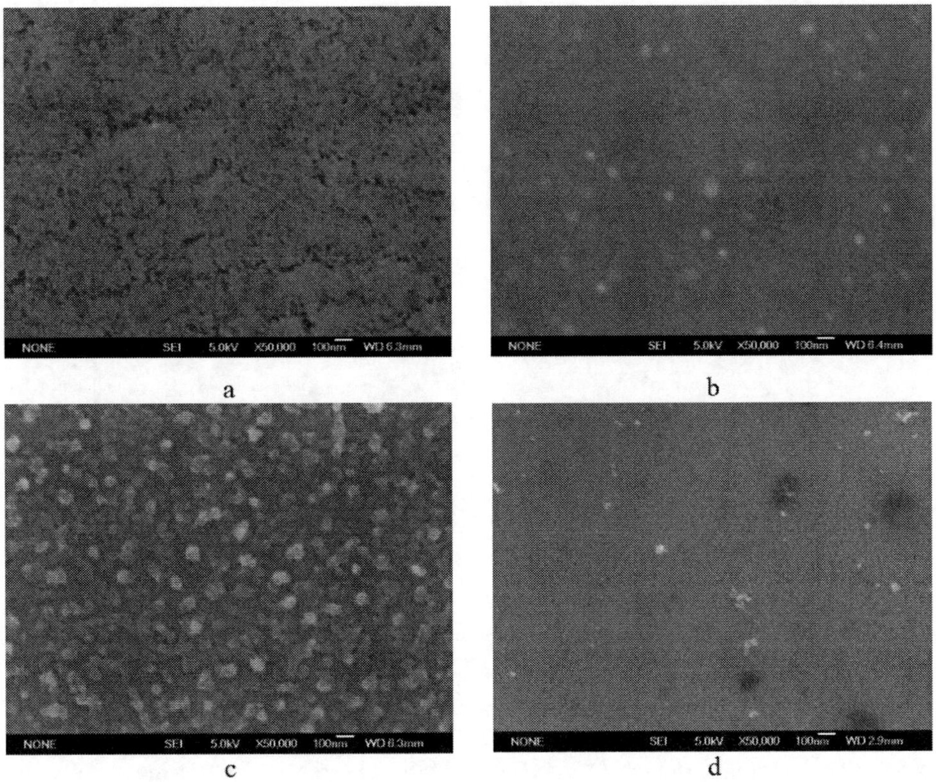

Fig. 2. SEM images of the fluoropolymer coatings surface obtained at different precursor gas pressure in the deposition process: a – 0.1 Torr; b – 0.5 Torr; c – 1 Torr; d – 5 Torr.

The model of formation process the fluoropolymer coating on the surface of the substrate is proposed to determine the influence of the main parameters on the deposition rate. The gas precursor molecules of the fluoropolymer passing through the mesh activated on its surface. The molecules C_2F_4O and active radicals $:CF_2$ [5] are formed. Further the radicals $:CF_2$ are polymerized on the substrate surface. For the considered steady flow the number of activated particles is equal

$$N_a = \frac{aP}{kT} \tag{1}$$

where a is the molecule activation coefficient. The formed active radicals can be deactivated by collisions. Let the coefficient of the deactivation on the free path λ is equal $\Gamma \geq 1$. Then the number of active radicals at a distance x from the surface activation will be equal

$$N_{ax} = \frac{aP/kT}{\Gamma^{\frac{\sqrt{2}\sigma Px}{kT}}} \tag{2}$$

where σ - section of collisions of molecules, P and T the pressure and temperature of the gas, correspondingly. The deposition rate obviously is proportional to the number of collisions of the active particles with the substrate surface. Thus, the deposition rate decreases exponentially with increasing distance x. The pressure dependence has a maximum at

$$P = \frac{kT}{\sqrt{2}\sigma x \ln \Gamma} \tag{3}$$

These simple estimates allow us to establish the qualitative dependence of the main parameters of the process on the deposition rate.

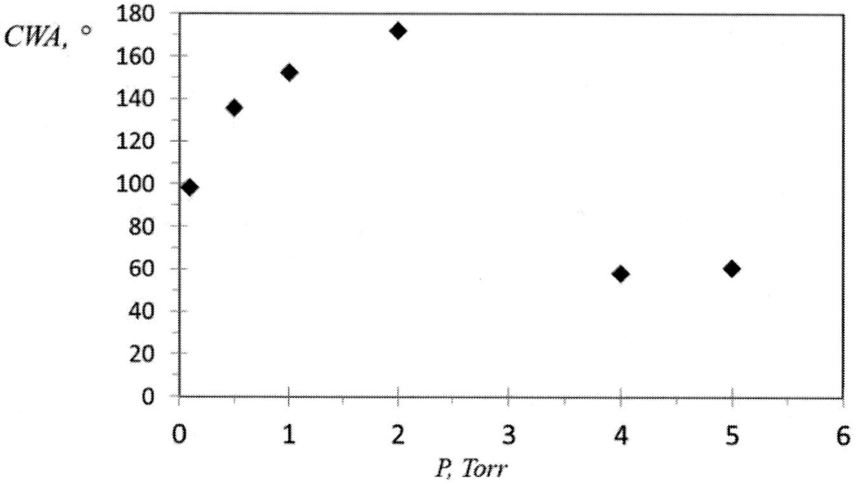

Fig. 3. The change of contact wetting angle depending on the precursor gas pressure in the deposition process.

The hydrophobic properties of the obtained fluoropolymer coatings was determined by measuring of the contact wetting angle (CWA) of its surface by a drop of water using a multifunctional modular device DSA 100 KRUSS. The results of the measurements are presented on the Fig. 3. It was found that the increasing of pressure in the range from 0.1 to 2 Torr lead to increasing the mean of contact wetting angle. The coatings had hydrophobic (90°< CWA <150°) and superhydrophobic (150°< CWA <180°) properties. Further increasing of precursor gas pressure leads to a significant decreasing in wetting angle of the obtained coatings. The coatings had hydrophilic (30°< CWA <90°) properties.

4 Conclusions

The change of deposition rate of fluoropolymer coatings depending on the precursor gas pressure has non-monotonic character with maximum at a defined mean of pressure. In the paper we propose a simple estimation of this relationship according to the number of deactivating the radical collisions.

The study has shown a significant effect of precursor gas pressure on the structure of formed fluoropolymer coating. The different structures of the obtained fluoropolymer coatings significantly affect its wetting properties. It leads to vary of the contact wetting angle of the obtained coatings from 60 to 170°.

This work was supported by the Grant of President of Russian Federation MK-5052.2016.8.

References

1. H. Yasuda. *Plasma Polymerization*. (Academic Press, 1985)
2. W.-C. Wang. Plasma Process. Polym. **4**, 88 (2007)
3. K. S. Lau, J. A. Caulfield, K. K. Gleason. Chem. Mater. **12**, 3032 (2000)
4. K. S. Lau, H. G. Lewis, S. J. Limb, M. C. Kwan, K. K. Gleason. Thin Solid Films **395**, 288 (2001)
5. A. K. Rebrov, A. I. Safonov, N. I. Timoshenko. Technical Physics Letters **35:5**, 395 (2009)
6. K. Lau, Y. Mao, H. Lewis, S. Murthy, B. Olsen, L. Loo, K. Gleason. Thin Solid Films **501**, 211 (2006)
7. A. Safonov, V. Sulyaeva, N. Timoshenko, E. Gatapova, O. Kabov, E. Kirichenko, A. Semenov. Matec **37**, 01047 (2015)

EPJ Web of Conferences **159**, 00043 (2017) DOI: 10.1051/epjconf/201715900043
AVTFG2016

The thermal expansion of hard magnetic materials of the Nd-Fe-B system

Igor Savchenko[1,*], *Yurii* Kozlovskii[1], *Dmitriy* Samoshkin[1], and *Oleg* Yatsuk[1]

[1]Kutateladze Institute of Thermal Physics, 630090 Novosibirsk, SB RAS, Russia

Abstract. The results of dilatometric measurement of the thermal expansion of hard magnetic materials brands N35M, N35H and N35SH containing as a main component the crystalline phase of $Nd_2Fe_{14}B$ type are presented. The temperature range from 200 to 750 K has been investigated by the method of dilatometry with an error of $1.5\text{-}2\times10^{-7}\ K^{-1}$. The approximation dependences of the linear thermal expansion coefficient have been obtained. The character of changes of the thermal coefficient of linear expansion in the region of the Curie point has been specified, its critical indices and critical amplitudes have been defined.

1 Introduction

A large number of experimental studies are devoted to measuring the thermophysical properties of substances in a magnetic field. It is known that under its influence curious thermal effects are observed in the magnetic materials. In particular there are magnetostrictive and magnetocaloric effects. The aim of the present work was to measure the linear thermal expansion coefficient (LTEC) of some technically important compounds of Nd-Fe-B system, as well as to clarify the question about the influence of the residual magnetization by density and thermal expansion of the samples. The materials which are used to create advanced permanent magnets with record values of the maximum magnetic product were investigated. Literature search revealed that the magnetic properties of iron compounds with neodymium and boron were studied in sufficient detail, which is impossible to tell of their thermal characteristics. Specifically it was succeeded to find only one paper [1], where the measurement results of thermal expansion of amorphous alloy of Nd-Fe-B system are provided; and it was not possible to find data for crystalline phases. The results presented in this paper allow to partially fill the gap.

2 Experimental technique

The thermal expansion was investigated on horizontal dilatometer DIL-402C [2] produced by the German company NETZSCH with holder and pushrod made of fused silica. Technique of carrying experiments was described in detail in [3]. The sample mounted on a stand was clamped between the holder and pushrod with a load of 45 cN, which was held

* Igor Savchenko: savchenko@itp.nsc.ru, savchenko.igor@gmail.com

 DOI: 10.1051/epjconf/201715900043

constant during the experiment. The sample length change was registered by an inductive linear variable displacement transducer (LVDT) with a resolution better than 1 nm, and the temperature was measured with Type E thermocouple, which hot junction was located in the vicinity to the lateral surface of the sample. The measurements were carried out in the temperature range 170 – 780 K during heating and cooling at a rate of 2 K/min and 30 min isothermal exposure at maximum and minimum temperatures. The temperature and the length change were registered every 2 s, which allowed to obtain the data with a step less than 0.3 K. Before the experiment, the setup is evacuated (up to 1 Pa) and filled with helium (99.995 vol. % pure), which was additionally purged by special device of cleaning and drying of inert gases – EPISHUR-A 11 SL [4]. The nonlinearity of the displacement transducer, differences in the temperatures and in LTECs of the holder and pushrod materials, etc. were taken into account by the measurement of the correction function of the dilatometer. The temperature dependence of the correction function was determined using certified reference samples of fused silica 25 mm long and 6 mm diameter certified by NETZSCH company in conditions which were identical to those of the main experiments. The correction function determined in this way was reproduced within 0.2 μm.

Apparatus, methods of measurement and processing of the results were tested in experiments with high-purity (99.99 mass. % pure) samples of aluminium and copper. Comparison of the results of measurements with the most credible literature data showed that the deviation in the LTEC values is no more than $(1\text{-}1.5)\times10^{-7}\ K^{-1}$. The control measurements of linear sizes, mass and magnetic field generated by the sample were carried out directly before and after the experiments. The electronic caliper Kraftool [5], electronic indicator Tesa Digico 10 [6], analytical balances AND GH-300 [7] and the magnetometer Aktakom ATE-8702 [8] were used for this measurements.

3 Investigated materials and sample preparation

The experiments were performed on samples of brands N35M, N35H and N35SH, this is hard magnetic materials containing as a main component the crystalline phase of $Nd_2Fe_{14}B$ type. The technology of their manufacture is described in [9], and their technical characteristics are listed in [10]. According with the information provided by the manufacturer the samples have the following chemical structure, namely Fe ~ 71%, Nd ~ 24–27%, Dy ~ 0.5–2.5%, Co ~ 1%, B ~ 1%. According to the composition various brands differ the content of dysprosium, which is 0.5, 1 and 2.5% for brands N35M, N35H and N35SH respectively. From the viewpoint of known properties the specimens of different grades distinguish range of the operating temperatures and the value of the coercive force. For our compounds we have that the more of dysprosium in the sample, the higher the coercive force and the upper limit of the permitted temperature interval. The specimens were the shape of cylinders 25 mm long and 6 mm diameter with plane base, their protective coating of zinc was mechanically removed before the beginning of the experiments.

4 Procedure of the experiments and data processing

Initially for the samples of each brand experiments to determine the temperature dependence of the residual magnetization and finding of maximum working temperature T_{max} were performed. The maximum working temperature was selected so that irreversible change of residual magnetization was not occurred after heating to this temperature and subsequent cooling down to 293 K. The experiments consisted of successive cycles of heating, isothermal holding and cooling of the initially magnetized sample in the dilatometer furnace. The maximum heating temperature was increased for each subsequent cycle, and the magnetic field of the sample was measured in gaps between cycles. Dilatometric experiments with the magnetized samples were carried out at temperatures that did not exceed the maximum working temperature. Thereafter the samples were demagnetized by heating above the Curie temperature (T_C), and measurements of thermal expansion were made in the entire temperature range.

Measurement results are given as the temperature dependence of relative expansion of the sample ε under heating or cooling:

$$\varepsilon(T) = \left(L(T) - L_{293}\right)/L_{293}\,, \tag{1}$$

where L_{293} and $L(T)$ are the sample length at 293.15 K and at T respectively. The LTEC α, and the true LTEC α^* are determined as follows:

$$\alpha(T) = \frac{1}{L_{293}}\left(\frac{\partial L}{\partial T}\right)_p = \left(\frac{\partial \varepsilon}{\partial T}\right)_p, \tag{2}$$

$$\alpha^*(T) = \frac{1}{L}\left(\frac{\partial L}{\partial T}\right)_p = \frac{\alpha(T)}{1+\varepsilon(T)}, \tag{3}$$

where p is the pressure.

The temperature dependence of the LTEC is found by direct numerical differentiation of the raw experimental data on relative expansion:

$$\alpha(T_i) = \frac{1}{2}\left(\frac{\varepsilon_{i+1} - \varepsilon_i}{T_{i+1} - T_i} + \frac{\varepsilon_i - \varepsilon_{i-1}}{T_i - T_{i-1}}\right), \tag{4}$$

where $\varepsilon_i = \varepsilon(T_i)$ is the relative expansion of the sample at the temperature T_i. The real interval of determination of the LTEC ($T_{i+1} - T_i$) is no more than 0.2 K. Such approach allows to get rid errors caused by the ambiguity of the choice depending on the type $\varepsilon(T)$, as well as to identify the features behavior of the linear coefficient of thermal expansion in the field of phase transformations.

In the field of magnetic phase transition a description of the results is carried out using scaling theory [11]. The magnetic contribution is stood out from the values of $\alpha^*(T)$ for this purpose:

$$\alpha^*_{\mathrm{mag}}(T) = \alpha^*(T) - \alpha^*_{\mathrm{para}}(T)\,, \tag{5}$$

where $\alpha^*_{\mathrm{para}}(T)$ is found by the approximation of experimental data in the region of paramagnetic state. The magnetic component of the true LTEC is written in the form:

$$\alpha^*_{\mathrm{mag}} = A|\tau|^a + B\,, \tag{6}$$

where A, B are constants, and a is the critical exponent of the true LTEC, and $\tau = \left(T - T_C\right)/T_C$ is reduced temperature. It is obvious that $B = \alpha^*_{\mathrm{mag}}(T_N)$. Then, introducing a new variable $Y_{mag} = \alpha^*_{\mathrm{mag}} - \alpha^*_{\mathrm{mag}}(T_N)$, we obtain from (6):

DOI: 10.1051/epjconf/201715900043

$$\ln\left(Y_{\mathrm{mag}}\right)=\ln(A)+a\ln\left(\left|\tau\right|\right). \tag{7}$$

As it is seen from (7), it is possible to calculate the values of the critical amplitude A and the critical exponent a when performing the linear approximation of $\ln\left(Y_{\mathrm{mag}}\right)$ by $\ln\left(\left|\tau\right|\right)$.

5 Results and discussion

The experiments to determine the maximum working temperature showed a good agreement of this parameter with values declared by the manufacturer [10]. According to the results of our measurements the maximum working temperature for brands N35M, N35H and N35SH was 373, 393 and 423 K respectively. The results of dilatometric measurements of the sample of brand N35H in magnetized and demagnetized states are presented in Fig. 1, at temperatures no greater than T_{max}. The measurement results of the relative expansion $\varepsilon(T)$ of demagnetized samples throughout the investigated temperature range are shown in Fig. 2, and the temperature dependences of the LTEC $\alpha(T)$ and the density $\rho(T)$ are illustrated in Fig. 3 and Fig. 4 respectively. As may be seen from the Figure 1, the residual magnetization does not have any influence on the temperature dependence of the relative expansion. The measurement results are practically identical and at the higher temperatures for magnetized and demagnetized samples. A similar pattern is typical for experiments with samples N35M and N35SH brands. The density of the samples under normal conditions are not changed after the demagnetization and takes the following values $\rho_{N35M}=7412$, $\rho_{N35H}=7332$ and $\rho_{N35SH}=7355$ kg/m^3. Thus it can be concluded, that the magnetostrictive effect is negligibly small for test materials at a temperature of 473 K to T_{max} and a magnetic field which is not exceeded 400 mT. The sharp minimum in the temperature dependence of $\alpha(T)$ is associated with the transition from the ferromagnetic to the paramagnetic state. This phase transition of the second order is clearly seen not only on the LTEC curve, but on the relation of the relative expansion (see Fig. 2).

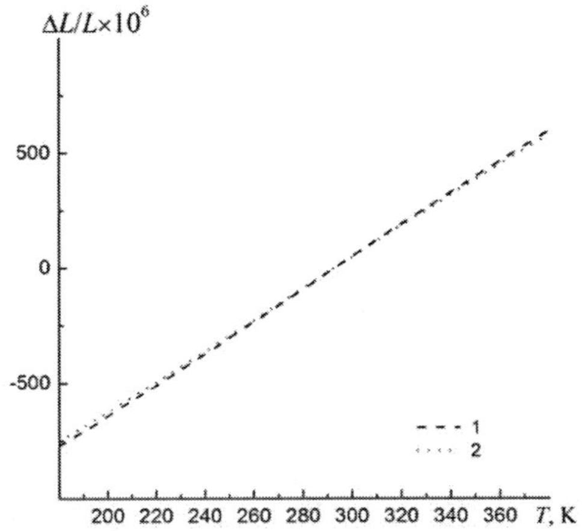

Fig. 1. The results of dilatometric measurements of the sample of brand N35H in magnetized and demagnetized states, at temperatures no greater than Tmax. 1 – the magnetic field on the surface of the sample is 400 mT; 2 – the magnetic field on the sample surface is 0 mT.

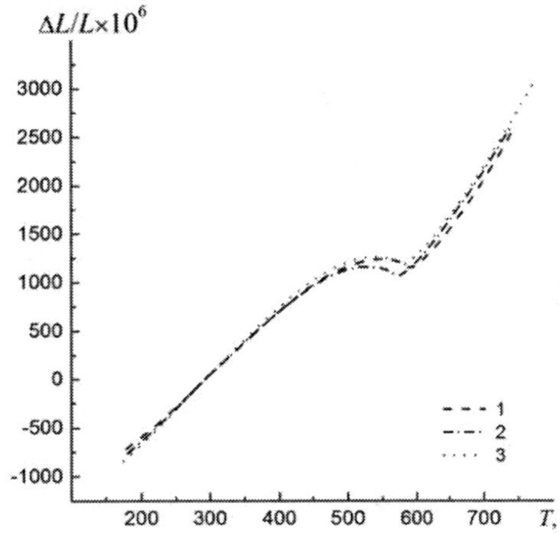

Fig. 2. The measurement results of the relative expansion of investigated materials. 1 – N35M, 2 – N35H, 3 – N35SH.

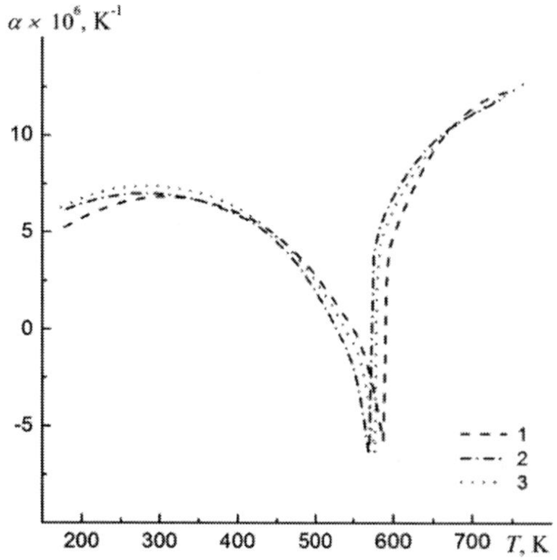

Fig. 3. The linear thermal expansion coefficient. 1 – N35M, 2 – N35H, 3 – N35SH.

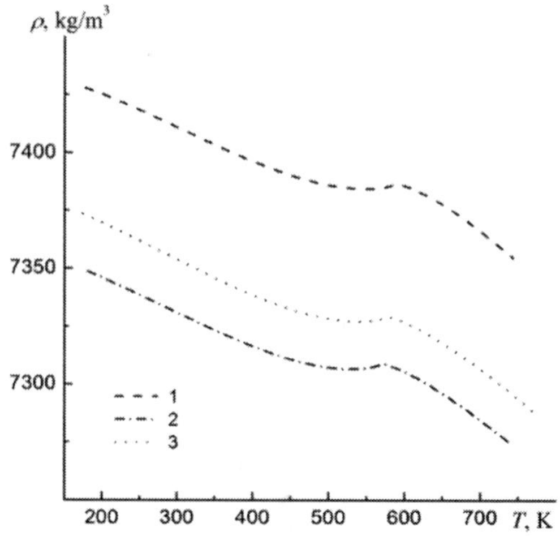

Fig. 4. The density of the samples. 1 – N35M, 2 – N35H, 3 – N35SH.

It can be seen from Fig. 3, that it is impossible to describe the dependence of the $\alpha(T)$ on a single relation. Therefore, the entire measurement interval was divided into areas in which the separate data processing was carried out. Outside the temperature intervals directly adjoined to the Curie point the primary data were approximated by the least squares method of the third degree polynomials. The approximations were made using scaling dependencies for the magnetic phase transition area. The variation of the magnitude of T_C was performed to clarify the Curie temperature so that in the region adjacent to the Curie point, the experimental data were approximated by the equation (7) with the lowest standard deviation. Finding all the points in the confidence limits of the random error of approximation (95% probability) was also the additional requirement. The values of the phase transition temperature determined in this way amounted to 586.3, 569.4 and 576.4 K for the brands N35M, N35H and N35SH respectively. Approximations of the LTEC are given in the Table. It was found that the critical indices for ferromagnetic and paramagnetic areas differed significantly, as well as were the same with good accuracy for all of studied materials.

Table. Approximations of the LTEC of hard magnetic materials brands N35M, N35H and N35SH.

Temperature interval, K	Approximation equation of $\alpha(T)$, $K^{-1}\times10^{-6}$
N35M	
200 – 525	$\alpha(T)=-3.65+0.07\times T-1.14\times10^{-4}\times T^2$
525 – 586.3	$\alpha(T)=-26.6+0.033\times T+33.3\times((586.3-T)/586.3)^{0.5}$
586.3 – 625	$\alpha(T)=-26.6+0.031\times T+19\times((T-586.3)/586.3)^{0.1}$
625 – 750	$\alpha(T)=-165.7+0.475\times T-3.17\times10^{-4}\times T^2$
N35H	
200 – 525	$\alpha(T)=-2+0.066\times T-1.17\times10^{-4}\times T^2$
525 – 569.4	$\alpha(T)=-22+0.026\times T+30.6\times((569.4-T)/569.4)^{0.5}$
569.4 – 625	$\alpha(T)=-22+0.026\times T+18\times((T-569.4)/569.4)^{0.1}$

Temperature interwal, K	Approximation equation of $\alpha(T)$, $K^{-1}\times10^{-6}$
625 – 750	$\alpha(T) = -75.1 + 0.224\times T - 1.44\times10^{-4}\times T^2$
N35SH	
200 – 525	$\alpha(T) = -1.9 + 0.066\times T - 1.15\times10^{-4}\times T^2$
525 – 576.4	$\alpha(T) = -22.8 + 0.028\times T + 30.8\times((576.4 - T)/576.4)^{0.5}$
576.4 – 625	$\alpha(T) = -22.8 + 0.026\times T + 18.7\times((T - 576.4)/576.4)^{0.1}$
625 – 750	$\alpha(T) = -72.2 + 0.208\times T - 1.27\times10^{-4}\times T^2$

6 Conclusion

For the first time, new experimental data on the thermal expansion of hard magnetic materials brands N35M, N35H and N35SH containing as a main component the crystalline phase of $Nd_2Fe_{14}B$ type were obtained. The measurements were carried out by dilatometric method with an error of $1.5\text{-}2\times10^{-7}$ K^{-1} at temperature range from 200 to 750 K. The approximation dependences of the linear thermal expansion coefficient were obtained. The character of changes of the thermal coefficient of linear expansion in the region of the Curie point was specified. It was shown that the critical indices for the ferromagnetic and paramagnetic regions vary significantly. The values of critical exponents for all investigated materials coincide. The values of the maximum working temperature of the permanent magnets created from the investigated ferromagnets were identified. It was shown that the presence of the constant magnetic field up to 400 mT did not lead to noticeable changes in the LTEC or the density of investigated materials.

This work supported by the Russian Foundation for Basic Research, project no. 15-38-20223.

References

1. K. Fukamichi, H. Komatsu, T. Masumoto, T. Kaneko, K. Shirakawa, Y. Satoh, High Pressure Research, **1**, 193-201 (1989)
2. http://www.paralab.pt/sites/default/files/pdf/DIL402C.pdf
3. Yu.M. Kozlovskiy, S.V. Stankus, High Temperature, **52**, 536-540 (2014)
4. http://granat-e.ru/stend_epishur-a_sl.html
5. http://kraftool-instrument.ru/SHtangencirkul-kraftool-elektronnyj-metallicheskij-150mm-p-624.html
6. http://technobearing.ru/tesa_digico-10
7. http://a-and-d.ru/product/gh-300
8. http://www.aktakom.ru/kio/index.php?ELEMENT_ID=39662
9. Patent No: US 8182618 B2
10. magnit54.com/biblioteka/harakteristiki_neodimovyh_magnitov
11. F.H. Spedding, B.J. Beaudry, J. Less-Common Metals, **25**, 61-73 (1971)

Experimental study of the evaporation of sessile nanoliter liquid drops

Andrey Semenov[1,2*] and *Dmitry* Zaitsev[1,2]

[1]Kutateladze Institute of Thermophysics, 630090, Lavrentiev ave. 1, Novosibirsk, Russia
[2]Novosibirsk State University, 630090, Pirogova str. 2, Novosibirsk, Russia

Abstract. Evaporation of a sessile nanoliter water droplet was investigated experimentally at a temperature difference between the solid substrate and surrounding atmosphere from 30 to 50 ^{0}C. The studies were performed on the float glass substrate with aluminum nanocoating of optical quality. The research has shown that the specific rate of evaporation (mass loss per unit of the drop surface area) increases with the decrease in droplet volume and at the last stage several times exceeds the initial value.

1 Introduction

The phenomenon of liquid drop evaporation, which takes place in a variety of technological systems in power engineering, agriculture, medicine, cooling systems, chemical and other industries [1], has been actively investigated during the last few years [2-5]. In most experimental and theoretical studies found in the literature, isothermal evaporation of water droplets has been studied. Publications on nonisothermal evaporation of liquid droplets with a sufficiently high temperature difference between the solid surface and the atmosphere are very limited.

In this paper, we study the process of evaporation of sessile liquid droplet with initial volume less than 1 μl. A drop of liquid is placed on a surface with controlled wettability. Then the liquid drop is heated until quasi-stationary heat transfer between the solid substrate and the drop. Distilled deionized nano-filtered water of Milli-Q company is used as the working fluid. The use of water with a relatively high boiling point as working fluid provides for a rather detailed investigation of the influence of temperature difference between the solid surface and the surrounding atmosphere on the rate of evaporation and heat transfer. The temperature difference between the solid and external atmosphere varies from 30 to 50 ° C.

2 Apparatus and technics

The experiments on water drops evaporation in the open atmosphere were made on the experimental stand the photograph of which is presented in Fig. 1. The work area was an

* Corresponding author: semenov.itp@gmail.com

 DOI: 10.1051/epjconf/201715900044

aluminum plate of 15 mm thickness, on whose bottom side the heater was located. The used heater was a Peltier element with dimensions 40x40 mm^2. On the upper side of the aluminum plate the investigated substrate was fixed. The substrate is a float glass disk (with diameter of 52 mm and thickness of 3.2 mm) with aluminum nanocoating of optical quality. Measurement of the surface temperature of the studied substrate was carried out using a thermocouple, fixed at a distance of 1-2 mm from the contact line of the liquid drop. The temperature under the substrate was measured by three thermocouples of K-type. Thermocouples were placed radially at different distances from the centre of the disc and served to measure the temperature distribution along the substrate radius. The experimental setup was covered with transparent box made of PMMA which has dimensions of 1000x800x900 mm^3. The relative humidity of air in the box during the experiments was 20-30 %.

Fig. 1. Photography of experimental stand.

Two optical techniques were used. The first one is a shadow method, consisting of the halogen light source Edmunds Optics MI-150, connected by a fiberoptic lightguide with a system of lenses of Edmunds Optics Telecentric backlight illuminator. This system generates a plane-parallel light beam of 50 mm diameter at the output of one side of the work area. High-speed video camera FastVideo 500M with a microscope lens 10X Mitutoyo Plan Apo Infinity-Corrected Long WD, connected through the focusing bellow Novoflex Balpro TS, on the other side of the droplet. The spatial resolution of this system was 1 μm/pixel, and the frequency of shooting was up to 500 frames per second. Using the shadow method we obtained shadow images of the liquid drop profile. The images were processed by the Drop Shape Analysis software of KRUSS company. Shadow photography was used to select the drop profile by the method of Young-Laplace. Further, the following geometric parameters of the liquid drop were determined: the base diameter, maximum diameter, height, volume, contact angles (right, left, middle), the surface area, and the basal area.

The second optical technique consisted of a digital video camera Point Grey Grasshopper3 (4.1 Mpix) with a microscope lens 20X Mitutoyo Telecentric Objective. It was placed above the liquid drop and allowed visualizing the evaporation from the top and controlling the symmetry of the droplet. Spatial resolution of the system was 0.5 μm/pixel, and the frequency of shooting was up to 90 frames per second. If the drop lost its symmetry, the experiment repeated.

3 Results

The experiment began with a preliminary preparation of the investigated substrate, which consisted in a thorough cleaning of the working surface. Then the substrate was placed in a reservoir with distilled water for a day. Thereafter, to remove residual water from the surface the substrate was flushed with clean compressed air designed for cleaning optical components. A liquid drop of pre-set size was placed on the surface, heated to the desired temperature, using a high-precision dispenser Lenpipet. The lens used in the shadow method was focused (depth of field was a few hundreds of μm); and after that the periodic shooting was performed until complete evaporation of the liquid droplet. The upper camera was switched on simultaneously with the shadow method.

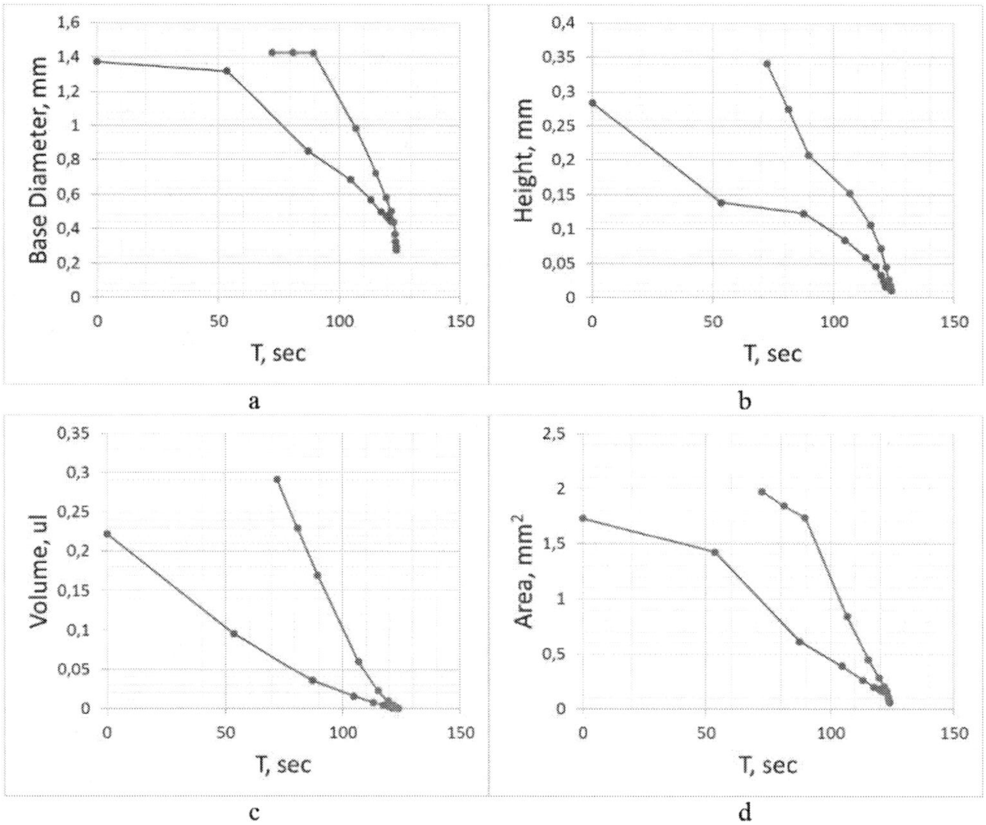

Fig. 2. Dependences of different parameters of droplet vs evaporation time. (a) – the diameter of droplet base, (b) – the height of droplet, (c) – the volume of droplet and (d) – the surface area of droplet. Blue lines and dots is data for substrate temperature T=30 ^{0}C, orange – T=50 ^{0}C. All data was synchronized for the moment of full droplet evaporation.

Dependences of various parameters of a sessile liquid drop during evaporation vs time are shown in Fig.2. The obtained data are synchronized in time of complete evaporation of the droplet. Fig. 2 shows the difference between the evaporation of a drop of liquid on a substrate with an initial temperature of 30 (blue dots) and 50 °C (orange dots).

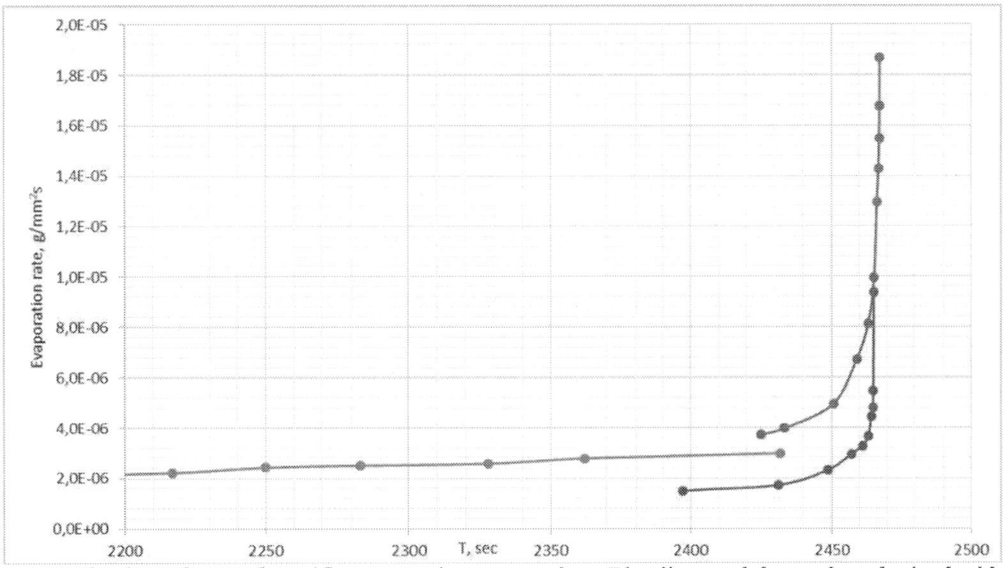

Fig. 3. The dependence of specific evaporation rate on time. Blue lines and dots – data obtained with the substrate temperature T=30°C, the initial drop volume V_0=220 nl; orange - T=50°C, V_0=290 nl; grey – data obtained in the work [2] at T=45°C, V_0=95.16 μl.

The dependence of the specific rate of liquid evaporation on time at different initial surface temperature is shown in Fig. 3. The evaporation rate was calculated as a loss in the droplet mass per unit of droplet surface area per unit of time. All built-in data were synchronized to the time of complete evaporation of liquid droplets. The research has shown that the specific rate of evaporation increases with the decrease in droplet volume and at the last stage several times exceeds the initial value. The obtained results are in good agreement with the results of [2, 3], studying the evaporation of drops with initial volume of about 100 μl.

This work was supported by the Russian Science Foundation (project 14-19-01755).

References

1. J. Kim, International Journal of Heat and Fluid Flow, **28**, 753-767 (2006).
2. E.Ya. Gatapova, A.A. Semenov, D.V. Zaitsev, O.A. Kabov, Colloids and Surfaces A: Physicochemical and Engineering Aspects, **441**, 776-785 (2014).
3. A.A. Semenov, D.V. Feoktistov, D.V. Zaitsev, G.V. Kuznetsov, and O.A. Kabov, Thermophysics and Aeromechanics, **22**, 771- 774 (2015).
4. Kabov O.A. and Zaitsev D.V. Doklady Physics, **58**, 292 (2013).
5. G.V. Kuznetsov, D.V. Feoktistov, E.G. Orlova, K.A. Batishcheva, Colloid Journal, **78**, 335-339 (2016).

EPJ Web of Conferences **159**, 00045 (2017) DOI: 10.1051/epjconf/201715900045
AVTFG2016

Identification of possible non-stationary effects in a new type of vortex furnace

Evgeniy Yu. Shadrin[1,*], *Igor S.* Anufriev[1], and *Anatoly P.* Papulov[1,2]

[1] Kutateladze Institute of Thermophysics SB RAS, Novosibirsk, Russia
[3] Novosibirsk State University, Novosibirsk, Russia

Abstract. The article presents the results of an experimental study of pressure and velocity pulsations in the model of improved vortex furnace with distributed air supply and vertically oriented nozzles of the secondary blast. Investigation of aerodynamic characteristics of a swirling flow with different regime parameters was conducted in an isothermal laboratory model (in 1:25 scale) of vortex furnace using laser Doppler measuring system and pressure pulsations analyzer. The obtained results have revealed a number of features of the flow structure, and the spectral analysis of pressure and velocity pulsations allows to speak about the absence of large-scale unsteady vortical structures in the studied design.

1 Introduction

Currently the most important source for producing electrical and thermal energy is coal combustion. One of the promising technologies is the pulverized coal combustion in a vortex flow. The flow swirling allows solving a number of problems: to increase the residence time of fuel particles in the combustion chamber, which reduces mechanical underburning; to reduce the level of harmful emissions to environmental standards; to optimize the design of boiler equipment; and to ensure effective management of the combustion process. One of the stages in the development of furnaces is the study of their internal aerodynamics in the laboratory models. A detailed study of the main features of the isothermal flow structure allows optimizing the structural and operating parameters of the furnace. It is known that the intensely swirling flows under certain conditions are characterized by the loss in stability of the stationary regime, which may be expressed in the appearance of the precessing vortex core (PVC). Intensive pressure pulsations associated with the PVC cause wear of power plants and deterioration of the performance of vortex devices and have a negative impact on the furnace processes. Therefore, to improve the efficiency and reliability of operation of the vortex combustion devices under development, one must examine possible unsteady effects that occur in the working areas.

In previous works [1, 2] authors visualized the vortex structure of the flow in the model of the improved vortex furnace with distributed fuel-air mixture supply and vertically oriented nozzles of the secondary blast. The obtained results were based on measuring the time-averaged flow characteristics and did not allow making an unambiguous conclusion about the vortex core dynamics. The aim of this work is to experimentally study the pulsation characteristics of the flow in this model of the vortex furnace.

2 Experimental setup and measurement techniques

The study of the flow structure and its pulsation characteristics was performed on an automated experimental stand. The scheme of the experimental setup with the mounted three-component (3D-) LDA system is presented on figure 1-a. The main elements of this setup are: the supply line of compressed air with control and regulating devices; the model of an improved vortex furnace; fog generator (for seeding the investigated flow with tracer-particles); measure devices and computer with specialized software. Air flows through the supply line is injected in the isothermal model of the vortex furnace. The study of this flow is based on a variety of modern optical methods.

The main elements of the vortex furnace are (Fig. 1-b): the combustion chamber; a diffuser; and a cooling chamber ending with a horizontal flue. Dimensions of the model are XYZ 320×1200×256 mm (scale 1:25), and the diameter of the vortex combustion chamber is 320 mm. A distinctive feature of this design compared to the previously studied one [3] (where the additional tangential inlet is located in the lower part of the combustion chamber), is the vertical arrangement of the nozzles of the secondary blast, and the presence of the "visor" inside the vortex chamber to prevent entrainment of fuel particles from the combustion chamber. The advantages of this design of the vortex furnace are described in [1, 2].

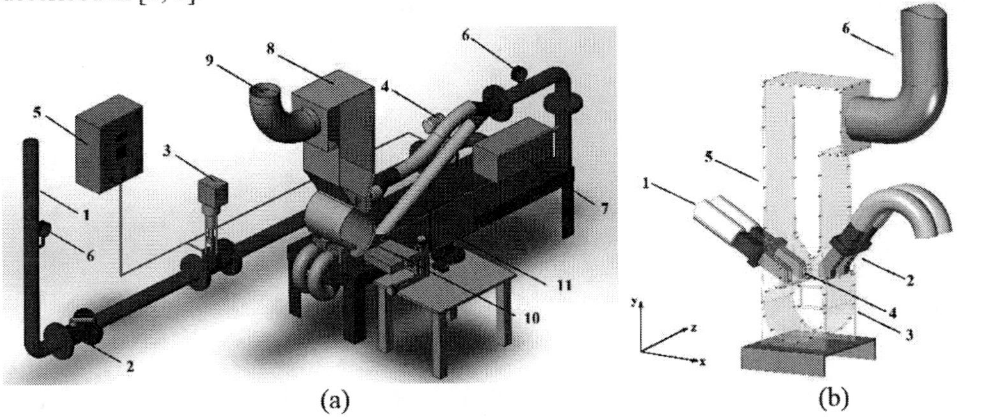

Fig. 1. (a) Experimental stand scheme: 1 – supply line of compressed air, 2 – stop valve, 3 – shut-off control valve, 4 – flow converter, 5 – control cabinet, 6 – manometer, 7 – fog generator, 8 – vortex furnace model, 9 – ventilation, 10 –measuring system, 11 – computer. (b) Laboratory model of the improved vortex furnace: 1 – main nozzles, 2 – secondary nozzle, 3 – combustion chamber, 4 – diffuser, 5 – cooling chamber, 6 – flue.

Measurements of pressure fluctuations were carried out using noise analyzer Bruel&Kjaer (pressure measurement up to 103.5 kPa, frequency measurement - 4.2 Hz÷20 kHz, sensitivity - 54.9 mV/PA). The sensor was placed inside the vortex combustion chamber with a metal sampler – thin-walled tube with a diameter of 2.2 mm and a length of 160 mm. The transfer function of the sampler is presented in [4], which shows its applicability without adjusting to frequencies of ~100 Hz. The measurements were carried out close to the conventional center of the vortex chamber ($x = y = 160$mm, $z = 126$ mm) at various operating parameters (γ=1÷4, where γ is the ratio of the flow rates through the main and additional nozzles). The signal digitized by the ADC (L-CARD E14-440), was expanded into a spectrum using fast Fourier transform.

To diagnose the flow velocity pulsations the two-component (2D-) laser Doppler anemometer LAD-05 (frequency up to 3 kHz), developed at IT SB RAS was used. The method of laser Doppler anemometry is based on measuring the displacement velocity of

 DOI: 10.1051/epjconf/201715900045

the particles (tracers) suspended in the flow. The measurements were performed in the range of Reynolds numbers $3\cdot10^5<Re<6\cdot10^5$, calculated on the diameter of the vortex chamber (320 mm) and the velocity module in the top burners (V_0=5÷25 m/s). These conditions (Re > 10^4) provide the self-similarity regime and the applicability of the results of physical modeling to the analysis of the structure of isothermal flow in a full-size furnace. The measurements were carried out in a period of time (about 1 min) sufficient to obtain 16384 measurements for each component (2^{14}) to facilitate further processing using the fast Fourier transform (FFT).

3 Measurement results and their analysis

Earlier, based on independent measurement techniques, using 3D-LDA [1] and Stereo PIV (Particle Image Velocimetry) systems [2], the authors studied the aerodynamics of a model of the studied vortex furnace with distributed tangential air supply. Based on the "minimum total pressure criterion" [5] the vortex flow structure was visualized. Figure 2-a presents the vector velocity field (a section in the "center of the nozzle"), obtained by PIV-measurements [1] as well as the isosurface of dynamic pressure (p_{dyn}=0.25 Pa, Fig 2-b), visualizing the vortex flow structure [2]. It has a W-shaped form, which is typical for the well-known vortex furnace of CBTI [6]. For comparison, the figure shows the isosurface of total pressure and the plotted distribution of the Q-criterion (Fig. 2-c), obtained by numerical simulation [7].

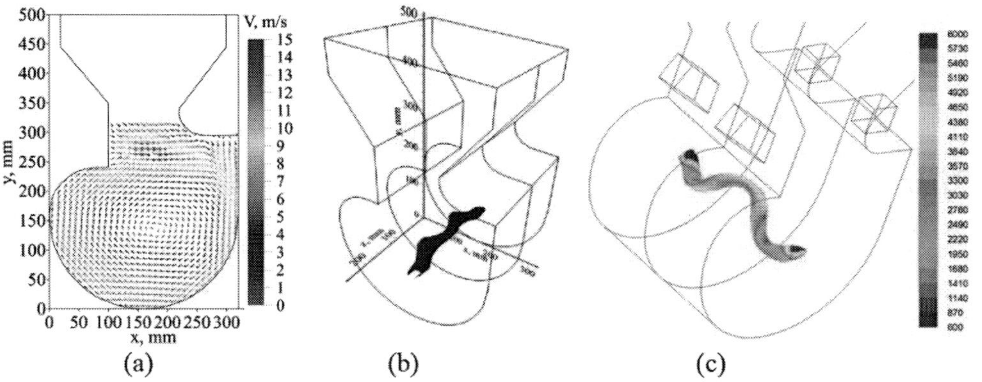

Fig. 2. The vector velocity field in the cross-section "over the nozzle center" (a); isosurface of dynamic pressure (b); isosurface of total pressure (c).

To analyze stationarity of the revealed vortex structure the pressure fluctuations were measured in the central part of combustion chamber. Spectra for the microphone signal (U, Volts) obtained for different ratios of flow rates through the main and additional nozzles (γ) are presented in figure 3. They have a complex form with multiple peaks. However, the position of these peaks neither depends on the flow rate ratio, nor on the magnitude of the total flow, which indicates the absence of the unsteady vortex structures (such as precession vortex core or other structures [8]). The presented peaks characterize the model as an acoustic resonator [6].

 DOI: 10.1051/epjconf/201715900045

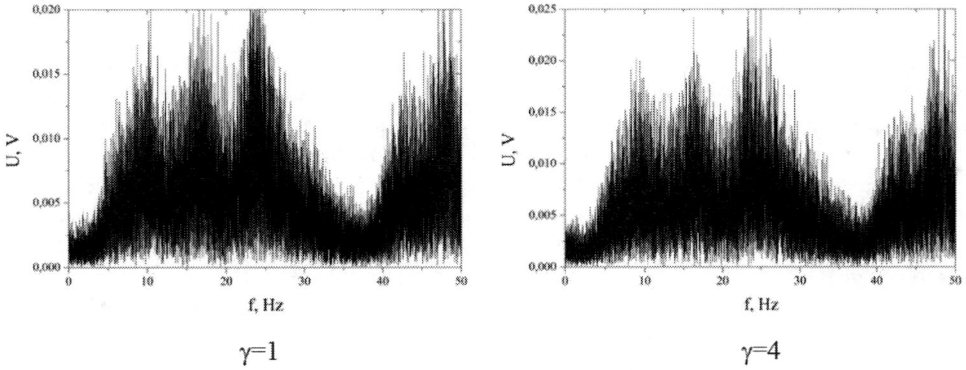

Fig. 3. Spectra of pressure pulsations in the center of the combustion chamber at different ratios of the flow rates through the primary and secondary nozzles (γ).

The noise analyzer used in the measurements is limited in frequency range. Therefore, the analysis of pulsations at low frequencies (<5 Hz) was performed with the use of the laser Doppler anemometer LAD-05. The measurements were carried out at different points of the model in the vicinity of the conventional axis of the curved vortex core and near the entrance nozzles, for different values of initial velocity. Figure 4 shows typical spectra of pulsations of the horizontal velocity component (normalized on the initial velocity V_0) in the point (x=150 mm; y=100 mm, z=64 mm) in a plane passing through the centers of the nozzles. The complete absence of peaks confirms the earlier conclusion about the PVC absence. These spectra are characteristic for all points of measurements.

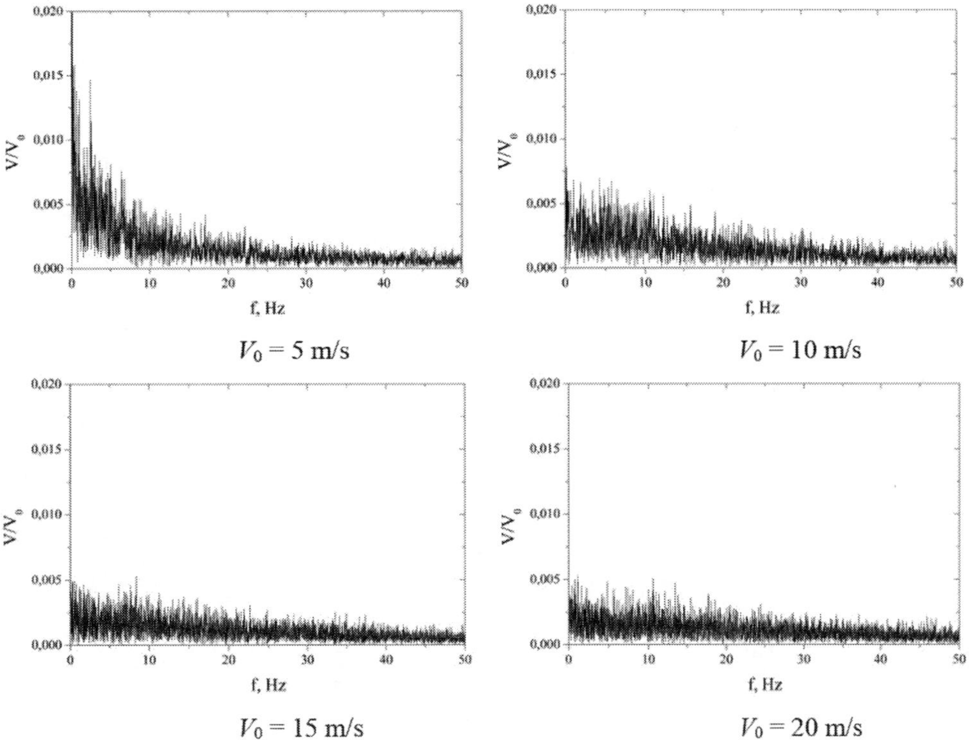

Fig. 4. Pulsation spectra of the x- velocity component for different values of initial velocity (γ=1).

4 Conclusions

Using the modern methods of measurements the pulsation characteristics of the swirling flow have been investigated in the improved model of the vortex furnace with distributed fuel-air supply and vertically oriented nozzles of the secondary blast. The results of the spectral analysis of pulsations of pressure and velocity of the turbulent swirling flow with different prove the stability of the stationary structure of the vortex core flow in the studied improved model of the vortex furnace. The absence of negative effects associated with the vortex core precession is one of the important practical advantages [9] of the studied furnace design.

Research was supported by the Russian Science Foundation (Project No. 14-19-00137).

References

1. Anufriev I. S., Strizhak P. A., Chernetskii M. Yu., Shadrin E. Yu., Sharypov O. V., Tech. Phys. Letters **41**, 727 (2015)
2. Anufriev I. S., Kuibin P. A., Shadrin E. Yu., Sharaborin D. K., Sharypov O. V., Thermophys. Aeromech. **6**, 621 (2016)
3. Krasinsky D. V., Salomatov V. V., Anufriev I. S., Sharypov O. V., Shadrin E. Yu., Anikin Yu. A., Thermal Engineering **62**, 117 (2015)
4. Litvinov I.V., Shtork S.I., Kuibin P.A., Alekseenko S.V., Hanjalic K., Int. J. Heat and Fluid Flow **42**, 251 (2013)
5. Dubief Y, Delcayre F., Journal of Turbulence **1**, 22 (2000)
6. Anikin Yu. A., Anufriev I. S., Shadrin E. Yu., Sharypov O. V., Thermophys. Aeromech. **21**, 775 (2014)
7. Krasinsky D. V., Sharypov O. V., Journal of Engineering Thermophysics **24**, 348 (2015)
8. Okulov V. L., Naumov I. V., Sorensen J. N., Tech. Phys. **52**, 583 (2007)
9. Chernetskiy M.Yu., Dekterev A.A., Burdukov A.P., Hanjalić K., Fuel. **135**, 443 (2014)

An experimental model of the evaporative cooling system of a single powerful LED

Evgeny Shatskiy

Kutateladze Institute of Thermophysics SB RAS, 630090, 1 Lavrentyev Ave., Novosibirsk, Russia
Novosibirsk state University, 630090, 2 Pirogova Str., Novosibirsk, Russia

Abstract. An experimental model of the evaporative cooling system of a single powerful LED with a natural circulation of the coolant capable of removing a heat flux density of more than 1 kW / cm2 is created. It is shown that on the finned surfaces the overheating relative to the saturation temperature in comparison with a smooth surface decreases up to three times for the heater with a diameter of 5 mm. There is up to two times increase in heat transfer coefficient on finned surfaces as compared to the smooth ones. For finned surfaces on the heater with a diameter of 1 mm the surface overheating relative to the saturation temperature decreases in four times. More than three times increase is observed for the heat transfer coefficient on finned surfaces as compared to the smooth ones.

1 Introduction

The future of a number of sectors of the economy is linked with the technology of solid-state light sources and one of the main characteristics of uptime is the removal of heat. Study of the evaporation mechanism on the surfaces of small size was started in [1]. The main objective was to study the temperature fields of the heating surface and liquid, and to study conditions for the formation and existence of a thin film of liquid at the base of the vapor bubble. In this work, as in the follow-up, dedicated to the study of a single center of vaporization [2], the heat flux limited with value at which there is a steady generation of a single center of evaporation, that is relatively modest values. A review of studies on heat transfer during vaporization at the surface of the small size is contained in [3].

Cooling of heat-emitting surfaces in different technological applications is an important problem, which can be solved by such processes as film flows [4, 5], etc. In our days the problem of creating an effective compact cooling system for powerful LEDs whose efficiency and durability are highly dependent on the efficiency of the power removal remains unsolved.

Study of evaporation mechanism on the surfaces of small size (SSS) was started in [6]. Heat transfer at boiling of dielectric liquids on the heaters of small size in the sunk vertical channels was studied in [7]. A review of works on heat transfer at evaporation on SSS may be found in [3]. In the literature, there is a dependencies for the critical heat flux density for each type of SSS, having the general form:

$$q_{cr}/q_{cr\infty} = k(d_s/l_\sigma)^n$$

but differing in correlation multipliers; and d_s is the diameter of the heating surface. Here

$$q_{cr\infty} = 0.13 r_{lv} \rho_v \left(\frac{\sigma(\rho - \rho_v) g}{\rho_v^2} \right)^{0.25}$$

r_{lv} is the latent heat of vaporization, ρ_v is the vapor density, ρ is the fluid density, σ is the surface tension, and g is the gravity acceleration.

For locally heated horizontal plates k=3.4, n=-0.5 [3], for horizontal cylinders k=1.4 [8], and for spheres k=2.5 [8]. The reason for such significant effect of the size of SSS on qcr is connected with the feature of hydrodynamic situation that occurs around the surface of small size (SSS).

The aim of the present work is studing of pool boiling of water on the surfaces with different geometry.

2 Experimental setup

The drawing of the working section is shown in Fig.1. The working chamber is formed by a radiator, a steam tube, a cover and a core body. Assembling, compression and sealing is performed by the means of fastening elements - flanges, studs and nuts. Before assembling, a cover is installed on the radiator. The casing consists of a cylindrical shell and two rings. The radiator has the compact form of a multilobal drum with a large heat exchange surface.

For evacuation of air, supply of liquid and installation of a thermocouple, an inclined branch with a union is glued into the working cavity to the cover. Through the cover, it is possible to visually observe the processes occurring in the layer of boiling liquid.

The core assembly is removable and consists of a core, core rings and nozzles. The core is firmly inserted into the ring, and the nozzle is mounted on the end of the core. The core assembly is firmly inserted into the core body. The tightness of the core assembly relative to the housing is provided by a rubber ring that is installed in the outer groove of the core ring. Cover of the core and screw are used for pressing the core to the end of the case. The cover is screwed to the core housing. The screw is screwed into the cover and presses the core assembly.

Two thermocouples are mounted on the core. One thermocouple measures the temperature of the core in the area of contact with the boiling liquid, another thermocouple measures the temperature at the base of the core. In the ring of the core there is a radial groove for laying the wires of the thermocouple. The wires of both thermocouples are discharged through the side openings located in the core body.

When the heat flux is applied to the core, the liquid is heated and boiled in a layer covering the upper surface of the core with a nozzle. The vapor of the liquid passes through the steam pipe and enters the upper part of the radiator. In the radiator, the steam condenses by cooling the outside walls with air. Condensate flows down and flows through the slots of the steam tube into the layer of boiling liquid. For vacuuming, two-stage rotary vane vacuum pump VEB MLW 2DS1 is used. To circulate the liquid, a COLE PARMER EW-74905-54 dispenser is used. Feed control is performed from the control panel. The transfer of heat to the core is due to the passage of current through the heating element - a wire wound around the core in a spiral and having a reliable thermal contact with it. The wire is made of a heat resistant material - nichrome, which has a high electrical resistance. The DC power source GW INSTEK GPD-74303S is connected to the ends of the wire. The change in the input power is due to the manual regulation of the voltage at the power supply. To monitor the input power in the power supply, indication and recording of current and voltage readings are provided.

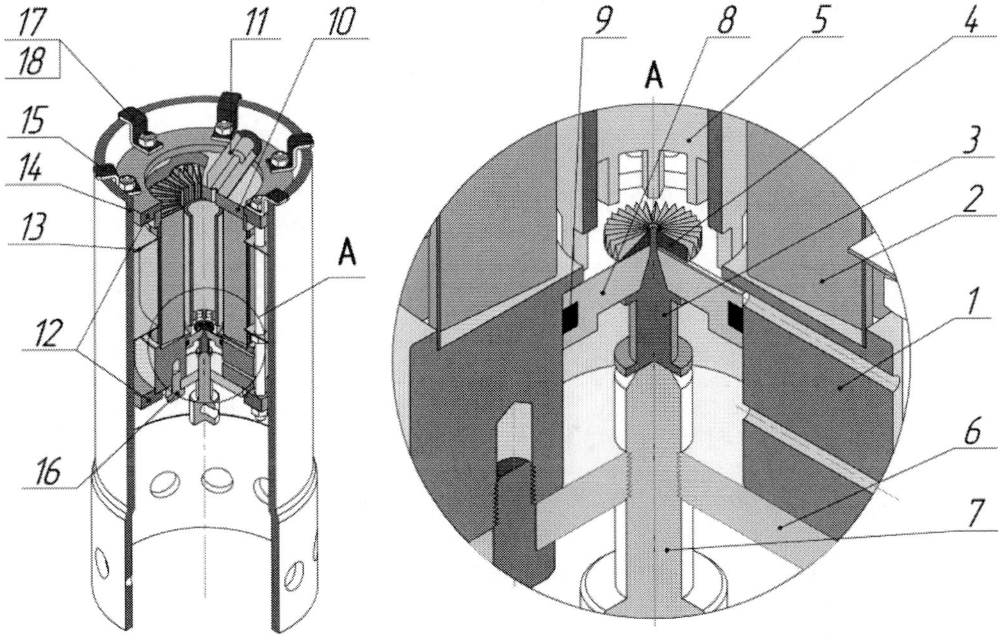

Fig.1 Test section: 1 - core body; 2 - radiator; 3-core; 4 - the core nozzle; 5 - a steam pipe; 6 - the cover of the core; 7 - core screw; 8 - core ring; 9 - a rubber ring; 10 - cover; 11 - withdrawal; 12 - flange; 13 - a casing of a radiator; 14 - blowing pipe; 15 - the bracket; 16-screw; 17, 18 - a hairpin and a nut.

Temperature measurement is performed using K type thermocouples. The thermocouples are connected to the NI TB 9214 measuring module via a 16-channel cDAQ 9171 switching unit. The measuring module is connected to a personal computer (PC) via a USB cable. The PC is equipped with the appropriate software, which allows you to continuously record the values of the corresponding temperatures, build graphical dependencies and archive data. The system uses an analogue absolute pressure sensor WIKA P-30 to measure the vapor pressure above the bed of boiling liquid. The sensor has a measuring range from 0 to 1.6 bar. The output signal of the sensor is 0-10V. The pressure sensor through the power supply TRACO POWER TXL 750-24S is connected to one of the inputs of the analog measuring module NI USB 6001, which is connected to the PC via a USB cable. The measuring module has 4 analog inputs. The corresponding software is installed on the PC, allowing to record the vapor pressure.

3 Results and conclusions

Heat transfer at boiling on local heaters with a diameter of 5 and 10 mm at the saturation temperature under conditions of a large volume of liquid was studied. It is shown that on the finned surfaces the overheating with respect to the saturation temperature in comparison with a smooth surface decreases to 1.5 times for a heater 10 mm in diameter and up to three times for a heater with a diameter of 5 mm. Up to two times the coefficient of heat transfer on the finned surface increases in comparison with the smooth one. For finned tips on a 1 mm diameter heater, surface overheating with respect to the saturation temperature is reduced to four times. The coefficient of heat transfer on the finned surface increases by more than three times in comparison with the smooth one.

A comparison of the results obtained for the heat transfer crisis with the literature data for small-sized surfaces is made. It is shown that the data for relatively large diameter heaters coincide with the calculated ones, and for the 1 mm diameter heater lie substantially higher. The effect of underheating of the liquid to the saturation temperature on the characteristics of the cooling system is studied. It is shown that when the temperature of the liquid surrounding the heat exchange surface decreases, the temperature of the LED crystal model also decreases.

As a result of the experimental work, it is established that heat flux densities in the system can be removed up to 2500 W/cm^2. Moreover, the optimal cooling when the LED semiconductor crystal is heated to a temperature not exceeding 120 °C is possible for heat flow densities up to 1000 W/cm^2. The expediency of using radial fins with a diameter of 1 mm is shown. It is recommended to maintain a level of filling the liquid above the heated surface in a thermosyphon of at least 10 mm.

The work was supported by the RF Ministry of Education and Science (Agreement No. 14.604.21.0053, project ID RFMEFI60414X0053).

References

1. Moor F. D., Mesler R. B. AIChE Journal **7** (1961)
2. Borisov V. Z., Kirillov P. L. Journal of Engineering Physics and Thermophysics **28,** 910 (1970)
3. Tolubinskiy, V.I., Antonenko, V.A., Kudritskiy, G.R., Ostrovskiy, Yu.N Industrial Heat Engineering **9** (1987)
4. Chinnov E. A. Journal of Heat Transfer **138** (2016)
5. Kabov O.A., Zaitsev D.V., Cheverda V.V. and Bar-Cohen A. Experimental Thermal and Fluid Science **35**, 825 (2011)
6. Moor F. D., Mesler R. B. AIChE Journal **7** (1961)
7. Chinnov E.A., Dyatlov A.V., Kravchenko D.N. Siberian Physico-Technical Journal **5** (1992)
8. Linard D., Dir B. Teploperedacha C **2** (1973).
9. Yu. V. Lyulin, S. E. Spesivtsev, I. V. Marchuk and O. A. Kabov. Technical Physics Letters **41** (2015)

EPJ Web of Conferences **159**, 00047 (2017) DOI: 10.1051/epjconf/201715900047
AVTFG2016

Numerical modelling of heat transfer in the layer of viscous incompressible liquid with free boundaries

Ekaterina Rezanova[1,2,*]

[1] Kutateladze Institute of Thermophysics SB RAS, 1, Lavrentyieva ave., 630090, Novosibirsk, Russia
[2] Altai State University, Faculty of Mathematics and Information Technologies, 61, Lenina, 656049 Barnaul, Russia

Abstract. The dynamics of a viscous incompressible liquid layer and the temperature distribution in it are investigated numerically in three-dimensional case. The planar layer with free boundaries under condition of zero gravity is studied on the basis of the special class of exact solutions of the Navier-Stokes equations. The thermocapillary forces and additional tangential stresses on the boundaries caused by the environment are taken into account. The influence of additional tangential stresses on the layer dynamics and heat distribution is studied.

1 Introduction

The problems of a nonstationary liquid flow in domains with free boundaries are rather difficult to study. Such problems may include the consideration of additional shear stresses caused by the external environment [1, 2]. Investigations of the flows in areas with free boundaries carried out with the use of exact solutions of special type are presented in [3, 4]. The numerical modelling of deformation of the viscous liquid layer with free boundaries and temperature distribution in it is given in [5-7].

In the present paper, the modelling of liquid flows which take place under the influence of thermocapillary forces and additional shear stresses on the free boundaries of infinite horizontal liquid layer is performed with the help of special class of exact solutions of the Navier-Stokes equations. The velocity field and temperature distribution in a layer with moving boundaries are determined by numerical algorithms. The results of numerical study of the dynamics of free boundaries and the temperature distribution within the layer depending on the direction of additional shear stresses are presented.

2 Statement of the problem and numerical algorithm

Consider the infinite plane-parallel layer of incompressible viscous thermally conducting fluid $\Omega = \{(x, y, z): -\infty < x < +\infty, -\infty < y < +\infty, -Z(t) < z < Z(t)\}$ under condition of zero gravity. Unknown function satisfies the Navier-Stokes and heat transfer equations:

* Corresponding author: katerezanova@mail.ru

$$\mathbf{v}_t + \mathbf{v} \cdot \nabla \mathbf{v} = -\nabla p + \frac{1}{Re} \mathbf{v} \Delta \mathbf{v},$$

$$\nabla \cdot \mathbf{v} = 0,$$

$$T_t + \mathbf{v} \cdot \nabla T = \frac{1}{RePr} \Delta T,$$

Here $\mathbf{v} = (u, v, w)$ is the velocity vector, p is the pressure, T is the temperature, Re is the Reynolds number ($Re = \upsilon_* \, l / \nu$), Pr is the Prandtl number ($Pr = \nu/\chi$), ν and χ are the kinematic viscosity and thermal diffusivity coefficients, l is the characteristic length of the flow domain, υ_* is the characteristic velocity, $t_* = l/\upsilon_*$ is the characteristic time.

Temperature and shear stresses $\tau_1(x, t)$ and $\tau_2(y, t)$ on free boundaries are the given functions of time and longitudinal coordinate:

$$T(x, y, \pm Z(t), t) = \frac{1}{2} A(t) x^2 + \frac{1}{2} B(t) y^2 + \Theta(t), \quad \tau_1(x,t) = x\tilde{\tau}(t), \quad \tau_2(y,t) = y\tilde{\tau}(t).$$

where $A(t)$, $B(t)$, $\Theta(t)$, $\tilde{\tau}(t)$ are the arbitrary functions that depend on time.

The kinematic and dynamic conditions on the free boundaries should be as follows:

$$\mathbf{v} \cdot \mathbf{n} = \pm w|_{z=\pm Z(t)} = \frac{dZ}{dt}, \quad -p + \frac{2}{Re} \mathbf{n} \cdot D(\mathbf{v}) \mathbf{n}|_{z=\pm Z(t)} = -P_g,$$

$$2\mathbf{s}_1 \cdot \mathbf{n} \cdot D(\mathbf{v}) \mathbf{n}|_{z=\pm Z(t)} = \tau_1(x,t) - \frac{Ma}{RePr} T_x, \quad 2\mathbf{s}_2 \cdot \mathbf{n} \cdot D(\mathbf{v}) \mathbf{n}|_{z=\pm Z(t)} = \tau_2(x,t) - \frac{Ma}{RePr} T_y.$$

Here $\mathbf{n} = (0, 0, \pm 1)$ is the unit vector of external normal, $\mathbf{s}_1 = (1, 0, 0)$ and $\mathbf{s}_2 = (0, 1, 0)$ are the unit tangential vectors, $\mathbf{D(v)}$ is the tensor of deformation rate, $P_g = \overline{P}_g - (2/Re)\overline{\rho}\overline{\nu}\mathbf{n} \cdot D(\mathbf{v}_g)\mathbf{n}|_{z=\pm Z(t)}$, $\overline{P}_g$ is the external pressure, $\overline{\rho}, \overline{\nu}$ is the relations of the densities and kinematic viscosity coefficients of gas and liquid, Ma is the Marangoni number ($Ma = \sigma_T T_* l/(\rho\nu\chi)$), T_* is the characteristic temperature, $p_* = \rho\upsilon_*^2$ is the characteristic pressure, ρ is the liquid density, $\mathbf{v_g}$ is the gas velocity, σ_T is the temperature coefficient of surface tension ($\sigma = \sigma_0 - \sigma_T(T - T_0)$, $\sigma_T > 0$). It is assumed that the action of normal stresses from the external environment may be neglected.

Let the solution that determines the layer dynamics have the form [3, 4]:

$$u(x,z,t) = (f+g)x, \quad \upsilon(y,z,t) = (f-g)y, \quad w(z,t) = -2\int_0^z f(\alpha,t)d\alpha.$$

The functions $f(z, t)$, $g(z, t)$ and the position of the free boundary $Z(t)$ are defined as the solutions of the integro-differential equations [3, 4].

Also the problem has been supplemented by conditions at infinity and initial position of the layer.

Functions $f(z, t)$, $g(z, t)$ and $Z(t)$ are determined numerically using the "predictor-corrector" algorithm of second-order accuracy [5].

The values of temperature distribution in the layer can be determined in the parallelepiped $\Omega = \{(x, y, z)$: $-N < x < N$, $-L < y < L$, $-Z < z < Z\}$. The "lateral walls" $x = \pm N$ and $y = \pm L$ of the parallelepiped (so-called the working boundaries) are artificially introduced. The boundary conditions at the "lateral walls" are the result of the heat transfer equation and the conditions at infinity. Such conditions may be called the "smooth" conditions (see. [1, 2]). The relations on the boundaries $x = \pm N$ and $y = \pm L$ of a parallelepiped can be specified as $T_{xx} = 0$ and $T_{yy} = 0$, respectivelly. These conditions are obtained as a result of estimated execution expressions $T \to T_\infty$ ($T_\infty = const$) at infinity and additional relations on the assumption $T_x \to 0$ on $x \to \pm\infty$, $T_y \to 0$ on $y \to \pm\infty$.

The temperature function is determined numerically. Finite-difference scheme of second order approximation for the solution of the heat transfer equation is based on the stabilizing corrections method [8, 9]:

$$\frac{T^{k+1/3}-T^{k}}{\Delta t}=K_1T^{k+1/3}+K_2T^{k}+K_3T^{k}+S^{k},\quad \frac{T^{k+2/3}-T^{k+1/3}}{\Delta t}=K_2\left(T^{k+2/3}-T^{k}\right),\quad (1)$$

$$\frac{T^{k+1}-T^{k+2/3}}{\Delta t}=K_3\left(T^{k+1}-T^{k}\right).$$

Here K_i are the finite-difference analogues of the corresponding (каких именно) differential operators: $K_1 \sim (1/(Re\ Pr))\partial^2/\partial z^2$, $K_2 \sim (1/(Re\ Pr))\partial^2/\partial x^2$, $K_3 \sim (1/(Re\ Pr))\partial^2/\partial y^2$; S_k is the finite-difference analog of the convective term: $S \sim - uT_x - \upsilon T_y - wT_z$; $T^k(x, y, z) = T(x, y, z, t^k)$, Δt is the time step. The finite-difference scheme (1) is implemented using the grid (x_n, y_l, z_m) which is movable along the vertical coordinate due to the mobility boundaries $z = \pm Z(t)$. Thus the grid along the coordinate z has the form: $z_m = (m - 1)h_z$ $\left(m=1,\ldots,\bar{M}+1\right)$, $h_z=\bar{Z}/\bar{M}, \bar{Z}=Z^{k+1}$ is the new position of the free boundary on the time layer $k + 1$. The Newton interpolation formula is used to transition to the new spatial grid at the new time step [7]. The sweep method is applied to implementation the scheme (1) at each time sublayer.

3 Numerical results

In the present paper, the results of numerical calculations of the dynamics of the liquid layer and the temperature distribution in it are performed. The modelling of the liquid layer spreading in cases where the action of the thermocapillary forces can be intensified or reduced by the additional tangential stresses is presented. The position of the free liquid boundary at the initial time is defined as $Z_0 = 0.5$, Θ is assumed to be 10. Functions $A(t)$, $B(t)$ and ~ have the form $A(t) = A_0 t exp(t)$, $B(t) = B_0 t exp(t)$, ~ $= A_\tau\ \tau_0 t exp(t)$, where $A_0 = -0.1$, $B_0 = -0.01$, $A_\tau = 400$. The dimensionless parameters are chosen as follows: $Pr = 17$, $Re = 1$, and $Ma = 7800$.

Figures 1-3 illustrates the temperature distribution in the parallelepiped (here $N = L = 5$). Figures 1a, 2a and 3a demonstrate temperature distribution in the spreading liquid layer in the XZ plane when the coordinate y is equal to 2.5. The "lateral wall" $x = 5$ (YZ plane) is shown on the figures 1b, 2b and 3b. The more intensive behavior of the liquid layer with

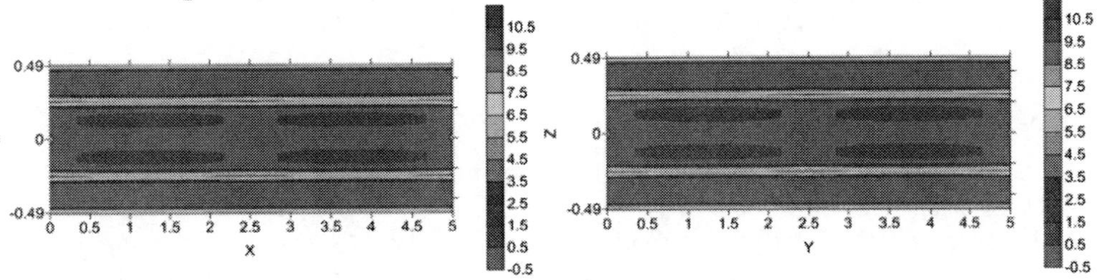

a b

Fig. 1. The temperature distribution in the XZ plane (y=2.5) (a) and YZ plane (x=5) (b), t=0.1, $\tau_0 = -0.1$.

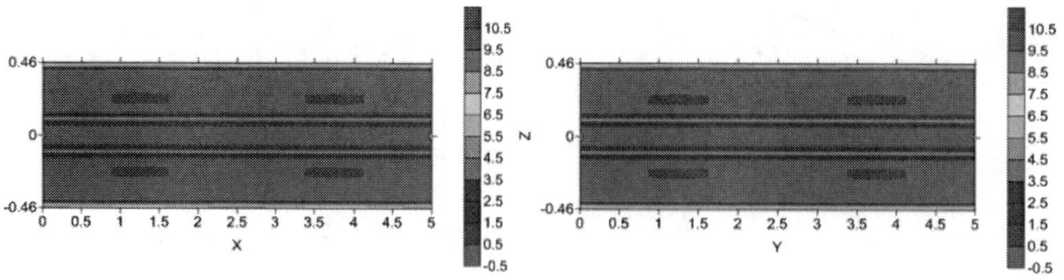

a b

Fig.2. The temperature distribution in the XZ plane (y=2.5) (a) and YZ plane (x=5) (b) , t=0.1, τ_0 = 0.1.

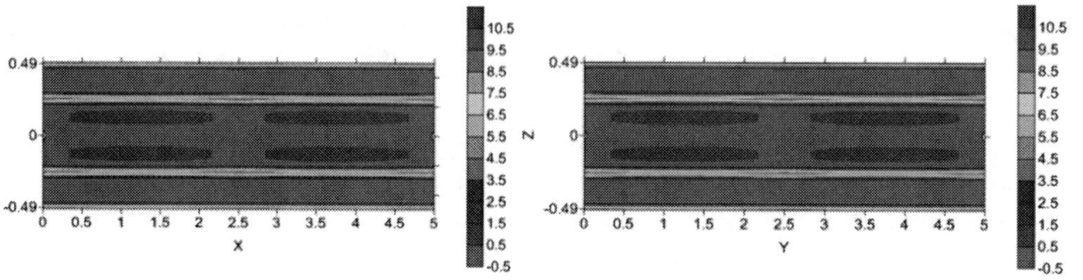

a b

Fig.3. The temperature distribution in the XZ plane (y=2.5) (a) and YZ plane (x=5) (b) , t=0.05, τ_0 = 0.1.

time occurs in the case when the parameter τ_0 = 0.1, i.e. when the additional tangential forces intensify the action of the capillary forces. The temperature distribution in the case τ_0 = 0.1 at time t = 0.05 qualitatively and quantitatively similar to the temperature distribution in the case τ_0 = -0.1 at time t = 0.1 (see figs. 1 and 3).

The less intense layer spreading is observed in the case when the gas-related tangential forces have the opposite direction in comparison with the capillary forces (τ_0 = -0.1, see fig. 1). The case when the additional tangential forces intensify the action of the capillary forces occurs for τ_0 = 0.1 (see fig. 2).

The research was supported by the Ministry of Education and Science of Russia (Agreement No 14.613.21.0011, project identifier RFMEFI61314X0011).

References

1. O. Goncharova, O. Kabov, Microgr. Sci. Technol., **22,** 407-414 (2010)
2. O. Goncharova, Yu. Kabova, O. Kabov, Comp. Thermal Sci., **3**, 5, 389-396 (2011)
3. Pukhnachov V *Model of a viscous layer deformation by the thermocapillary forces* (Max-Planck-Institut fuer die Mathematik in den Naturwissenschaften, Leipzig. Preprint, **50**, 2000)
4. V. Pukhnachov, European J. of Applied Math., **13**, 2, 205-224 (2002)
5. T. Pukhnachova Symmetry and diff. equations (Krasnoyarsk) 183-186 (2000)
6. O. Goncharova, O. Kondratenko, Izv. AGU, **1-2**, 69, 23-31 (2011)
7. A. Voevodin, O. Goncharova, O. Kondratenko, Izv. AGU, **1-2**, 77, 16-21 (2013)
8. O. Goncharova, Izv. AGU, **1**, 53, 39-44 (2007)
9. E. Rezanova, JPSC, **754** (2016)

DOI: 10.1051/epjconf/201715900048

Vortex rope instabilities in a model of conical draft tube

Sergey Skripkin[1,*], *Mikhail* Tsoy[1], *Pavel* Kuibin [1] and *Sergey* Shtork[1]

[1] Kutateladze Institute of Thermophysics, Lavrentiev avenue 1, 630090, Novosibirsk, Russia

Abstract. We report on experimental studies of the formation of vortex ropes in a laboratory simplified model of hydroturbine draft tube. Work is focused on the observation of various flow patterns at the different rotational speed of turbine runner at fixed flow rate. The measurements involve high-speed visualization and pressure pulsations recordings. Draft tube wall pressure pulsations are registered by pressure transducer for different flow regimes. Vortex rope precession frequency were calculated using FFT transform. The experiments showed interesting features of precessing vortex rope like twin spiral and formation of vortex ring.

1 Introduction

Vortex ropes are often formed in draft tubes of hydraulic turbines operating under partial load or overload conditions. The precessing vortex core (PVC) is associated with strong pressure pulsation commonly called draft-tube surge. Interaction of the PVC with draft tube elbow has negative impact on whole hydrodynamic system. When precession frequency approaches to acoustic eigenfrequency of hydraulic unit, the resonance phenomenon is appeared. It may lead to serious negative consequences as mechanical damage, efficiency and safety decreasing. Thus, the study of the phenomenon of vortex rope formation in hydraulic turbines is an important task.

Draft tube surge has been investigated for a long time both numerically and experimentally, usually on scaled hydro turbine models. Numerous publications related to investigations of flow regimes in draft tube can be found in literature [1-3]. It is well known that the pattern, precession frequency and intensity of the vortex rope dependent on the operating mode of hydraulic turbine. However, a large number of swirl devices used as a model of the runner constantly expand and complement the information on a variety of PVC patterns. The main vortex can be replaced by twin spiral vortex, whereby the frequency of the pressure pulsations in hydraulic unit becomes unpredictable in certain part load conditions [4, 5]. In turbine regime, close to the optimum, the pattern of vortex rope becomes unstable and constantly changes [6].

In this study, experimental modelling of swirling flow under the hydraulic turbine runner have been conducted. Experiments include high-speed visualization and pressure measurements at a fixed flow rate of fluid and varying the speed of the runner.

* Corresponding author: skripkin.s.g@gmail.com

2 Experimental setup and measurement equipment

The swirl generator modelling of turbine runner consists of two swirlers: the first with fixed vanes and the second is a forcedly rotating runner. We used magnetic coupling between the motor on the dry side of turbine model and the runner in the water on the other side which transfers torque from the electric motor to runner. Thus, we manage the runner by controlling the motor rotational speed. This swirl device was designed similarly the swirl generator designed at Politehnical University of Timisoara [7]. The main idea of using the simplified draft tube model is not a complete reproduction of whole hydro turbine unit but the reconstruction of velocity profile close to distribution under nature turbine runner. Also conical draft tube model excludes the effect of interaction between PVC and draft tube elbow. Because the flow regime in this system dependent on the flow rate Q and the runner rotational speed N, all experiments were conducted at the fixed flow rate $Q = 80$ m^3/h (it is correspond $Re \approx 10^5$) and rotational speed from 0 to 1000 rpm. Changing one parameter allows to completely cover the whole map of the flow regimes at fixed Reynolds number.

Pressure fluctuations measurements and vortex rope visualizations are carried out on a simplified draft tube model. We used high-speed cam PCO 1200 hs and Canon 7D for detail investigations the main patterns of the PVC in different regimes. Quantitative data on the pressure pulsation and the precession frequency of vortex rope were obtained after signal processing from a pressure sensor Keller PA-4LC mounted in the wall of the draft tube.

3 Visualisation and pressure measurements

The swirl flow in a conical draft tube model is divided into several main regimes depending on the runner rotational speed at fixed discharge. Regime map is shown in fig. 1, where each vortex rope is illustrated at different N. As it is shown in work [3], the shape of vortex rope and its behaviour depend on the stagnation point which moves along the centre axis in a conical part.

Fig. 1. Vortex rope visualisation at constant flowrate $Q = 80$ m^3/h by varying runner rotational speed

Thus, at the rotational speed from 0 to 450 rpm flow is swirled counter clockwise. In region about 450 rpm the swirl of flow is close to zero. In this regime, there is straight axially symmetric vortex rope, at the end of which the spiral vortex breakdown is observed (fig. 2). At rotational speed, more than 450 the flow is swirled clockwise and with increasing runner rotational speed (N > 500 rpm) swirl of flow increases and thin spiral vortex visualized cavitation is formed. This vortex pattern is interesting in that the shape of the PVC is not stable and formation of vortex rings due to process of self-reconnection is observed (fig 2).

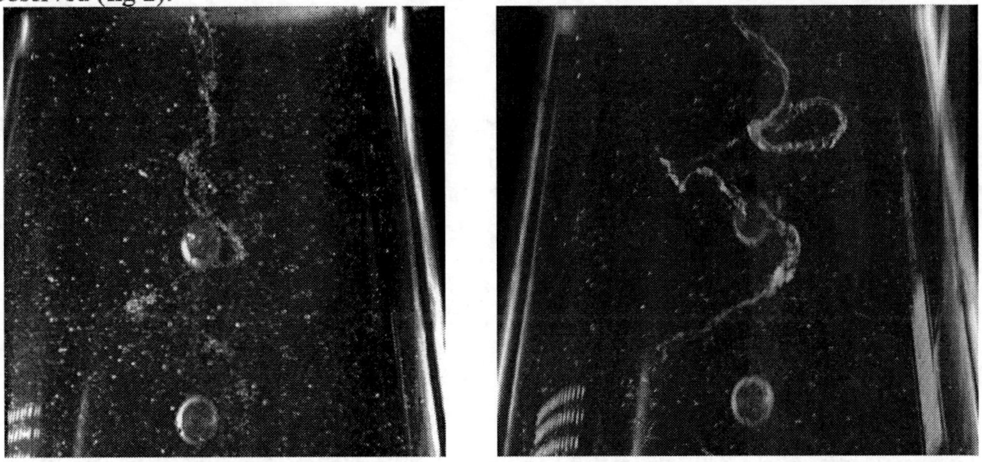

Fig. 2. The view of axially symmetric vortex rope (left) and formation of isolated vortex ring (right)

Vortex ring moves close to the wall of the draft tube, and presumably can cause strong pressure pulsations. With a further increase of speed of the runner vortex rope is intensified, cavitation cavity increases. At rotation frequency about 600 rpm there is an unstable formation of the twin vortex, with a predominance of single mode. With further increasing rotational frequency (N > 700 rpm) vortex rope replaced by a large number of small vortices rotating as one.

Also, experiments on study the draft tube wall pressure pulsation were conducted. The figure 3 shows the graphs obtained by fast Fourier transform.

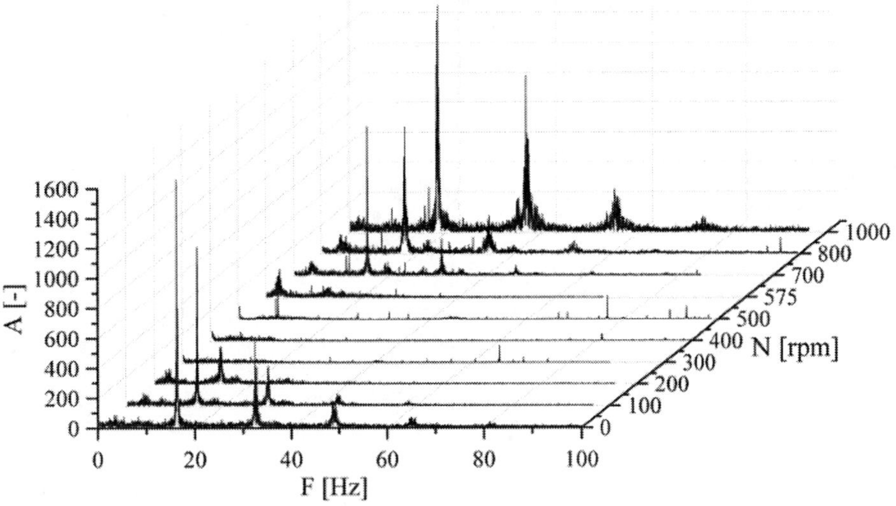

Fig. 3. Pressure spectra diagram at different rotational speed

It can be seen that the amplitude of fluctuations under the regime with minimal residual swirl falls. Despite the presence of vortex rope at the speed of 500 rpm PVC amplitude is very weak and draft tube pressure pulsation is very small, because the vortex precession radius is much smaller than in the regime of 600 rpm or more. The precession frequency is also reduced, but there is an interesting feature at N = 500 rpm. Precession frequency increases significantly up to 40-50Hz, these data were obtained through the analysis of high-speed visualisation. After that, the precession frequency falls to 10Hz and then it increases monotonically with runner rotational speed.

4 Conclusion

Experimental modeling of various flow regimes in a model of hydraulic turbine have been conducted. PVC formation was observed in various forms in a wide range of runner rotational speed. Flow visualization revealed some interesting features of the flow pattern, these are twin vortex breakdown and formation of vortex ring due to reconnection of vortex rope.

Earlier the regime with formation of vortex rings was obtained only on the freely-rotating runner. Here, we defined the position of these phenomena on the regime map [8]. It is shown that the precession frequency of the PVC is dependent on the rotational speed of the runner. The effect of a sharp rise in the precession frequency was observed at a small residual swirl of flow ($N \approx 500$ rpm).

This research was supported by the Russian Foundation for Basic Research, grant No 16-08-01191.

References

1. Skotak, J. Mikulasek, and P. Troubil, Proceedings of the 10th international meeting of the work group on the behavior of hydraulic machinery under steady oscillatory conditions (Trondheim, Norway, 2001)
2. Wen-Tao Su, Xiao-Bin Li, Feng-Chen Li, Xian-Zhu Wei, Wen-Fu Han, and Shu-Hong Liu, Advances in Mechanical Engineering, **2014**, 16, (2014)
3. M. Nishi, S. Liu, International Journal of Fluid Machinery and Systems, **6**, 1 (2013)
4. S. Skripkin, M. Tsoy, S. Shtork and K. Hanjalić, J. Hydraul. Res., **54**, 4 (2016)
5. T. L. Wahl, , Hydro Review, **13**, 1 1994
6. Nicolet, A. Zobeiri, P. Maruzewski, and F. Avellan, Int. J. Fluid Mach. Syst., **4**, 1 (2011)
7. Javadi, A. Bosioc, H. Nilsson, S. Muntean, and R. Susan-Resiga, ASME J. Fluids Eng., **138**, 8 2016
8. S. Alekseenko, P. Kuibin, S. Shtork, S. Skripkin and M. Tsoy, JETP Letters, **103**, 7 (2016)

EPJ Web of Conferences **159**, 00049 (2017) DOI: 10.1051/epjconf/201715900049
AVTFG2016

Breakdown dynamics of a horizontal evaporating liquid layer when heated locally

Serafim Spesivtsev[1,2,*]

[1]Kutateladze Institute of Thermophysics SB RAS, 630090, Novosibirsk, Russia
[2]Novosibirsk State University, 630090, Novosibirsk, Russia

Abstract. Breakdown of liquid layer when heated from a localized hot spot was investigated experimentally. Water and ethanol were used as working liquids with a layer thickness of 300 μm. Basic steps of the breakdown process were found and mean velocities of the dry spot formation were determined. The formation of residual layer over the hot-spot before the breakdown has been found for both liquids. The creation of a droplet cluster near the heating region is observed when using water as a working fluid. It was shown that evaporation is one of the general factors influencing the process of layer breakdown and dry spot formation as well as thermocapillary effect.

1 Introduction

Study of heat transfer from a local heat source becomes one of the most important and complicated problems in modern thermophysics. The problem is closely connected to the problem of microelectronic equipment cooling [1]. Permanent consumption of electricity by the device inevitably leads to increasing the microprocessor temperature and degradation of its performance and reliability. The average heat flux density on the surface of chips of commercially available computers and other electronic devices is currently known to reach 100 W/cm^2. Continuous development and complication of microchip structure produce nonuniform heat flux distribution on the chip surface. This effect takes place due to the design features of computer chips, where the processor cores cause the formation of "hot" spots. Heat flux density in some regions is much higher than the chip average [2], of the order of 1 kW/cm^2. These specific regions called "hot spots" could have size from several hundred microns to 1-2 millimeters.

However, using special localized cooling it is possible to produce large performance gains in microprocessors. Nowadays, there are several effective techniques for cooling of local hot spots such as spray cooling [3], boiling in microchannels [4], thermoelectric coolers [5]. One of the promising methods for removing such high heat fluxes from a spotted heat source is technology based on evaporation of a thin liquid layer. Dynamics of evaporation essentially depend on the conditions in the layer [6]. In particular, the breakdown of liquid layer leads to dramatic decreasing of heat transfer from a spotted heat source [7]. Processes of liquid layer rupture are actively studied experimentally [7-9] and

[*] Corresponding author: simafir94@gmail.com

theoretically [10-12] in the present time. The goal of the present work is to study using schlieren technique the breakdown dynamics of a horizontal evaporating liquid layer when heated from a localized hot spot.

2 Experimental rig

Experiments were conducted on the rig shown in fig. 1. The working area consists of a caprolon base, a metal substrate and a heating element. Fluid from the syringe pump enters the work area, forming a horizontal liquid layer opened to the atmosphere. Spot heating of the horizontal liquid layer takes place in the center of the substrate. The test cell consists of caprolon base, metallic substrate and the heating element. The caprolon base has a special cut on the upper side for installation of the substrate and a central through hole with a diameter of 1.6 mm. The substrate is made of stainless steel and has a diameter of 50 mm and a thickness of 1 mm. In the center of the substrate is a closed hole with a diameter of 1.6 mm and a height of 0.8 mm. The heating element is made of brass and has a round tip with a diameter of 1.6 mm and a height of 3 mm. It is tightly inserted into the closed hole of the substrate through the caprolon base. Thermal paste is used for better thermal contact between the heater tip and the substrate. The distance between the tip and the upper side of substrate is 0.2 mm. The heat source of the heater is a nichrome wire wound on the core rod. Heating is controlled by the power source. The heater core was placed in the cutout of the base on its lower side, and between the heater and the base there was an air gap of 2 mm. From the bottom of the heater there was an insulating material to minimize heat loss. Temperature in the test cell is measured by thermocouples (type K) with an accuracy of 0.1°C. Location of the thermocouples is shown in Fig. 1. Relative humidity and atmosphere temperature are measured using the thermohygrometer Testo 645 with an accuracy of 2% and 0.1°C, respectively.

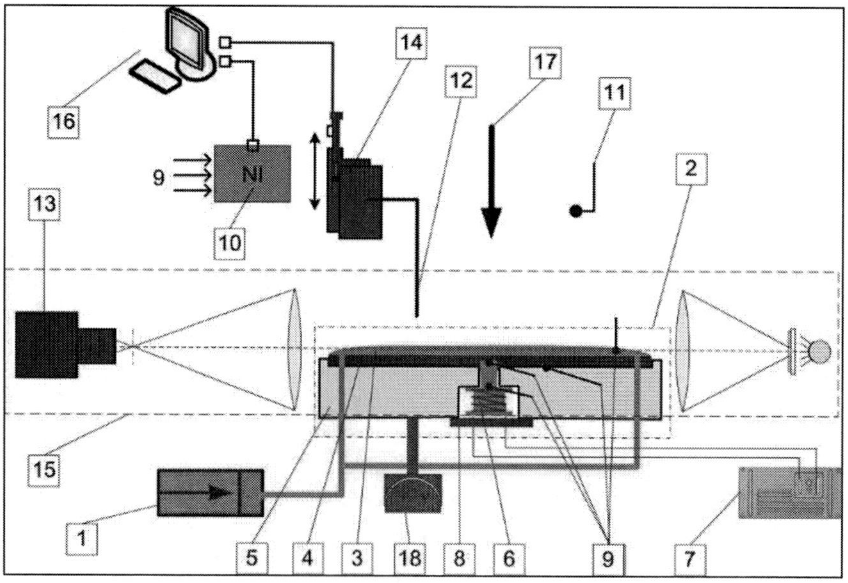

Fig. 1. Scheme of the experimental rig. 1 – syringe pump; 2 – test cell; 3 – liquid layer; 4 – substrate; 5 – base; 6 – heating element; 7 – power source; 8 – insulating material; 9 – thermocouples; 10 – measuring system; 11 – thermohygrometer; 12 – probe; 13 – camera; 14 – linear actuator; 15 – shadowgraphy technique; 16 – PC; 17 – schlieren technique; 18 – goniometer.

The heat flux density is determined by measuring the temperature difference between two different sections along the heater tip.:

$$q=\lambda\,\Delta T\,/\,l, \tag{1}$$

where λ is the thermal conductivity of the heater material, W/cm·K; l is the distance between the two sections of the heater tip, cm; and ΔT is the temperature difference in two cross sections along the length of the heater tip, K.

The height of the horizontal liquid layer is maintained in constant position during the all experiment. The probe with a diameter of 100 μm is installed at a required distance from the substrate surface. Precise linear actuator and shadowgraphy technique are used in order to determine the position of the probe. The probe moves in range of 5 mm by steps of 1 μm. With the help of high-precision syringe pump the flow rate is chosen for providing a constant layer thickness, taking into account the evaporation. The horizontal layer surface and the probe are observed using video camera Imaging Source DFK 23GP031 with a resolution of 2592x1944 pixels. To visualize surface deformations and register the breakdown an optical schlieren system is used with high-speed video camera Photron FASTCAM 675K-M3 (speed of 5000 fps at a resolution of 640x640 pixels and a scale of 25 μm/pix). The test section is mounted in horizontal position using a goniometer. The surface roughness of the substrate was determined by profilometer "Micro Measure 3D station" and the average roughness value is equal to R_a=0.327 μm. The contact angle on the working surface in the heating area was determined by the sessile drop method (Young-Laplace) [13] at a room temperature of 25±2°C and equals to θ_1=6±1° for ethanol and θ_2=76±1° for water.

3 Results and discussion

Experiments were conducted at atmospheric pressure, temperature and relative humidity of 28±2°C and 25±3%, respectively. The used working liquids were ethanol (95% (mass.), GOST R 51723-2001) and ultrapure water. For water purification the system Merck Millipore Direct-Q 3 UV is used which allows providing the water of type I (ultrapure water). The height of the liquid layer was 300 μm. Injection liquid flow rate was up to 200 μl/min. During the experiment heat flux is increased up to a critical value at which the liquid layer ruptures. At this moment heating process is stopped to prevent the failure of heating element. For both working fluids the critical heat flux density when the breakdown of the liquid layer occurs is measured. The value of the critical heat flux density for ethanol equals to 12.6 W/cm^2 at the substrate temperature in the heating area of 37.1°C, for water it is 117 W/cm^2 at the substrate temperature of 133°C, respectively.

It was found out that for both working liquids the layer breakdown occurs according to one scenario [7, 14, 15]. First, a thermocapillary deformation of the layer above the point heating area appears (Fig. 2a, b). Further thinning leads to the formation of a residual liquid layer in the area of the point heating, Fig. 2c [8]. Then, the residual liquid layer evaporates to a critical thickness, at which the layer breakdown takes place, Fig. 2d. After the breakdown the entire heating area is intensely dried, and the round dry spot is formed, Fig. 2e. It should be noted that local heating of the water layer is followed by the formation of a droplet cluster [6, 14, 16] above the heating area and the breakdown takes place at a distance of about 1 mm from the substrate center. When using ethanol as a working liquid residual layer formation and rupture occur directly above the heating element in the center of the substrate.

The average rate of dry spot formation was measured for two working fluids and was determined as the ratio of the characteristic radius of the resulting dry spot to the time of its formation in the heating area. The time of dry spot formation is counted from the moment of the residual layer breakdown to its complete evaporation. The average velocity of dry spot formation for ethanol is 0.06 mm/s, and for water it is 5.15 mm/s, respectively. The time of dry spot formation for ethanol as a working fluid is 7.85 seconds, and for water - 0.13 seconds. First of all the difference in the velocity of dry spot formation is connected with different evaporation rate of the residual layer. The evaporation rate directly depends on the heat flux density and the substrate temperature. In addition, it is influenced by the difference in properties and hydrodynamic parameters of working fluids: contact angles and surface tension.

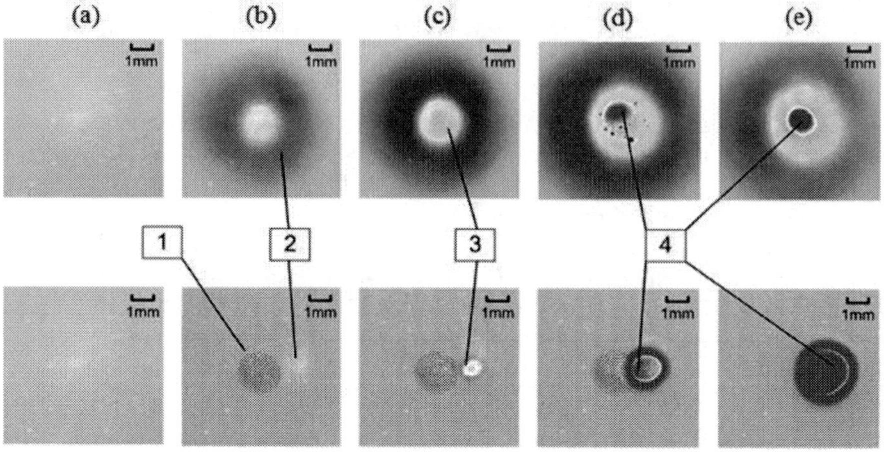

Fig. 2. The dynamics of the liquid layer breakdown in case of point heating. Liquids are ethanol and water, the layer depth is 300 μm. 1 – droplet cluster; 2 – thermocapillary deformation; 3 – residual layer; 4 dry spots.

In the study of thermocapillary breakdown of the water layer with local heating the existence of the droplet cluster was found (Fig. 3), followed by its falling and formation of capillary waves [16]. Droplets in a cluster are held above the layer surface due to intensive evaporation in the heating area and falling of individual droplets begins after breakdown of residual layer (Fig. 3 c, d). The phenomenon of droplet cluster was investigated in detail in [16-18].

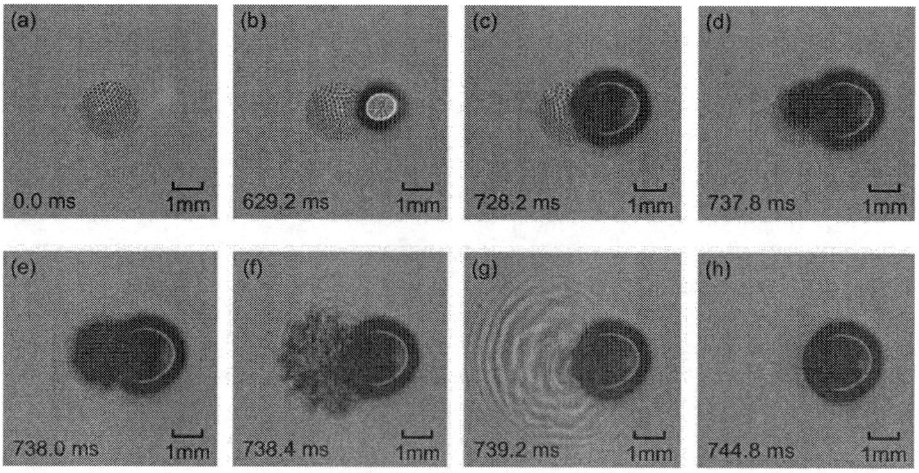

Fig.3. Visualization of the droplet cluster and the breakdown dynamics. The liquid is water and the layer depth is 300 μm.

4 Conclusions

Influence of the working liquid on the breakdown dynamics was studied. Visualization of the breakdown dynamics has been performed for thin layers of ethanol and water using schlieren technique. It is found that for rupture of the water layer the density of heat flux required to be by an order of magnitude higher than the critical heat flux density for ethanol layer at the same thickness. At the same time typical velocities of breakdown of water and ethanol differ by two orders of magnitude. This fact is directly related to the difference in the critical heat fluxes and intensity of evaporation. It is proved that before the breakdown the residual layer appears in the area of local heating. When using water as the working liquid the formation of a droplet cluster may be observed near of the heating area. Along with the thermocapillary effect, evaporation is one of the main factors that influence the breakdown of residual liquid layer and dry spot formation in the heating area.

The study was financially supported by the Russian Science Foundation (Project 14-19-01755).

References

1. A. Bar-Cohen, P. Wang, J. Heat Transfer **134**, 5 (2012)
2. R. Mahajan, C. Chiu, G. Chrysler, Proceedings of the IEEE **94**, 8 (2006)
3. E.A. Silk, E.L. Golliher, R.P. Selvam, Energy Convers. Manag. **49**, 453–468 (2008)
4. P. Di Marco, W. Grassi, Int. J. Therm. Sci. **41**, 567–585 (2002)
5. A. Bar-Cohen, P. Wang, Microgravity Science and Technology **21**, 351-359 (2009)
6. A.A. Fedorets, I.V. Marchuk, O.A. Kabov, Interfacial Phenomena and Heat Transfer **1**, 1 (2013)
7. Yu.V. Lyulin, S.E. Spesivtsev, I.V. Marchuk, O.A. Kabov, Technical Physics Letters **41**, 11 (2015)
8. D.V. Zaitsev, D.A. Rodionov, O.A. Kabov, Microgravity Science and Technology **19**, (2007)
9. D.V. Zaitsev, O.A. Kabov, Microgravity Science and Technology **XIX-3/4**, (2007)
10. V.S. Ajaev, Interfacial Phenomena and Heat Transfer **1**, 1 (2013)
11. M. B. Williams and S. H. Davis, J. Colloid Interface Sci. **90** (1982)
12. J. P. Burelbach, S. G. Bankoff and S. H. Davis, Phys. Fluids A **2**, 322–333 (1990)
13. I.V. Marchuk, V.V. Cheverda, P.A. Strizhak, and O.A. Kabov, Thermophysics and Aeromechanics **22**, 3 (2015)
14. S.E. Spesivtsev, Yu.V. Lyulin, MATEC Web of Conferences **72**, 01107 (2016);
15. S.E. Spesivtsev, Yu.V. Lyulin, I.V. Marchuk, O.A. Kabov, Journal of Physics: Conference Series **754** (2016);
16. A.A. Fedorets, I.V. Marchuk, P.A. Strizhak and O.A. Kabov, Thermophysics and Aeromechanics **4**, 535-538 (2015)
17. A.A. Fedorets, I.V. Marchuk and O.A. Kabov, Technical Physics Letters **37**, 45 (2011).
18. A.A. Fedorets, I.V. Marchuk and O.A. Kabov, JETP Letters **99**, 5 (2014)

An experimental model of evaporative cooling system with forced circulation of coolant for high-performance electronic components

Konstantin Stepanov[1], *Dmitry* Mukhin[1], *Karapet* Eloyan[1], *Egor* Tkachenko[1,2]

[1]Institute of Thermophysics SB RAS, Novosibirsk, Russia
[2]Novosibirsk State University, Novosibirsk, Russia

Abstract. A promising way to remove heat fluxes from the surfaces of electronic devices with high heat generation is the use of evaporating thin liquid film, moving under the action of gas flow in a channel. On the basis of the studies conducted previously, we have developed a pilot model of the experimental-industrial prototype for removing heat fluxes with densities of up to 1000 W/cm^2 from the surface of the heat-stressed element with dimensions 10x10 mm^2. Testing of the model has proved its efficiency in continuous operation.

1 Introduction

Currently, the global industry is ready to produce high-performance electronic components where the heat flux density at individual sites can reach 1000 W/cm^2 and higher [1]. However, the introduction and use of these devices faces challenges of removing such high specific heat fluxes into the ambient medium. One of the promising ways of removing large heat fluxes from the surface of heat-stressed elements of electronic devices is the use of evaporating thin layer of liquid film, moving under the action of the gas flow in a flat channel [2,3,4].

Recent experimental investigations conducted in works [5,6] proved possible removal of heat fluxes with density of up to 870 W/cm^2 from the heating area of 1x1 cm^2 using this method. The works [7,8] showed that such a system can operate stably in a wide range of the channel heights (0.17-2.00 mm) and angles of the channel inclination to the horizon (0-360°). In [9] a 3D non-stationary mathematical model of joint motion of evaporating liquid film and co-current gas flow in a microchannel with local heating has been developed, taking into account a deformable gas-liquid interface, convective heat transfer in the liquid and gas phases as well as temperature dependence of surface tension and liquid viscosity. The performed investigations have resulted in the development of a promising model of evaporative cooling system for electronic devices with high heat flux removal and forced circulation of coolants. The model is an experimental-industrial prototype of cooling system with closed loops of coolants.

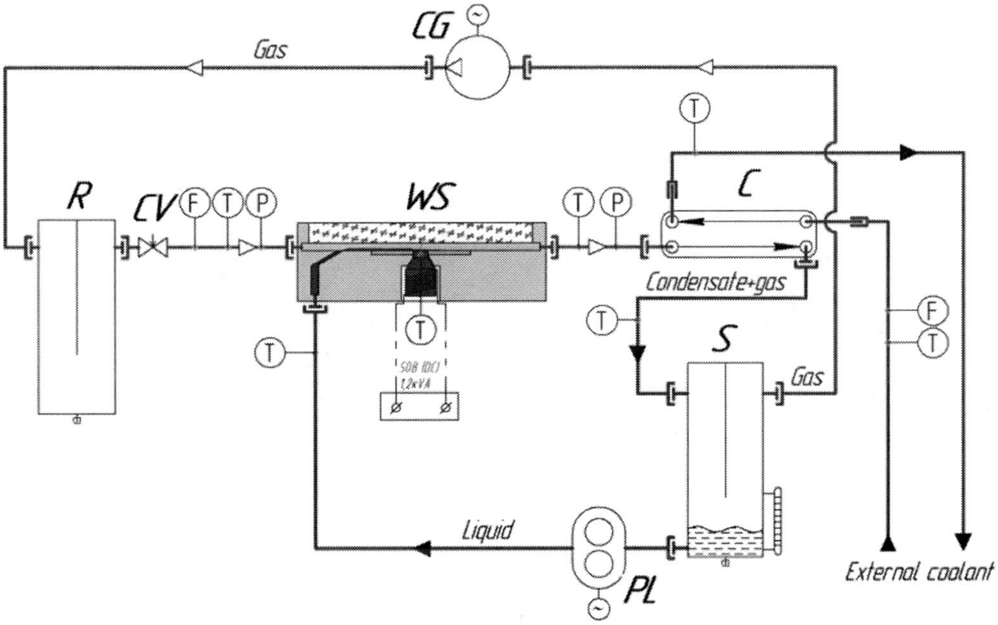

Fig. 1. Schematic diagram of the experimental model of evaporative cooling system: WS – working section; C – condenser; S – separator; R – receiver; CV – control valve; PL – liquid pump; CG – gas vacuum pump-compressor.

2 Experimental model

A schematic diagram of the experimental model of evaporative cooling system is shown in Fig. 1. The principle of heat removal from the surface simulating the heated wall of heat-stressed electronic component is based on the use of intensively evaporating thin liquid film, moving under the action of cocurrent gas flow or vapor-gas mixture in a slot mini- or microchannel. This method of cooling allows acheaving of substantially increased critical heat fluxes needed for a liquid film to rupture and also removing high heat fluxes from surfaces with small area.

To ensure continuous operation of the experimental model gas and liquid circulate in closed loops. In the loops different pressures can be maintained: atmospheric, vacuum or excess pressure. In the gas loop the gas is purified from impurity – air or nitrogen with a small amount of liquid vapor. To enable gas circulation in the loop, vacuum pump-compressor CG is used. In the fluid circuit the purified liquid is used, i.e. water, ethanol or HFE-7100. Fluid is circulated using high-precision gear pump PL. After entering into the working section WS the liquid moves under the action of the cocurrent gas flow.

In the working section, gas and liquid move in a slot channel with the height of 1.0 mm and the width of 11 mm. The slot channel is formed between the working surface and a viewing glass, and the channel height is set by a spacer plate (the channel height can be changed from 0.1 to 2.0 mm). The fluid flows through the channel in the form of a film, entrained by the gas flow. To form a liquid film with certain thickness in the working section a slit nozzle is used; it is formed by a wedge-shaped knife and the working plate, over which the film is distributed (see Fig. 2).

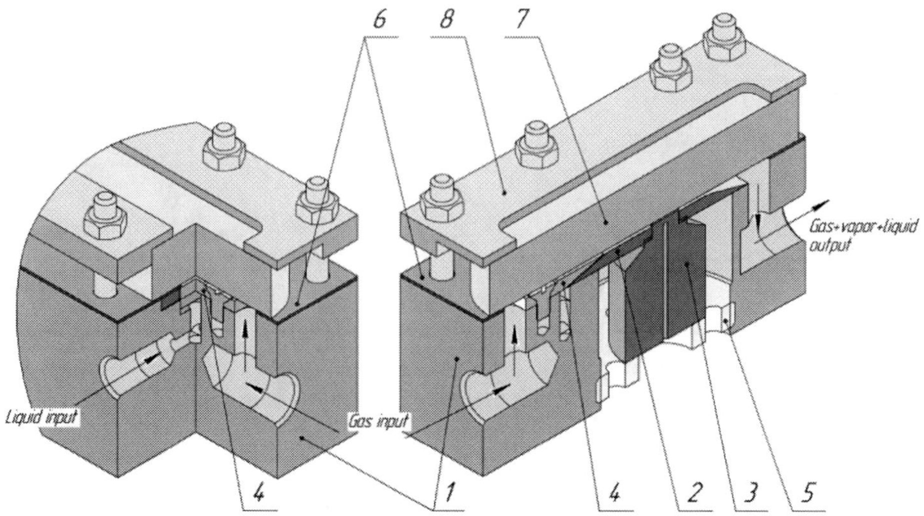

Fig. 2. Design of the working section WS: *1* – housing; *2* – working plate; *3* – the local heater (copper rod); *4* – a wedge-shaped knife; *5* – holder; *6* – spacer plate; *7* – viewing glass; *8* – flange.

The formed liquid film is supplied to the center of the slot channel, to the zone of the heat-stressed electronic component simulator – the local heater, where it is evaporating, due to heat flux supply to the local heater. The resulting vapor is carried away by the gas passing over the liquid film. At the end of the slot channel the vapor-gas mixture together with the liquid is supplied through the loop to the condenser C (Fig. 1), where vapor condensation of the evaporated liquid occurs. After the condenser C the gas-liquid mixture goes to a separator S, where the separation of liquid and gas takes place. The gas then enters the suction line of the pump-compressor CG, and the liquid accumulates in the lower part of the separator and enters the pump PL.

The local heater is a copper rod, heated by a nichrome wire (not shown in Fig. 2), by means of passed electric current. The wire is helically wound around the core of the local heater. The dimensions of the flat surface of the local heater are 10x10 mm^2. Power is supplied to the heating element using a direct current source. The heat flux on the local heater is adjusted by changing the current strength.

The removal of condensation heat from the condenser C is realized by external coolant – water (or other liquid). The flow of circulating gas is regulated by the frequency converter of pump-compressor CG, and the control valve CV, mounted at the inlet of the working section. The flow rate of the circulating fluid is regulated by means of the control unit of the liquid pump PL.

Measurements of temperatures, pressures and flow rates are performed in a centralized measurement system, consisting of measurement modules and a personal computer.

3 Test results

The pilot model of the evaporative cooling system was used to test the removal of the heat flux with density of up to 1000 W/cm^2 from the local heater of the working section. The used working fluid was distilled water, and the used working gas was purified air. The used external coolant was tap water. The parameters of the pilot model are given in Table 1.

Table 1. The results of the test of experimental model.

Parameter	Value
Density of the heat flux, W/cm^2	0÷1000
Volumetric flow rate of circulating fluid, ml/min	100÷120
Volumetric flow rate of circulating gas, l/min	45÷50
Temperature of liquid at the inlet the working section, °C	28÷30
Gas temperature at the inlet of the working section, °C	40÷45
Temperature of the gas mixture at the output from the working section, °C	55÷60
Absolute gas pressure at the inlet of the working area, Bar	1.3÷1.4
Absolute pressure of the gas mixture at the output of the working section, Bar	1.0÷1.1
Temperature of the external coolant at the inlet of the condenser, °C	23÷25

Tests have confirmed the efficiency of the experimental model in a continuous operation at the maximum heat flux for 50 minutes. The local surface temperature of the heater in the zone of evaporative cooling did not exceed 140 °C. Visual observations of the local heater surface have revealed no formation of stationary dry spots in the range of heat fluxes up to 1000 W/cm^2.

4 Conclusions

Initial tests of the experimental model of evaporative cooling system with forced circulation of coolants for electronic devices with high heat emission have confirmed the possibility of continuous discharge of the heat flux with density of up to 1000 W/cm^2 from a flat surface of heat-stressed element with the size of 1x1 cm^2 at relatively low flow rates of liquid and gas.

The experimental setup realized in a closed cycle of coolants has the potential of providing even higher heat fluxes. Further stages of model testing will be realized to develop this potential.

This work was supported by the Ministry of Education and Science of Russia (Agreement No 14.604.21.0053, project identifier RFMEFI60414X0053).

References

1. A. Bar-Cohen, C. Holloway, Journal of Physics: Conference Series, **745**, 022002 (2016)
2. O.A. Kabov and D.V. Zaitsev, Multiphase Science and Technology, **21**, 249-266 (2009)
3. D.V. Zaitsev, D.A. Rodionov, O.A. Kabov, Technical Physics Letters, **35**, 680-682 (2009)
4. O.A. Kabov, D.V. Zaitsev, V.V. Cheverda, A. Bar-Cohen, Experimental Thermal and Fluid Science, **35**, 825 (2011)
5. D. Zaitsev, E. Tkachenko, E. Orlik, O. Kabov, MATEC Web of Conferences, **92**, 01037 (2016)
6. E.M. Tkachenko, D.V. Zaitsev, E.V. Orlik, O.A. Kabov, Journal of Physics: Conference Series, **754**, 032019 (2016)
7. D. Zaitsev, O. Kabov, MATEC Web of Conferences, **84**, 00043 (2016)
8. E.M. Tkachenko, D.V. Zaitsev, MATEC Web of Conferences, **72**, 01114 (2016)
9. Y. Kabova, V.V. Kuznetsov, O. Kabov, T. Gambaryan-Roisman, P. Stephan, International Journal of Heat and Mass Transfer, **68**, 527-541 (2014)

Study of thermal behavior of microlayer under vapor bubble at liquid boiling

Anton Surtaev*, *Vladimir* Serdyukov, and *Andrey* Chernyavskiy

Kutateladze Institute of Thermophysics, Novosibirsk, 1 Akad. Lavrentiev ave., Russia

Abstract. The results of experimental study of evolution of temperature fields under single vapor bubble obtained by high-speed infrared thermometry with high spatial resolution (13 μm) are presented in this paper. The data were obtained at pool boiling of saturated ethanol and deionized water at atmospheric pressure. Reconstruction of local instantaneous heat flux distribution on the heater surface was carried out with the use of numerical simulation. It is shown that maximal local heat flux was observed in the microlayer region on the bubble growth stage and reached the value an order of magnitude greater than the input heat flux. Based on the results of experimental and numerical researches the estimations of the microlayer thickness were carried out at pool boiling of water and ethanol, which are in good agreement with the experimental data presented in the literature and obtained using laser interferometry.

1 Introduction

Nucleate boiling is common and one of the most effective heat transfer mechanisms, it is widely used in various industrial applications: energy and chemical industry, nuclear power plants, electronics cooling system, etc. Despite of large number of papers published since the middle of the last century and devoted to the study of different aspects of the liquid boiling, many questions remain open and there is no generally accepted boiling theory [1]. This is associated with the fact, that boiling is very complicated process by the need to consider phenomena that occur over multiple length scales, ranging from centimeters, down to few nanometers. In particular, the importance of the evaporation from a liquid microlayer with a thickness of several microns, developing underneath a bubble during its initial growth, still debated in scientific society.

In last decade, new experimental techniques with high temporal and spatial resolution, including infrared thermometry, laser interferometry, PIV, etc., that allow to discover the boiling process on microscale are developed. For example laser interferometric method is widely used in recent years to determine the thickness of the microlayer under single vapor bubbles at different time [2-4]. With the use of PIV technique authors [5] analyzed the hydrodynamics of microflows at bubble appearance, growth and detachment from the surface. In recent years, a general method to study the evolution of temperature field under single vapor bubble is the high-speed infrared thermometry (IR) [3-9]. This technique

* Corresponding author: surtaev@itp.nsc.ru

allows to investigate not only integral heat transfer but also the main local characteristics at boiling such as nucleation site density, nucleation frequency, local heat transfer under vapor bubble, etc. For reconstruction of local instantaneous heat flux distribution on the thin heaters some authors [3, 4] used commercially available computational fluid dynamics (CFD) software package. Also it should be noted that many of the experimental data on the evolution of the temperature field obtained using the IR method, are presented in the literature only for certain values of the heat fluxes and only for water. Therefore, for a deeper analysis of local heat transfer beneath a single vapor bubbles the further research in a wide range of heat fluxes and the use of fluids with different physical properties is needed.

The aim of this research is the experimental investigation of evolution a temperature field under single vapor bubbles using IR method with high temporal (1000 fps) and spatial (13 μm) resolutions at pool boiling of water and ethanol at saturation line in wide range of heat fluxes. The second task is to develop numerical algorithm for reconstruction of local instantaneous heat flux distribution on the heater with the use of obtained experimental results of unsteady temperature field.

2 Experimental techniques

Experiments were carried out at pool boiling of saturated ethanol and deionized water at atmospheric pressure. Detailed description of the experimental setup and measurement techniques is presented in [9]. Thin (1 μm) conductive indium tin oxide (ITO) with exposed area of 30 × 20 mm2, vacuum deposited onto 400 μm thick sapphire substrate was used as the heater. The main advantage of usage ITO film as a heater is its property to be opaque in the mid-IR spectrum (3-5 μm) and to be transparent for visible spectrum (380-750 nm). When it deposited onto thin sapphire substrates, which is transparent to both IR and visible spectrum, it became possible to investigate unsteady temperature field of ITO film by infrared recording and to observe the dynamics of vapor bubbles and contact line with the use of high-speed video camera.

Measurements of the surface temperature field were performed by high speed infrared thermographic recording (IR) with the use of FLIR Titanium HD 570M camera (spectral range 3.7-4.8 μm). Thermographic recording was made with frame rate of 1000 fps at resolution up to 13 μm/px. High-speed visualization (HSV) was performed using Vision Research Phantom v. 7.0 camera with 2000 fps and resolution of 640 × 480 (1 px ≈ 0.07 mm). A quick-response incandescent lamp located under the work area was used to synchronize IR and HSV recordings in time. For correct signal conversation of the IR camera in temperature in order to account for the unknown surface emissivity the calibration experiments were carried out.

3 Numerical simulations

Reconstruction of the temperature fields and calculation of the heat fluxes based on experimental IR-thermometry data have been performed by cooperative solving of unsteady heat conductivity equation in the ITO film and sapphire substrate.

Because of small thickness of the ITO film local temperature was assumed equal throughout the thickness of the heater and solution of heat conductivity equation has been performed in 1D calculation area. With the presented assumptions heat conductivity equation looks like the following:

$$(\rho C_p)_{ITO} \frac{\partial T}{\partial t} = \lambda_{ITO} \frac{\partial^2 T}{\partial x^2} + \sigma \cdot \quad (1)$$

Volume heat sources σ could be presented as the superposition of three components:

$$\sigma = \frac{q_{in} - q_L - q_S}{\delta_{ITO}}, \qquad (2)$$

where q_{in} – power of electric heating generated in ITO film per unit surface area, q_L – heat flux density taken by the liquid, q_S – heat flux density taken by the sapphire substrate.
The heat flux density taken by the liquid could be obtained from (1), (2) and looks like the following:

$$q_L = q_{in} + \delta_{ITO}\lambda_{ITO}\frac{\partial^2 T}{\partial x^2} - \delta_{ITO}(\rho C_p)_{ITO}\frac{\partial T}{\partial t} - q_S, \qquad (3)$$

which is consistent with the mathematical formulation used in [7]. In presented work dimensional and time derivatives were calculated from experimentally obtained temperature fields. The heat flux density from ITO film to sapphire substrate was calculated using (4) from the temperature field in the sapphire which was obtained as a solution of the heat conductivity equation (5) in 2D calculation area (x,y):

$$q_s = -\lambda_S \left.\frac{\partial T}{\partial y}\right|_{y=0} \qquad (4)$$

$$(\rho C_p)_S \frac{\partial T}{\partial t} = \lambda_S\left(\frac{\partial^2 T}{\partial x^2} + \frac{\partial^2 T}{\partial y^2}\right) \qquad (5)$$

Temperature of the sapphire substrate on the interface with the ITO film ($y = 0$) was assumed to be equal to the ITO film temperature for all the calculated moments of time. On the opposite surface of the sapphire substrate ($y = \delta_S$) first derivative of y-axis was assumed to be zero. In order to smooth the edge effects temperature on the side surfaces of the substrate was assumed to be equal to the ITO film temperature. Temperature of ITO film and sapphire substrate were assumed to be equal to T_0 at the initial moment of time.

4 Results and discussion

With the use of synchronized high speed IR thermography and video recording there were obtained the new experimental data of heat transfer coefficients, nucleation site density, frequency and temporal charactestics of nucleation process, dynamics of vapor bubble growth and detachment. The detailed description and analysis of these data are presented in [9, 10]. In this paper we focus on the study of local heat transfer under single vapor bubble at liquid boiling obtained by macro IR thermometry and numerical simulation.

The frames of high-speed macro IR thermography of temperature field evolution under active nucleation site at boiling of water at $q = 50$ kW/m^2 are presented in Fig.1. Data of wall temperature distribution under isolated bubble in different times at boiling of water and ethanol are shown in Fig.2. As it is shown from the analysis of experimental data when vapor bubble appears on the surface, the temperature in the area of nucleation site sharply decreases (1 ms). In the growth stage of vapor bubble the regions with different heat transfer mechanism (microlayer region and area of dry spot in the center of nucleation site) appear. The boundary between different heat transfer modes is the triple contact line [2-6]. The area with a minimal temperature of the heater surface corresponds to microlayer region. The data analysis presented in [3, 10] shows, that after the bubble growth stage, where microlayer evaporation plays a dominant role in heat transfer, the stage of vapor bubble detachment begins. At the departure stage the equivalent bubble diameter almost unchanged with a sharp reduction of the area limited by triple contact line. Experimental data of thickness of microlayer under individual vapor bubble [3] showed that on this stage the microlayer was completely depleted.

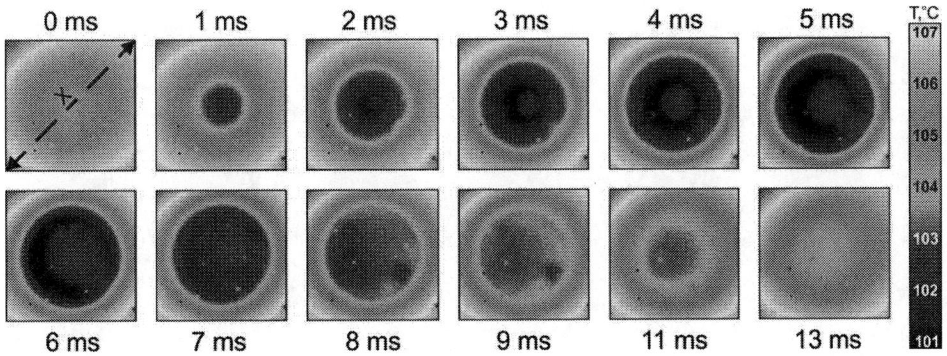

Fig. 1. High-speed IR camera images of evolution of temperature field under single vapour bubble at water boiling (q = 50 kW/m^2). Spatial and temporal resolutions are 13 μm and 1 ms, respectively.

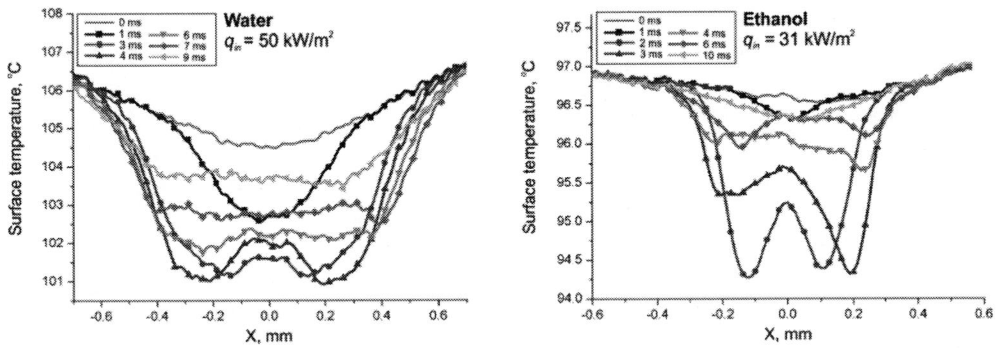

Fig. 2. The temperature distribution on the heater surface under individual vapor bubble at pool boiling of water and ethanol. x is coordinate along the bubble base (see Fig.1).

As it was shown in chapter 1 some authors for reconstruction of local instantaneous heat flux distribution on the thin heaters have used commercially available CFD software package. The aim of this research was numerical simulation of local heat transfer under individual vapor bubble at boiling with the use of obtained temperature fields and own-designed software based on the finite differences. The local heat flux distributions on the wall at different times at boiling of water and ethanol are presented on the Fig. 3 at low input heat fluxes. The graph shows that the maximum heat flux density is observed in microlayer and reaches about 1.2 MW/m^2 for water and ethanol. The resulting heat flux distributions at the region of active nucleation site at boiling of water are consistent with the data of [3]. With the use of heat flux maps and temperature field under individual vapor bubble the microlayer thickness can be estimated using the following equation:

$$q_{cond} = \lambda_{fl} \frac{(T_{ITO} - T_{sat})}{\delta_{ml}}. \tag{6}$$

Estimations obtained by this equation show that microlayer thickness on the stage of vapor bubble growth at boiling of water and ethanol ranged from 0.5 to 5 microns, which are in good agreement with the experimental data presented in the literature and obtained using laser interferometry [2, 3].

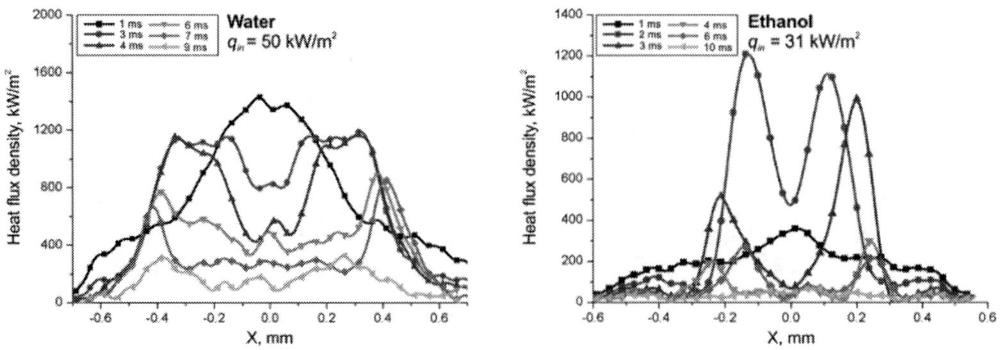

Fig. 3. The local heat flux distribution on the heater surface under individual vapor bubble at pool boiling of water and ethanol.

In the future, on the basis of experimental data of the evolution of the temperature field under the individual vapor bubbles at boiling of water and ethanol, obtained in a wide range of heat fluxes using the high-speed infrared thermography with max. resolution of 13 μm, and using the developed numerical algorithm, a detailed theoretical analysis of heat transfer in the vapor bubble will be held. This analysis will allow to define the influence of the input heat flux and liquid properties on the nucleation frequency, dynamic characteristics and the local heat transfer in the microlayer region, formed under the vapor bubbles, in order to determine the basic mechanism of highly effective heat transfer at liquid boiling and to improve the macroscopic boiling models.

This work was supported by the Russian Foundation for Basic Research (Grant No. 17-08-01342).

References

1. V. V. Yagov, *Heat transfer in one-phase media and at phase transitions* (Publishing House MPEI, 2014) (in Russian)
2. M. Gao, L. Zhang, P. Cheng, X. Quan, Int. J. Heat Mass Transfer **57**, 1 (2013)
3. S. Jung, H. Kim, Int. J. Heat Mass Transfer **73** (2014)
4. S. Jung, H. Kim, Heat Transfer Engineering **36**, 12 (2015)
5. X. Duan, B. Phillips, T. McKrell, J. Buongiorno, Exp. Heat Transfer **26**, 2-3 (2013)
6. C. Gerardi, J. Buongiorno, L. W. Hu, T. McKrell, Int. J. Heat Mass Transfer **53**, 19 (2010)
7. I. Golobic, J. Petkovsek, D. B. R. Kenning, Int. J. Heat Mass Transfer **55**, 4 (2012)
8. J. Petkovsek, Y. Heng, M. Zupancic, H. Gjerkes, F. Cimerman, I. Golobic, Int. J. Refrigeration **61** (2016)
9. A. S. Surtaev, V. S. Serdyukov, M. I. Moiseev, Instrum. Exp. Tech. **59**, 4 (2016)
10. V. S. Serdyukov, A. S. Surtaev, JPSC (to be published)

EPJ Web of Conferences **159**, 00052 (2017) DOI: 10.1051/epjconf/201715900052
AVTFG2016

Isothermal modeling of aerodynamic structure of the swirling flow in a two-stage burner

Roman Yusupov[1,2,*], *Sergey* Shtork[1,2] and *Sergey* Alekseenko[1,2]

[1] Institute of Thermophysics SB RAS, Novosibirsk, Russia
[2] Novosibirsk State University, Novosibirsk, Russia

Abstract. The work deals with the experimental study of the aerodynamic structure of a swirling flow in the isothermal model of two-stage vortex combustion chamber. The main attention is focused on the process of flow mixing of two successively connected tangential swirlers of the first and second stages of the working section. Data on flow visualization are presented for two patterns of flow swirling. Time-averaged profiles of the axial and tangential velocity components are obtained with the help of laser-Doppler anemometer. In the case of flow co-swirling between two stages of the working section, instability of a secondary flow in the form of precessing vortex was distinguished. For the regime with counter flow swirling, effective mixing of the swirl flows was found; this was reflected by formation of the flow with uniform distribution of axial velocity over the cross-section.

1 Introduction

Investigation of the vortex flows is a significant part of the modern fluid dynamics. Flow swirling can make a considerable contribution to the transport processes. Therefore, the swirling flows are widely used in many industrial processes to sustain atomization of liquid fuel, liquid mixing, formation of aerosols and stabilization of flames. Intensive heat and mass transfer occurring in the swirling flows and increased level of flow turbulence are used for cooling/heating and cleaning the working liquids, suspension separating, separation of materials and many other technological purposes. The devices, based on working medium swirling, are used in hydropower engineering, thermal power engineering, petroleum and chemical industries and many other fields [1].

Currently, one of the most interesting problems in the field of swirling flows is investigation of characteristics of behavior of the interacting swirling flows [2-3]. The need for such research is firstly caused by the use of the counter-swirling flows in energy dissipation devices and vortex combustion chambers. To optimize the process of flow mixing in these devices, it is necessary to understand the flow structure and mixing mechanisms. These studies are specifically relevant for modernization and improvement of the vortex combustion chamber [4]. In particular, this relates to the problems of the most efficient combustion of fuel, reducing pollutant emissions and increasing the operational lifetime of power plants [5-8].

2 Experimental methods

* Corresponding author: romansemsk@yandex.ru

Isothermal modeling of a two-stage burner was carried out using water as a working liquid. To study experimentally the aerodynamic structure of a swirling flow, the transparent model of a vortex chamber with two-stage flow swirling was made. This chamber (6) was connected to the hydrodynamic circuit (Fig.1), consisting of vessel (1) with tankage of 0.5 m^3, centrifugal pump X100-80-160T153 (2) with maximal flow rate and head of 100 m^3/h and 4 atm., respectively, lines of liquid supply to the first and second stages with the flow meters in every line (4 – metering orifice, 5 - vortex-acoustic flow meter Metran-300PR). The total flow rate was regulated by the frequency converter Vesper E2-8300-030N (3), and flows through each line were controlled by the regulating valves. Together with the flow rates between the stages, it was possible to measure flow swirling on the second stage of the model.

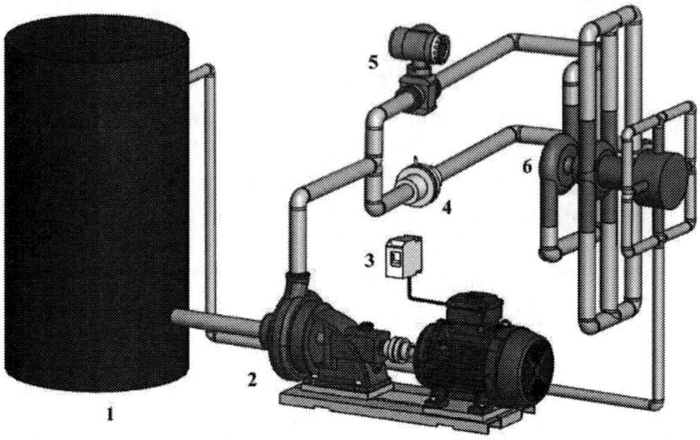

Fig.1. Scheme of hydrodynamic setup for modeling the aerodynamic structure of the swirling flow in the two-stage burner. 1 – vessel, 2 – pump with electromotor of 22-kW power, 3 – frequency converter, 4 – metering orifice, 5 – vortex flow meter, 6 – working section.

The hydrodynamic model of the two-stage burner is used as the working section (Fig. 2). This chamber consists of two successive swirlers with the diameter of 185 mm, where the flow is swirled by means of tangential supply through the pipe sockets with the inner diameter of 40 mm. The liquid flow is swirled by swirler (1), then, passing through a nozzle, it is fed to the zone of sudden expansion. Then, in this zone, it interacts with the flow, swirled by second swirler (3). Depending on the method of liquid supply to the second swirler, co-swirling (white arrows) and counter-swirling (dark arrows) of two flows are achieved between the swirlers of the first and second stages of the working section. The resultant flow from the cylindrical zone of 104-mm diameter (4) is fed to the outlet section (5), and it leaves it through four symmetrical pipes with the inner diameter of 20 mm.

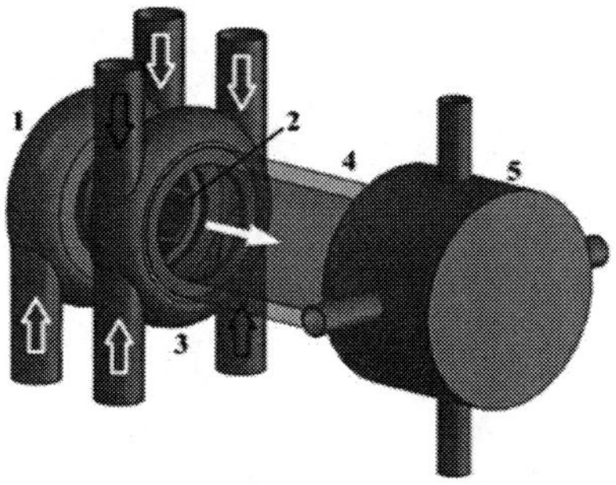

Fig.2. Scheme of hydrodynamic model of two-stage burner. 1 and 3 – swirlers of the first and second stages of the burner, respectively, 2 – shaped nozzle, 4 – zone of sudden expansion, 5 – outlet section with outlets and deswirler blades.

3 Experimental results

3.1 Visualization

The flow was visualized during the experiments on the isothermal model. Small air bubbles were used as the flow markers. These bubbles were illuminated by a light source (diode floodlight), to obtain the general flow pattern, or by a narrow light "sheet" formed via sweeping a laser beam on a cylindrical lens, to visualize the flow structure in a certain cross-section of the working section.

The flows with co-swirling and counter-swirling were visualized for different ratios of liquid flow rates between the stages of the working section. According to analysis of visualizations, the regime characterized by the aerodynamic flow structure, optimal from the point of efficient flow mixing at the first and second stages, was determined. Results of visualization of the flows with co- and counter swirling in both longitudinal and transverse cross-sections of the vortex chamber are presented in Fig. 3. The experiments were carried out for the liquid flow rates of 4 m^3/h at the first and 6 m^3/h at the second stages of flow swirling. For the cases of flow co-swirling, the high-speed recording (Fig.3 (c, d)) demonstrates the presence of secondary flow instability with formation of a precessing vortex at the boundary of the central stagnation zone, that causes induction of powerful low-frequency flow pulsations of a discrete frequency. In contrast to the flow with co-swirling, the version with counter-swirling showed efficient interaction of the swirling flows of the first and second stages. The resultant flow was characterized by uniform distribution of the axial velocity along the cross-section in combination with relatively expressed total rotation of the flow in the working section. The absence of large-scale non-stationary structures and, consequently, intense flow pulsations is the feature of this flow regime.

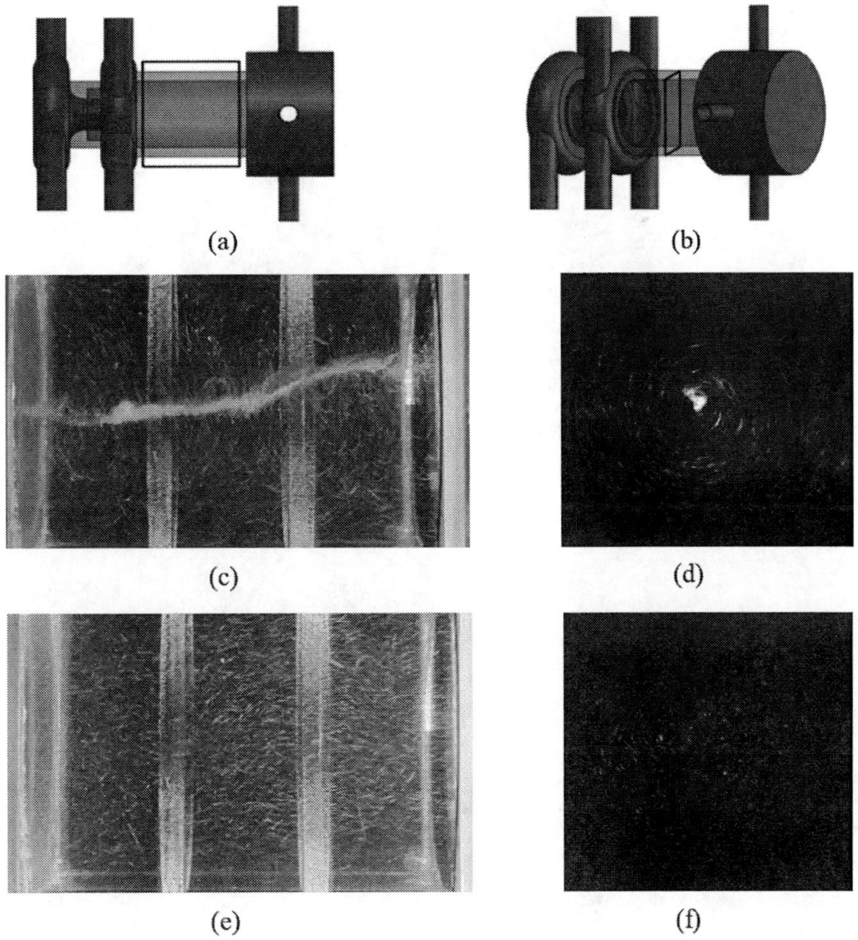

Fig.3. Flow visualization under different conditions of vortex chamber operation. Position of visualization area: a –side view, b – transversal cross-section. Regime with flow co-swirling: c – side view, d – transversal cross-section. Regime with flow counter-swirling: e – side view, f – transversal cross-section.

3.2 LDA measurements

To get quantitative information about the flow field, the two-component laser Doppler anemometer (LDA) LAD-06i was used. Time-averaged axial and tangential velocity components were measured along three lines in the cross-section of the cylindrical zone of sudden expansion (Fig. 4). The characteristic distances from the second stage of flow swirling to the measured areas are shown in Fig. 4. The spherical polystyrene particles with neutral buoyancy were used as the flow tracers in LDA measurements. The presented results were obtained for the liquid flow rate of 4 m^3/h at the first and 6 m^3/h at the second stages of flow swirling. The distance from the central axis r is normalized to radius R_0 of the cylindrical area of sudden expansion (Fig. 2 (4)).

In the case with flow co-swirling (Fig. 5), we can see formation of the axial flow, non-uniform over the cross-section, with intense flow along the sidewalls and extended area of a stagnant flow along the chamber axis. In the case of flow counter-swirling (Fig. 6), the resultant flow structure is characterized by uniform distribution of axial velocity over the cross-section in combination with relatively expressed rotation of the flow in the working section.

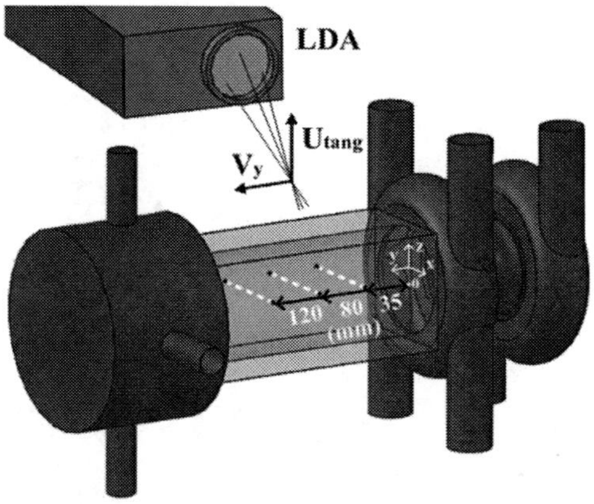

Fig.4. Layout of measurement sections

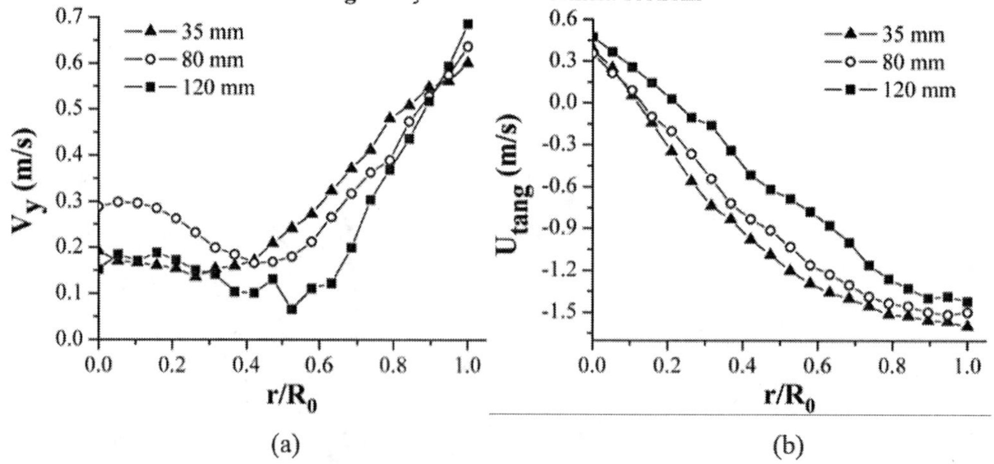

Fig.5. Velocity profiles for different conditions of vortex chamber operation. a, b – axial and tangential velocity components, for the regime of flow co-swirling.

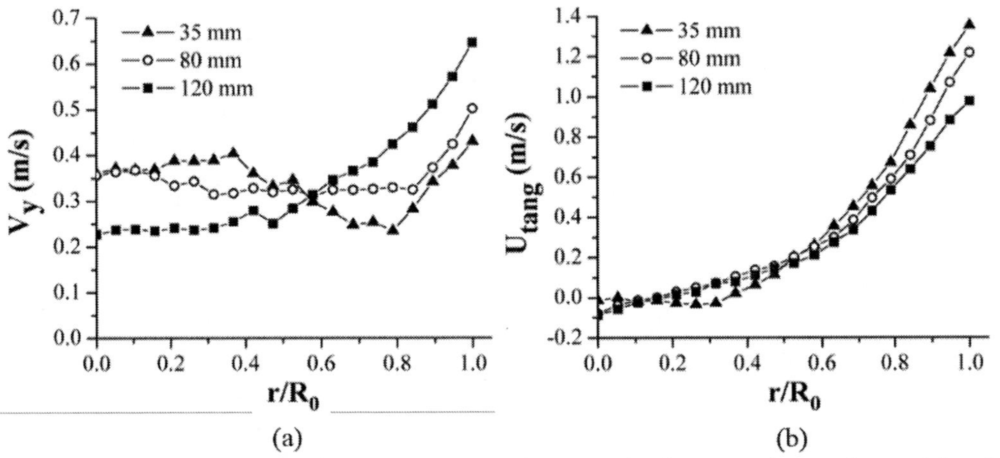

Fig.6. Velocity profiles for different conditions of vortex chamber operation. a, b – axial and tangential velocity components, for the regime with flow counter-swirling.

4 Conclusion

During this study, the high-speed recording of the flow was carried out for two types of flow swirling. When the flows were swirled in one direction, formation of a precessing vortex, generating low-frequency pressure pulsations in the flow, was revealed. The flow with swirling in different directions showed uniform filling of the working section in combination with moderate flow swirling and flow stability. Time-averaged profiles of axial and tangential velocity components, obtained using the two-component LDA, demonstrated that in the case of co-swirled flows the non-uniform distribution of axial velocity is observed in the cross-section with a strong flow along the sidewalls. In turn, the characteristic profiles of axial velocity component in the case of counter-swirled flows showed uniform distribution, which characterizes the effective mixing of two flows.

The work was financially supported by the Russian Ministry of Education and Science, according to Subsidiary Agreement No. 14.613.21.0005 (unique project identifier RFMEFI61314X0005).

References

1. V.V. Volshanik, G.V. Orekhov, Application fields for interacting swirling flows of liquids and gases, Vestnik MGSU, No. 7 (2015)
2. Chao Y.C., Recirculation Structure of the Co-annular Swirling Jets in a Combustor, AIAA Journal, **26**, pp. 623—625, No. 5 (1988)
3. Escudier M. P. and Keller J. J., Recirculation in Swirling Flow: A Manifestation of Vortex Breakdown, AIAA Journal, **23**, pp. 111-116 (Jan. 1985)
4. H. Altgeld, W. P. Jones and J. Wilhelmi., Velocity Measurements in a Confined Swirl Driven Recirculating Flow, Experiments in Fluids 1, 73-78 (1983)
5. H. Toral, J. H. Wihtelaw., Velocity and Scalar Characteristics of the Isothermal and Combusting Flows in a Combustor Sector Rig, Combustion and Flame, **45**, 251-272 (1982)
6. Bach T. Vu, F. C. Gouldint., Flow Measurements in a Model Swirl Combustor, AIAA, **20**, No. 5 (1982)
7. Theodore Providakis, Laurent Zimmer, Philippe Scouflaire, Sebastien Ducruix., Characterization of the Acoustic Interactions in a Two-Stage Multi-Injection Combustor Fed With Liquid Fuel, Journal of Engineering for Gas Turbines and Power, **134** (2012)
8. Yonmo Sung, Gyungmin Choi., Non-intrusive optical diagnostics of co- and counter-swirling flames in a dual swirl pulverized coal combustion burner, Fuel, **174**, 76–88 (2016)

EPJ Web of Conferences **159**, 00053 (2017) DOI: 10.1051/epjconf/201715900053
AVTFG2016

Testing of numerical simulation technique of concentrated suspensions agitation

Vladimir Zhigarev[1,*], *Andrey* Minakov[1,2], *Dmitry* Platonov[1,2], *Dmitry* Dekterev [1,2] and *Maxim* Pryazhnikov[1,2]

[1]Siberian Federal University, 79 Svobodny pr., Krasnoyarsk, 660041, Russia
[2]Institute of Thermophysics SB RAS, 1 Acad. Lavrentiev pr., Novosibirsk, 630090, Russia

Abstract. The paper presents testing results of the numerical technique to simulate the agitation process of concentrated suspensions. The simulating technique is based on Eulerian two-fluid model of granular media, k-w SST URANS turbulence model, and sliding mesh approach. The results obtained show a good qualitative and quantitative agreement between calculation and experiment in terms of the shape and location of the phase interface and the distribution of the solid phase along the height.

1 Introduction

Solid–liquid systems occur in a wide range of industrial applications such as mineral processing, precipitation, biochemical crystallization processes, and drilling of the wells. A considerable amount of work has been done to investigate the influence of hydrodynamics and mixing of solid suspensions using impellers. Mixing is the one of the most common ways to enhance many engineering processes [1]. Apparatuses with mixing devices are used in processes such as crystallization, absorption, extraction, homogeneous and heterogeneous chemical reactions, etc. The most efficient way of mixing is creation of swirling flow inside the reactor volume. Swirling flow can be created using a variety of mixing devices (agitators). Apparatuses with mechanical impellers are preferably used in processes with prevailing liquid phase. Up to now, 80% of application processes consist of blending, and at the same time 50% of them deal with solid–liquid suspensions [2]. These operations require employment of axial high speed impellers, because their flow characteristics significantly affect both off-bottom suspension [3] and the rate of homogenization of liquid phase in agitated charges [4].

2 Testing of numerical methods

The objective of the present work was testing the computational technique to simulate the suspensions agitation processes by means of devices with a swirling flow (agitators). Calculation of suspension agitation process is implemented using a hydrodynamic approach based on the solution of Reynolds averaged Navier-Stokes equations with a two-fluid Eulerian model for granular media [5].

* Corresponding author: zhigarev.vladimir@yandex.ru

The agitation of silicon carbide particles suspension in a cylindrical container was considered as the first task to test the computational technique. A detailed description of this experiment can be found in [6].

The computational domain was a cylindrical volume with a diameter of 240 mm and a height of 225 mm. The inner wall of the cylinder had 4 ribs with the dimensions of 25x10x225 mm. Agitation of the pulp in the cylinder was carried out by installed flat four-blade impeller with the blade angle of 45 degrees. The impeller had the following dimensions: diameter – 105 mm, blade width – 14 mm, and blade thickness – 2 mm. The rotation speed of the impeller was 300 rpm. Photograph of the experimental setup [6] is shown in figure 1. The experiment was conducted for volumetric concentration of silicon carbide particles equal to 5%, particle size was equal to 500 μm. Water was used as an actuating medium. As a result of the experiments we measured the distribution of the bulk concentration of particles along the height of the cylinder.

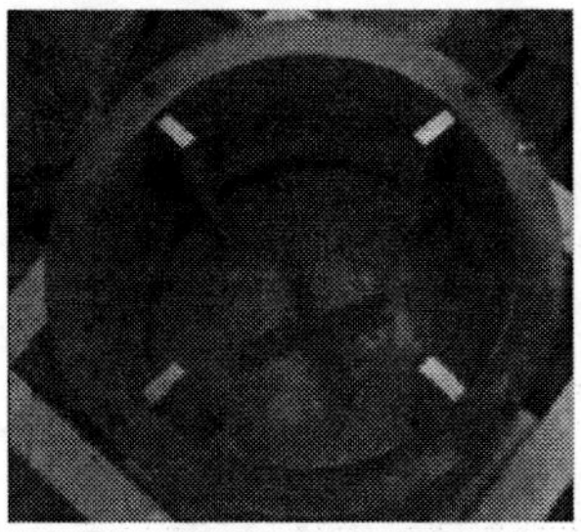

Fig. 1. Photograph of the experimental setup taken from [6].

Place the figure as close as possible after the point where it is first referenced in the text. If there is a large number of figures and tables it might be necessary to place some before their text citation.

In calculations we used unstructured mesh consisting of 950 000 cells with refinement in their number towards actuator walls. The impeller rotation was simulated using the "frozen rotor" approach. The problem was solved for spatial stationary conditions. For simulation we used k-w SST (RANS) turbulence model. For simulation of the pulp agitation we used Eulerian two-fluid model of granular media. The coefficient of hydraulic resistance of the particles was set according to the Syamlal-Obrien model, the lifting force was simulated according to the Saffman-Mei lift model, turbulent dispersion was simulated employing Simonin model, granular viscosity was calculated according to Gidaspow, while stresses in the solid phase – by the model of Lun at al.

Figure 2 shows the distribution of the particles volume fraction over the walls of the cylindrical area. As is obvious, the distribution of particles is rather uneven. Due to centrifugal force, particles are concentrated at the cylinder walls and in the corners of the ribs.

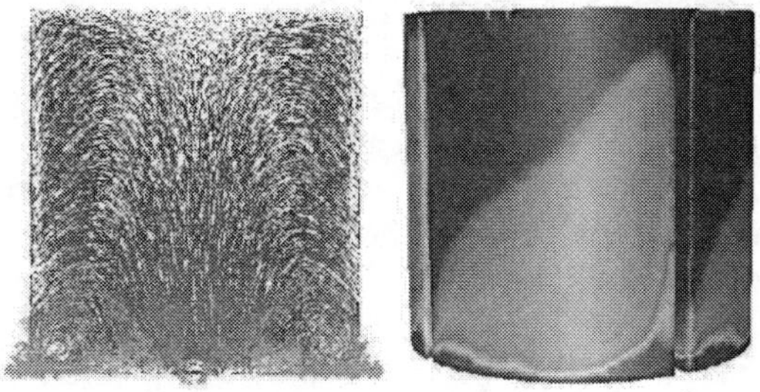

Fig. 2. The velocity field in the central plane (left) and the distribution of the particles volume fraction over the walls (right).

The results of quantitative comparison with experiment are shown in Fig. 3. The figure shows the distribution of particles volume fraction over the cylinder height along the line lying on the inner wall of the cylinder midway between the ribs. The diagram shows the results obtained using Eulerian two-fluid model without taking into account turbulent diffusion, though considering the model of Simonin [7]. As is obvious, not taking into account turbulent dispersion leads to a significant underestimation of the particles lifting height. This parameter is very important in the chemical heterophase devices. The Simonin model of turbulent dispersion gives satisfactory agreement with experiment on the particle concentration distribution over the height of the computational domain.

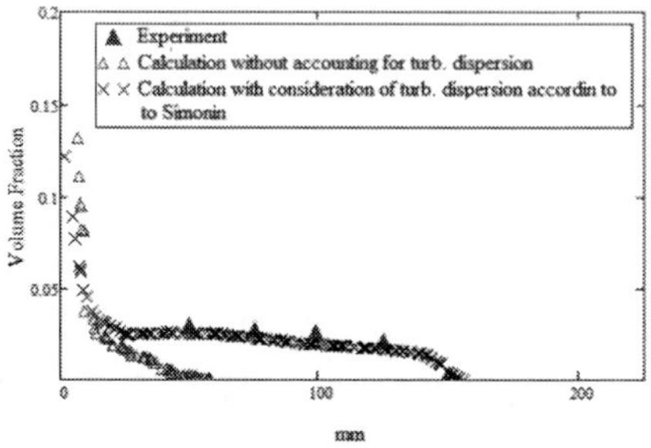

Fig. 3. The distribution of the volume fraction of solid phase over the height of the computational domain.

3 The experimental results and numerical calculation

A special experimental setup was designed for a more detailed study of the pulp agitation process. Photograph of the setup is shown in Fig. 4. The experimental setup is a glass cylindrical container served as a stirred tank with a diameter of 20 cm and a height of 40 cm.

Fig. 4. Photograph of the experimental setup.

The rotation of the agitator was carried out using a servomotor. The rotation speed of the agitator was controlled within the range from 0 to 3000 rpm. The distribution of particles in a fluid in the course of the experiment was recorded using high-speed camera. For testing mathematical model we conducted simulation of agitation process of 500 μm sand particles at a volume concentration of 10%. The agitation of two-phase medium was carried out by four-blade flat impeller with the blade angle of 45 degrees. The agitator had the following dimensions: diameter – 80 mm, blade width – 10 mm, and blade thickness – 2 mm. The agitator was mounted at a height of 20 mm from the bottom of the container and in this case had no guide ribs. The rotation speed of the agitator was varied from 100 to 800 rpm. In our calculations we used unstructured mesh consisting of 1 250 000 cells, with refinement towards the agitator walls. The simulation was performed under unsteady conditions. In our simulation we used k-w SST (URANS) turbulence model. For simulation

DOI: 10.1051/epjconf/201715900053

of the pulp agitation we used Eulerian two-fluid model with the sub-models described above. Simulation of the agitator rotation was carried out by a "sliding mesh" model. Examples of comparison are presented in Figs. 5.

a b

Fig. 5. A comparison of the volume fraction of the liquid with the experimental image at a rotation speed of 200 rpm without solid phase. a) Numerical simulation, b) experiment.

First, the experiment was carried out at zero concentration of particles. Figure 6b shows a photograph of such an experiment. As can be seen, the absence of the solid phase and guide ribs leads to the appearance of a large funnel. For the simulation of wave generation on a free surface, VOF method was used in simulation [7-9]. The presence of a funnel during the agitation process has negative effect, since at high rotation speeds the air can reach the agitator blades and mixing efficiency will be sharply reduced. This effect is called "flooding" of agitator. As can be seen from Fig. 6a, the calculated shape of the free surface very well corresponds to the funnel observed experimentally.

Next, we conducted a series of experiments with solid particles. The results of the experiments and simulations are shown in Figs.6-7. As is obvious, in the first case (200 rpm) no pulp agitation was observed. The phase interface remains almost flat. At the agitator rotation speed of 300 rpm there is strong enough wave generation on the free surface of the fluid and the interface between liquid and solid phase. In this mode, the pulp is moving wavily (see Fig.7). However, agitation in this mode cannot be called effective either. Homogeneous agitation of the particles can only be achieved at the rotation speeds of the agitator exceeding 600 rpm. Qualitative and quantitative comparison of calculation and experiment on the lift height of the particles has shown that for all cases the calculation very well describes the behavior of the pulp in the experiment. The results are shown in Fig. 5.

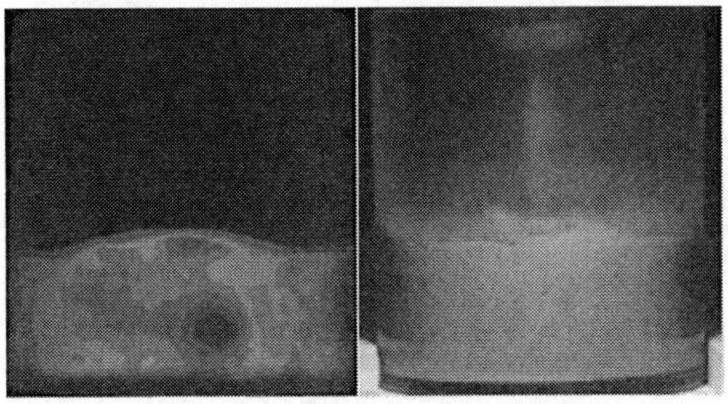

a b

Fig. 6. A comparison of the particles volume fraction at the wall with experiment at rotation speed of 200 rpm. a) Numerical simulation, b) experiment.

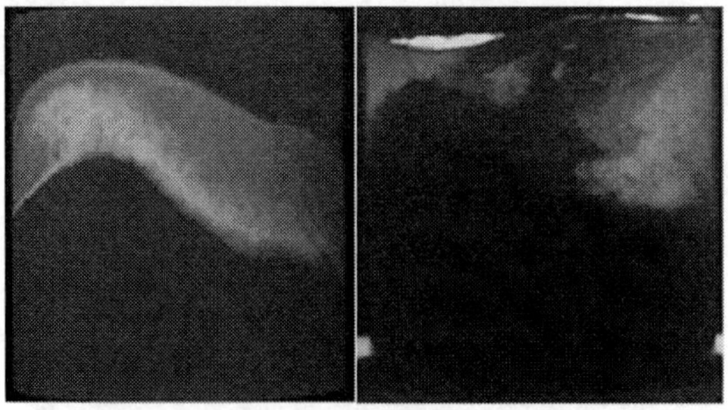

a b

Fig. 7. A comparison of the particles volume fraction at the wall with experiment at rotation speed of 300 rpm. a) Numerical simulation, b) experiment.

4 Conclusion

We have conducted a review and analysis of up-to-date types and methods of agitators for chemical heterogeneous processes. A model setup was designed and the experimental study of the mixing process of the 500 μm sand particles suspension was carried out. The pulp mixing efficiency at various rotation speeds of the agitator was determined by means of a high-speed camera. The mathematical model and numerical technique was developed to simulate suspension agitation process. The developed numerical technique was tested by comparison of calculations with the own experimental data as well as the data obtained by other authors. Good qualitative and quantitative agreement between calculation and experimental data was achieved in terms of the shape and location of the phase interface and the distribution of the solid phase over the reactor height. In the future we plan to perform a comparative analysis of the mixing efficiency for the agitators with different deign shapes.

The reported study was funded by Russian Foundation for Basic Research, Government of Krasnoyarsk Territory, Krasnoyarsk Region Science and Technology Support Fund to the research project № 16-41-243034.

References

1. Starback, P. Tausk. L., Chimizdat,(1963).
2. I. Fořt, P. Seichter, L. Pešl, F. Rieger, T. Jirout, Chem. Process Eng. Vol.34 (2013).
3. J. WU, L. PULLUM, J. AICHE, Vol.46 (2000)
4. I. FOŘT, Vol. 3, Academic Press,New York, (1986).
5. ANSYS FLUENT Tutorial Guide. Validation.
6. Lalit Upadhayay, Dissertation PhD (2010)
7. C.W Hirt, B.D Nichols. J. Comput.Phys. Vol. 39 (1981).
8. D.V. Guzei, A.V. Minakov, M.I. Pryazhnikov, A.A. Dekterev. THERMOPHYS AEROMECH+ Vol.22 (2015)
9. A.V. Minakov. COMP MATH MATH PHYS+. Vol. 54 (2014)

An experimental study of high heat flux removal by shear-driven liquid films

Dmitry Zaitsev[1,2], *Egor* Tkachenko[1,2] and *Oleg* Kabov[1,2]

[1]Institute of Thermophysics SB RAS, Novosibirsk, Russia
[2]Novosibirsk State University, Novosibirsk, Russia

Abstract. Intensively evaporating liquid films, moving under the friction of a co-current gas flow in a mini-channel (shear-driven liquid films), are promising for the use in cooling systems of modern semiconductor devices with high local heat release. In this work, the effect of various parameters, such as the liquid and gas flow rates and channel height, on the critical heat flux in the locally heated shear-driven water film has been studied. A record value of the critical heat flux of 1200 W/cm^2 has been achieved in experiments. Heat leaks to the substrate and heat losses to the atmosphere in total do not exceed 25% for the heat flux above 400 W/cm^2. Comparison of the critical heat fluxes for the shear-driven liquid film and for flow boiling in a minichannel shows that the critical heat flux is an order of magnitude higher for the shear-driven liquid film. This confirms the prospect of using shear-driven liquid films in the modern high-efficient cooling systems.

1 Introduction

Development of the modern microelectronic equipment requires the effective cooling solutions because it is necessary to remove high heat fluxes of up to 1 kW/cm^2 and higher from the local hot spots of the processor [1]. Thin and ultra-thin (less than 10 μm in thickness) liquid films, moving under the action of a forced gas flow in a mini-channel, are promising for the use in the temperature control systems of modern semiconductor devices [2]. The numerical and analytical studies of the heat and mass transfer processes and hydrodynamics of joint motion of intensively evaporating liquid film and gas flow in a mini-channel have been performed in [3-4]. The authors of [5-7] have determined the basic laws of the flow and crisis phenomena in the liquid film, moving under the influence of a gas flow in a horizontal channel, under moderate heat fluxes. The first experiments with locally heated liquid films, moving under the action of gas, showed that they are less subject to the breakdown than the liquid films moving under action of gravity. This is because the film breakdown and formation of dry spots can be controlled by the forced gas flow. Liquid film breakdown and formation of dry spots are important for studying the crisis phenomenon in the liquid films [8-9]. The main goal of this work is to study the breakdown and critical heat flux in the locally heated shear-driven films of water. Experimental data on removal of the heat fluxes higher than 1 kW/cm^2 from a heat source of 1x1 cm^2 are presented in the paper. Data on heat leakages into the substrate and heat losses into the atmosphere are also presented here.

2 Experimental setup

The working section construction is shown in Fig. 1. Its main part is a stainless-steel plate with an embedded copper rod. The rod has a shape of a square head of 1x1 cm^2. The rod is heated by a spiral, wound on its bottom part (not shown in Fig. 1). This construction of the heater ensures a constant temperature along the rod surface (this is proved by the measurements of thermocouples). The working section is covered by a transparent optical glass (not shown in Fig. 1) to form a flat channel. The channel width is 40 mm, while the channel height can be varied from 0.15 to 2 mm.

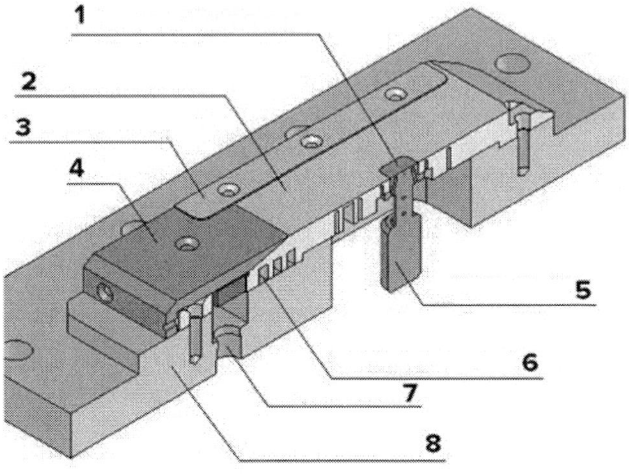

Fig. 1. Design of the working section: 1- source of the local heating; 2- stainless steel plate; 3- channel height control; 4- knife; 5- copper rod; 6- thermostabilizer; 7- liquid inlet; 8- textollite substrate.

Gas is fed into the working section by a compressor. Liquid is supplied from the thermostat to the channel through the liquid nozzle and moves under the influence of gas friction over the stainless-steel plate as a film. After passing through the working area, gas flows out to the atmosphere, while liquid returns to the thermostat. Distilled water with the initial temperature of 24°C is used as the working liquid. Air with the temperature of 24-27°C and relative humidity of 15-30 % is used as the working gas.

Thermocouples embedded into the stainless-steel plate and copper rod allow to determine the temperature of the working surface. The heat flux is determined by electrical power, released by the heating spiral. Thermal conductivity of copper is 400 W/mK, and this is more than an order higher than that of stainless steel (15W/mK); this ensures moderate heat leakage from the heater into the steel plate. According to the estimate using the measurements of thermocouples embedded into the steel plate, the heat leakages are about 15% at $q > 200$ W/cm^2. To reduce heat losses to the atmosphere, the bottom part of the copper rod was wrapped by a layer of insulating material (not shown in Fig. 1). According to the estimate using the measurements of thermocouples embedded into the rod, heat losses to the atmosphere do not exceed 10% at $q > 400$ W/cm^2. Consequently, the heat leakages and losses in total do not exceed 25% at $q > 400$ W/cm^2.

The flow and breakdown of the liquid film is visualized using a high-resolution infrared camera «Titanium HD 570M» (resolution of 640x512 pixels with the frame rate of 115 Hz).

The infrared camera works in the spectral range of 3.7 - 4.8 μm, with minimal resolvable temperature difference equivalent to noise (NETD) of 18 mK at 25°C.

3 Experimental results

The behavior of the liquid film moving under the influence of gas was studied in this paper with the values of heat flux q = 0-1200 W/cm², liquid Reynolds number Re_l = 8.5-193 and superficial gas velocity U_{sg} = 3.8-93 m/s (gas Reynolds number Re_g = 150 – 5960). Figure 2 shows infra-red images illustrating the process of film breaking with increased heat flux at Re_l = 30, U_{sg} = 22 m/s. At a certain value of the heat flux, the film breakdown occurs with dry spot formation on the substrate. As a rule, the first dry regions are formed below the heater (not shown in Fig. 2). With an increase of the heat flux, the dry areas reach the heater, and the film on the heater divides into 2-3 nonstationary rivulets (Fig. 2a). With further increase in the heat flux, the heater is covered by a metastable evaporating thin liquid film, with rapidly appearing and disappearing small dry spots (Fig. 2b). When the critical heat flux is achieved, the heater suddenly becomes dry, and the heater temperature starts increasing sharply (Fig. 2c-d).

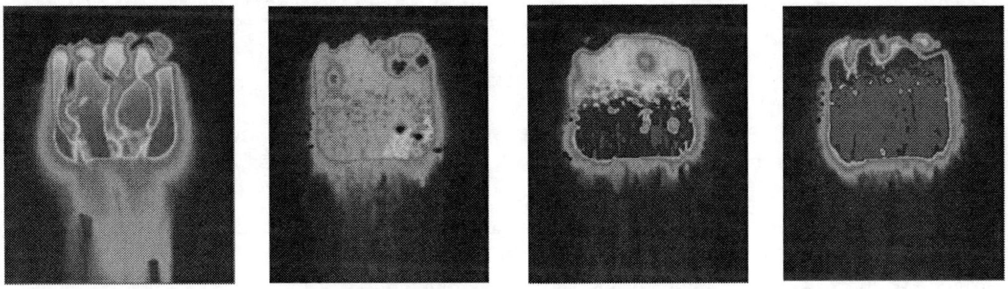

a) q = 125 W/cm², rivulet flow | b) q = 200 W/cm², thin film evaporation | c) q = 290 W/cm², drying out | d) q = 290 W/cm², crisis

Fig. 2. Breakdown and crisis in the liquid film moving under the action of gas, heater of 1x1 см², Re_l = 30, U_{sg} = 22 m/s, the flow is directed from top to bottom.

Data on the effect of liquid Reynolds number and superficial gas velocity on the critical heat flux are shown in Fig. 3. It is seen that with an increase in the liquid Reynolds number and gas velocity, the critical heat flux increases. The record value of the critical heat flux of 1.2 kW/cm² was achieved in experiments (Figure 3, point Re_l = 193, U_{sg} = 71.3 m/s). This critical heat flux for the water film flowing under the action of gas was achieved for the mass water flow rate of 175 kg/m²s. At that, as it is shown in [11], the critical heat flux for subcooled water flow boiling in a minichannel (at the same mass flow rate of liquid, at atmospheric pressure and subcooling to the saturation temperature of 75 K) reaches 170 W/cm² (Fig. 3, line 8). According to [11], to achieve the heat flux of 1.2 kW/cm² for water flow boiling in a minichannel, the mass flow rate of liquid should be about 7000 kg/m²s, which is 40 times higher than the mass flow rate in our experiment.

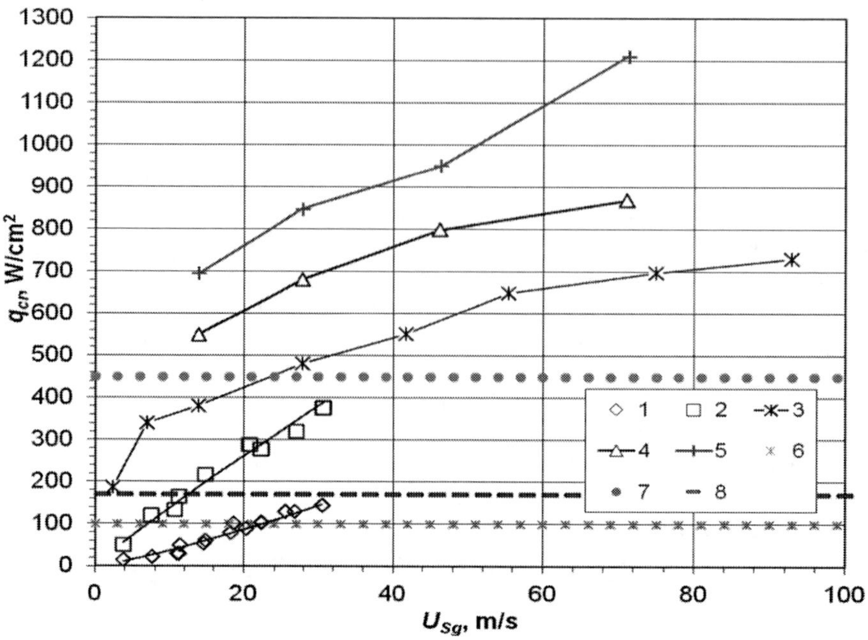

Fig. 3. Critical heat flux vs. superficial gas velocity, U_{Sg}, liquid Reynolds number, Re_l, and channel height, H. 1 – Re_l = 8.5, H = 1.2 mm; 2 – Re_l = 30, H = 1.2 mm; 3 – Re_l = 62, H = 1 mm; 4 – Re_l = 113, H = 1 mm; 5 - Re_l = 193, H = 1 mm; 6 – water pool boiling (formula of Kutateladze $q_{cr} = 0.14 r \sqrt{\rho_v} \sqrt[4]{\sigma g(\rho_l - \rho_v)}$, [10]); 7 – water pool boiling at subcooling to the saturation temperature of 75 K [10]; 8 – water flow boiling in a mini-channel for mass flow rate 175 kg/m²s, at atmospheric pressure and subcooling to the saturation temperature of 75 K [11].

4 Conclusions

The critical heat flux for the water films moving under the influence of gas flow friction in a minichannel can reach 1.2 W/cm², which is an order of magnitude higher than the critical heat flux for water flow boiling (for the same mass flow rate, same subcooling to the saturation temperature, and same pressure). Heat losses to the atmosphere and heat leakages to the substrate in our experiments in total do not exceed 25% at heat fluxes higher than 400 W/cm². The results obtained confirm the prospects of the use of shear-driven liquid films in the modern cooling systems for devices with high local heat release.

The work was performed under the support of Russian Science Foundation (project № 14-19-01755).

References

1. A. Bar-Cohen, C. Holloway. Journal of Physics: Conference Series, **745**, 022002 (2016)
2. O.A. Kabov, Yu.V. Lyulin, I.V. Marchuk and D.V. Zaitsev. International Journal of Heat and Fluid Flow, **28**, 103-112 (2007)
3. E. Ya. Gatapova and O. A. Kabov. International Journal of Heat and Mass Transfer, **51**, 4797-4810 (2008)
4. Y. Kabova, O. Kabov, T. Gambaryan-Roisman, P. Stephan and V. V. Kuznetsov. International Journal of Heat and Mass Transfer, **68**, 527-541 (2014)

5. D. V. Zaitsev, D. A. Rodionov and O. A. Kabov. Technical Physics Letters, **35**, 680-682 (2009)
6. O. A. Kabov and D. V. Zaitsev. Multiphase Science and Technology, **21**, 249-266 (2009)
7. O.A. Kabov, D.V. Zaitsev, V.V. Cheverda, A. Bar-Cohen. Experimental Thermal and Fluid Science, **35**, 825 (2011)
8. V. S. Ajaev. Interfacial Phenomena and Heat Transfer, **1**, 81-92 (2013)
9. M. Kadoura and S. Chandra. Experiments in Fluids, **54**, 1-11 (2013)
10. S.S. Kutateladze, *Heat Transfer in Condensation and Boiling* (Mashgiz, Moscow, 1952)
11. W. Zhang, T. Hibiki, K. Mishima, Y. Mi, International Journal of Heat and Mass Transfer, **49**, 1058-1072 (2006)

EPJ Web of Conferences **159**, 00055 (2017) DOI: 10.1051/epjconf/201715900055
AVTFG2016

Thermocapillary rupture in falling liquid films at moderate Reynolds numbers

Evgeny Shatskiy, *Evgeny* Chinnov

Kutateladze Institute of Thermophysics SB RAS, 630090, 1 Lavrentyev Ave., Novosibirsk, Russia
Novosibirsk state University, 630090, 2 Pirogova Str., Novosibirsk, Russia

Abstract. An experimental study of the flow of a water film over a heated surface for Re = 15-50 was performed. The influence of the development of thermocapillary instability on the wave amplitudes, the deformation of the surface of the liquid film, and the formation of the first stable dry spot on the heater are investigated. It is shown that the interaction of waves with thermocapillary structures can lead to an increase in the critical heat flux corresponding to the rupture of the liquid film, as compared with the data known in the literature.

1 Introduction

Film flows are widely used in various industrial technologies and apparatuses, and understanding the processes occurring in flowing films is an important task for the implementation of such devices. An important factor affecting the intensity of heat transfer and the stability of films to rupture is the interaction of waves on the interphase surface with thermocapillary structures.

Hydrodynamic two-dimensional waves in isothermal fluid films are unstable to three-dimensional perturbations. The length of the wave instability to transverse three-dimensional perturbations decreases with increasing Reynolds number [1]. It is established that the transition from regular two-dimensional structures to three-dimensional flow is accompanied by a significant redistribution of liquid in the longitudinal direction [2, 3]. Characteristic forms of three-dimensional structures developing during the transition are described.

While flowing along heated surfaces, in addition to the hydrodynamic instability leading to the development of the wave flow, there are also thermocapillary instabilities of various types associated with fluid transport along the interface due to the appearance of a surface tension gradient [4].

An investigation of the Marangoni effect in a heated liquid film with a developed wave motion is described in [5]. The film temperature fields were visualized by infrared thermography, and the film thickness was measured by confocal microscopy.

Investigations of the natural evolution of developed three-dimensional waves into thermocapillary-wave rivulets during heating of a vertically falling water film under conditions of small temperature gradients were performed in [6] with Re = 10. The deformation of horseshoe-shaped hydrodynamic waves as they pass through the heated region is recorded. It is shown that at the front of the three-dimensional wave, when moving along the heater, temperature inhomogeneities appeared.That led to the deformation of the liquid film and the formation of rivulets due to the action of thermocapillary forces.

An investigation of the dynamics of the rupture by means of a fiber-optic sensor was carried out in [7]. The influence of wettability of the substrate on the thermocapillary rupture of a liquid film was carried out in [8]. In [9] generalizing dependences on thermocapillary rupture of a film in a wide range of Reynolds numbers and sizes of the heater are given.

The aim of this work is an experimental study of the effect of the interaction of hydrodynamic waves with thermocapillary structures on the stability of the flow of a heated water film to rupture.

2 Setup and experimental results

The stand was a closed circulation circuit, including a reservoir with a pump, a working section, a filter, rotameters, pipelines and shut-off valves. The working area consisted of a working plate, on which a film former, a heat stabilizer and a heater were installed. The working liquid (water with a dye) was fed into the film former, which included a storage chamber, a dispenser and a nozzle with a calibrated flat slot. The liquid flowed down the plate and through the connecting channels under the influence of gravity, returning to the reservoir with the electric pump. As a heating element, a copper flat heat exchanger 150 mm wide and 100 mm long was used. A detailed description of the installation can be found in [10]. The critical heat flux corresponding to the rupture of the film was determined by the appearance of the first stable dry spot in the center of the heater.

A synchronous measurement of the surface temperature and a thickness distribution in the heating region at Re = 15-50 are performed. Re = W / ρν, where W is the specific mass flow rate of the liquid, ρ is the density of the liquid, and ν is the kinematic viscosity of the liquid. The method of investigation is described in detail in [10].

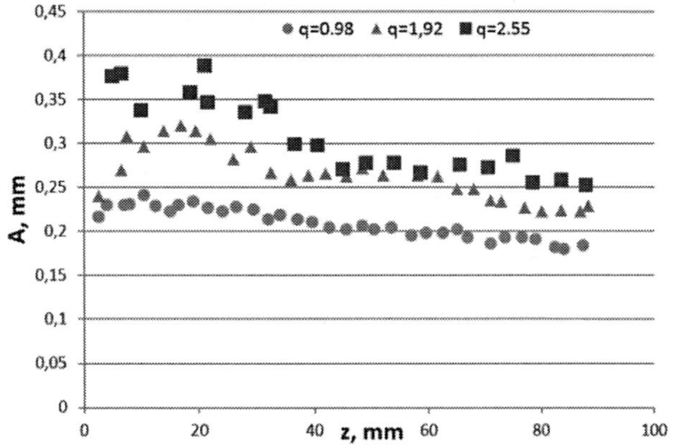

Fig. 1. Amplitude of waves in the interrivulet region for different values of the heat flux, Re = 15.

The growth of wave amplitudes was observed. Fig. 1 presents data on the amplitude of the waves in the rivulets and in the interrivulet region as a function of the film path along the heater (A is the wave amplitude, Z is the coordinate along the heater, and Z = 0 is the beginning of the heater). It can be seen from the graphs that the amplitude of the waves in the upper part of the heater is higher than in the lower part and increases with increasing density of the heat flux.

The front of the wave, during the movement along the heater, was deformed under the action of thermocapillary forces and in the lower part of the heater the flow passed into an rivulet with a thin film between the rivulets. It was found for Re = 15, that for a heat flux density q> 1.5 W/cm², the motion, amplitudes and deformation of the wave front differ sharply from the case q <1.5 W/cm². We see a significant change in the character of the dependence of the amplitude of the waves on the density of the heat flux, Fig. 1.

When high longitudinal gradient values are reached, which occurs at a heat flux density from 1.5 to 2 W/cm², the formation of thermocapillary structures begins. In the region of the leading edge of the heater Z <10-15 mm, on the residual layer of liquid film, after passing the wave front, thermocapillary structures are formed in the form of a series of rivulets with a thin film between them. The distance between the structures is $\Lambda = 10$ mm, which corresponds to the distance between the earlier investigated structures in regime A at other Reynolds numbers. When a new front is flowing onto such structures, a periodic inhomogeneity of the wave front along the thickness in the transverse direction appears. Such perturbation of the front leads to the formation of three-dimensional waves with a transverse distance between the ridges equal to 10 mm. However, at Z > 23 mm the liquid film temperatures close to the temperature of the heater, and the temperature gradients on its surface decrease sharply. In this region three-dimensional waves merge to the lower part of the heater and the formation of thermocapillary structures in regime B appears with an average rivulet spacing of 15 mm.

A similar picture was observed for other Reynolds numbers. So, for Re = 50, when the heat flux density was q = 4 W/cm², the temperature gradients increased to 10 K / mm, which is typical for the appearance of structures of type A. Periodically, these structures were clearly visible. They were formed in the residual layer before the wave front near the leading edge of the heater. The average distance between the structures in the transverse direction corresponded to the average instability wavelength in regime A at lower values of Re. It was found that the thickness of the residual film layer in the region between the rivulets can be greatly reduced. Three-dimensional hydrodynamic waves were destroyed in the upper part of the heater. The determining influence on the process was provided by thermocapillary instability in the upper part of the heater.

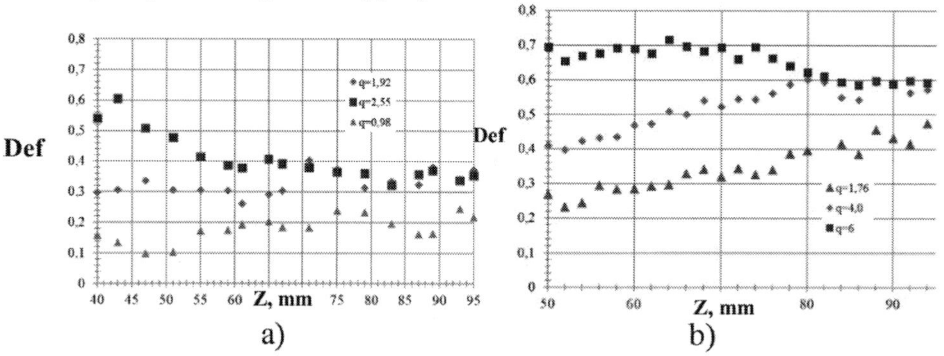

Fig.2. Dependence of the film surface deformation on the longitudinal coordinate along the heater for different heat fluxes a) Re = 15, b) Re = 50.

In Fig. 2. The dependences of the deformation of the film surface on the longitudinal coordinate along the heater for different values of the heat flux are presented. For the quantitative description of transverse deformations in a fluid film, the criterion $Def = (h_{riv}-h_{val})/h_0$ is used, defined as the ratio of the difference between the average film thicknesses on the rivulet crest and in the inter rivulet region to the initial film thickness. As can be seen from the graphs, the dependence character changes at high heat flux densities. At this density the thermocapillary structures in the form of rivulets with a thin film between them form in the residual layer of liquid between the wave fronts. The interaction of waves with these structures leads to an increase in deformation at the beginning of the heater, an increase in the amplitude of the waves by their transformation into a rivulets and the appearance of the rivulet movement (deviation from strictly vertical flow). Further, with an increase of path of the film along the heater, the deformations decrease. At the same time, a rivulet stream with a spacing of about 15 mm is formed in the lower part of the heater, as well as for smaller Re.

In Fig. 3 the comparison of the density of the critical heat flux obtained in this paper with data for heaters with the boundary condition q = const is presented. The initial temperature of the film was maintained equal to 23 °C. It can be seen that the values of the critical heat fluxes obtained in this work lie above the data for heaters, where only structures of type B are formed. The growth of wave amplitudes and deformation of the film surface after the formation of structures of type A, as well as the occurrence of the motion of rivulets in the direction across the flow leads to a substantial increase in the stability of the liquid film to rupture.

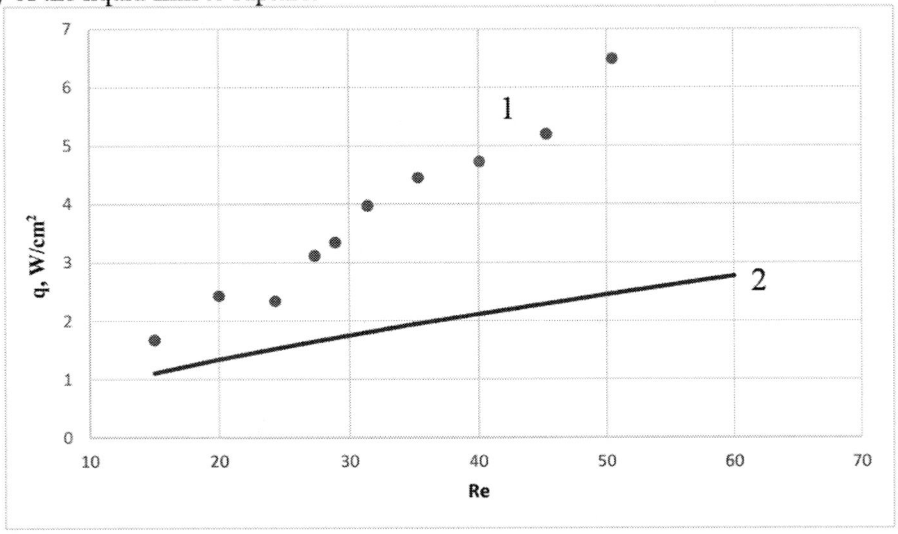

Fig.3. Dependence of the critical heat flux corresponding to the rupture of the liquid film on the Reynolds number. 1 - the critical heat flux density corresponding to the appearance of the first stable dry spot in the center of the heater, after the formation of structures of type A (the result of this work), 2 - the critical heat flux density for heaters with the boundary condition q = const, where only structures of the type B were formed [9].

3 Conclusions

Thus, it has been shown that the formation of metastable thermocapillary structures of type A in the upper part of the heater affects on the magnitude of the critical heat flux corresponding to rupture of the liquid film. The interaction of three-dimensional waves with thermocapillary structures leads to an increase in the amplitudes of the waves by their transformation into a jet and the appearance of the rivulet motion (deviation from strictly vertical flow). In this case, the deformation of the liquid film in the middle part of the heater increases. It is shown that the magnitude of the critical heat flux increases with the Reynolds number of the film and, at high values of Re, significantly exceeds the rupture data in the absence of thermocapillary structures of the type A. It can be concluded that for the first time a new method of influencing the film flow, leading to an increase in film stability was found. The result is important, since the discovered method of increasing the stability of liquid films can be used in industrial installations.

The work was supported by the RF Ministry of Education and Science (Agreement No. 14.604.21.0053, project ID RFMEFI60414X0053).

References

1. C.D. Park, T. Nosoko, AIChE Journal. **49**, 2715 (2003)
2. S.V. Alekseenko, V.V. Guzanov, D.M. Markovich, S.M. Kharlamov, Tech. Phys. Lett. **38**, 739 (2012)
3. S.V. Alekseenko, A.V. Bobylev, V.V. Guzanov, D.M. Markovich, S.M. Kharlamov, Tech. Phys. Lett. **40**, 1031 (2014)
4. Chinnov E.A., Kabov O.A., J. Appl. Mech. Tech. Ph. **44**, 128 (2003)
5. V.V. Lel, A. Kellermann, G. Dietze, R. Kneer, A.N. Pavlenko, Exp. Fluids, **44**, 341 (2008)
6. E.A. Chinnov, E.N. Shatskiy, Tech. Phys. Lett. **42**, 997 (2016)
7. Zaitsev D. V., Rodionov D.A., Kabov O.A., Microgravity science and technology, **19**, 100 (2007)
8. Zaitsev D.V., Kirichenko D.P., Kabov O.A., Technical Physics Letters, **41**, 551 (2015)
9. Zaitsev, D.V., Semenov, A.A., Kabov, O.A., Thermophysics and Aeromechanics, **23**, 625 (2016)
10. E.N. Shatskiy, E.A. Chinnov, High Temp., **54**, 965 (2016)

Mathematical modelling of liquid meniscus shape in cylindrical micro-channel for normal and micro gravity conditions

Igor Marchuk and *Yuriy* Lyulin*

Institute of Thermophysics, 1, Lavrentiev Ave., 630090, Novosibirsk, Russia
Novosibirsk State University, 2, Pirogov Ave., 630090, Novosibirsk, Russia

Abstract. Mathematical model of liquid meniscus shape in cylindrical micro-channel of the separator unit of condensing/separating system is presented. Moving liquid meniscus in the 10 μm cylindrical microchannel is used as a liquid lock to recover the liquid obtained by condensation from the separators. The main goal of the liquid locks to prevent penetration of a gas phase in the liquid line at the small flow rate of the condensate and because of pressure fluctuations in the vapor-gas-liquid loop. Calculation of the meniscus shape has been performed for liquid FC-72 at different values of pressure difference gas - liquid and under normal and micro gravity conditions.

1 Introduction

In recent decades, the use of two-phase flows for cooling of electronic components, such as computer chips, power electronics, converter chips and inverters in hybrid cars, powerful lasers, light-emitting diodes, etc., has developed significantly. One of the promising solutions allowing to remove high heat fluxes is technology using processes with phase change, for example evaporation of a thin liquid film moving in a flat micro-channel under the action of a gas flow [1, 2]. Often it is of fundamental importance for the film to use an inert (non-condensable) gas. The vapour produced during evaporation is mixed with non-condensable gas. To operate in a closed loop, the vapour should be condensed and the resulting liquid separated from the gas. Further liquid and gas has to be separately sent again to the evaporator for cooling. Thus, the creation of a condensing/separating system is one and the most important task for improving the efficiency of the two-phase cooling system. Specifically it plays a central role in thermal control systems for space application [3, 4].

Concept of condenser with liquid separators is shown in Fig. 1. To separate a condensate from a vapor-gas flow two separators are used. Liquid flows from condenser to the slot of the first separator by the tangential stress on the interface and is continuously pumped out by the pump. The gas phase is not completely free of vapour and liquid at this stage. The inert gas flow that contains some residuals of vapour and liquid droplets passes

* Corresponding author: lyulin@itp.nsc.ru

through a second separator. This device is used to extract from the gas phase a very small liquid droplets created in the condenser but transported by the inert gas flow beyond the first separator. The so-called "liquid lock" is used to recover the liquid obtained by condensation from the separators.

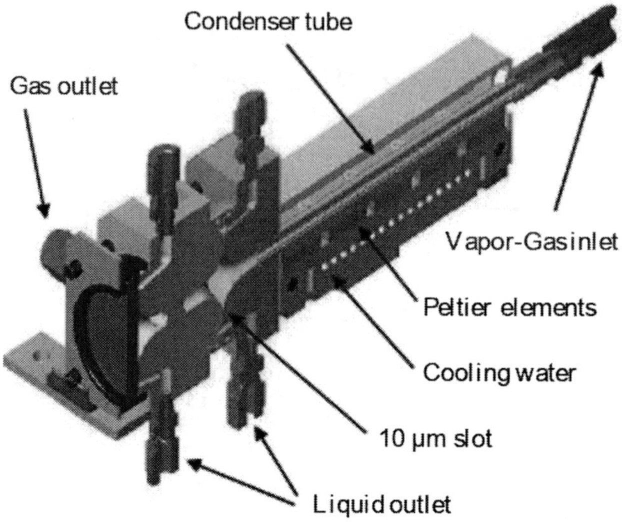

Fig. 1. Cross section of the condensing/separating system.

The main goal of this work is development of a mathematical model of liquid meniscus shape in cylindrical microchannel of the separator unit of condensing/separating system. Moving liquid meniscus is used as the liquid locks to prevent penetration of a gas phase in the liquid line at the small flow rate of the condensate and because of pressure fluctuations in the vapour-gas-liquid loop.

2 Mathematical model

The motion of a two-phase flow in a separator is described by the Navier-Stokes equations. The liquid-gas interface is assumed to be stable, since the process is stationary. Assuming that the flow rate of liquid pumped from the separator is relatively small, it can be considered that the velocities of liquid and gas are zero.

Under gravity meniscus shape is not round and it is deformed by hydrostatic pressure. Let find meniscus shape in this case. Wetting contact angle is equal γ. It means that angle between tangent plane to liquid – gas interface and tangent plane to solid surface in contact point is equal γ.

Pressure difference between liquid and gas is given by Laplas equation:

$$\Delta p_s = \sigma\left(\frac{1}{r_m} + \frac{1}{R_2}\right)$$

In this case $r_m << R_2$. It allows to neglect the radial curvature of the interface and pass on two dimensional problem.

Let x_m is distance from the beginning the channel widening to meniscus, r_m is the meniscus radius, R_s is the radius of separator wall, h is a half of separator gap thickness, R_{in} - inner radius of separator chamber. Real sizes in separator are following: R_s=10 mm, R_{in} =13.35 mm and h=0.005 mm, Fig.2.

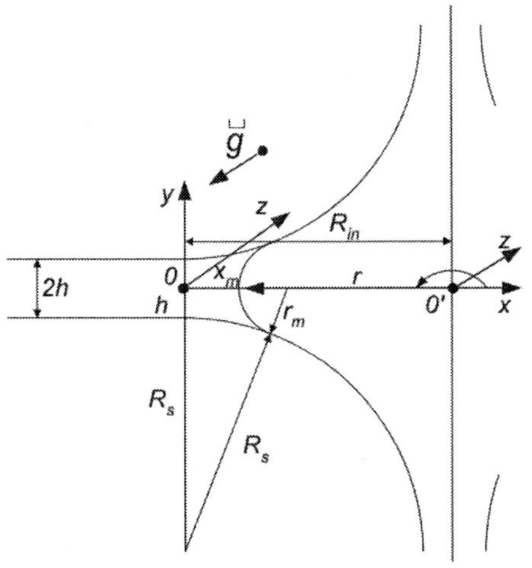

Fig. 2. Geometrical scheme of the meniscus position in separator, axis *z* directed opposite to the vector of gravitation.

The meniscus radius r_m obey the following system of equations:

$$\begin{cases} x^2 + (y + R_s + h)^2 = R_s^2 & \text{solid surface of separator} \\ (x - (r_m + x_m))^2 + y = r_m^2 & \text{liquid gas interface (meniscus)} \\ x(x - (r_m + x_m)) + y(y + R_s + h) = -r_m R_s \cos\gamma & \text{wetting contact angle condition} \end{cases} \quad (1)$$

Excluding from this system coordinates of the contact point *x*, *y* one obtain an expression for radius r_m of the meniscus:

$$r_m = \frac{x_m^2 + 2R_s h + h^2}{2\left(R_s \cos\gamma - x_m\right)} \quad (2)$$

Let pressure difference Δp is known. At gravity condition it satisfy to the expression

$$\Delta p = -\frac{\sigma}{r_m} + \rho g z - \rho g Z_{dp} \quad (3)$$

here $\Delta p = p_l - p_g$ is measured value with detector pressure difference between liquid and gas, Z_{dp} is coordinate *z* of the pressure difference detector sensor location. Hydrostatic pressure in the gas phase is neglected.

Value of coordinate *z* of the interface can be expressed through polar angle ϕ, counting from the positive direction of the axis *x* in the plane *zx*,

$$z = r \sin\phi \quad (4)$$

where *r* is a radius vector from separator centre

$$r = R_{in} - x_m \quad (5)$$

Substitution (2) and (4) into expression for pressure difference (3) gives following equation:

$$\sin\phi = \frac{\Delta p + \rho g Z_{dp}}{\rho g \left(R_{in} - x_m\right)} + \frac{\sigma}{\rho g} \frac{2\left(R_s \cos\gamma - x_m\right)}{\left(x_m^2 + 2R_s h + h^2\right)\left(R_{in} - x_m\right)}$$

Finally, with taking into account (5)

$$\sin\phi = \frac{\Delta p + \rho g Z_{dp}}{\rho g r} + \frac{\sigma}{\rho g}\frac{2\left(r-\left(R_{in}-R_s\cos\gamma\right)\right)}{\left(\left(R_{in}-r\right)^2+2R_sh+h^2\right)r} \qquad (6)$$

This is the equation of the meniscus shape at gravity condition.

At microgravity (g=0) meniscus shape equation (6) takes the form

$$\Delta p + \frac{2\sigma\left(r-\left(R_{in}-R_s\cos\gamma\right)\right)}{\left(\left(R_{in}-r\right)^2+2R_sh+h^2\right)} = 0 \qquad (7)$$

Hence

$$r = R_{in} - \frac{\sigma}{\Delta p} - \sqrt{\left(\frac{\sigma}{\Delta p}\right)^2 - 2hR_s - h^2 - 2\frac{\sigma}{\Delta p}R_s\cos\gamma} \qquad (8)$$

Here r is constant and meniscus has the shape of circumference.

3 Results

Meniscus shape at different values of pressure difference liquid – gas, given at the level of separator centre is shown in Fig. 3. Solid circumference is interior boundary of separator gap. Calculation of equation (8) is performed for liquid FC72 and Z_{dp}=0. One can use dimensionless criterion $K_{dp}=\rho g R_{in}/\Delta p$ for evaluation of gravitation influence on meniscus shape. At $K_{dp}<1$ the gravitation influence on meniscus shape is weak and meniscus shape is close to circumference. At K_{dp} of order of unity and higher meniscus is flattened from below by the gravity, Fig 3.

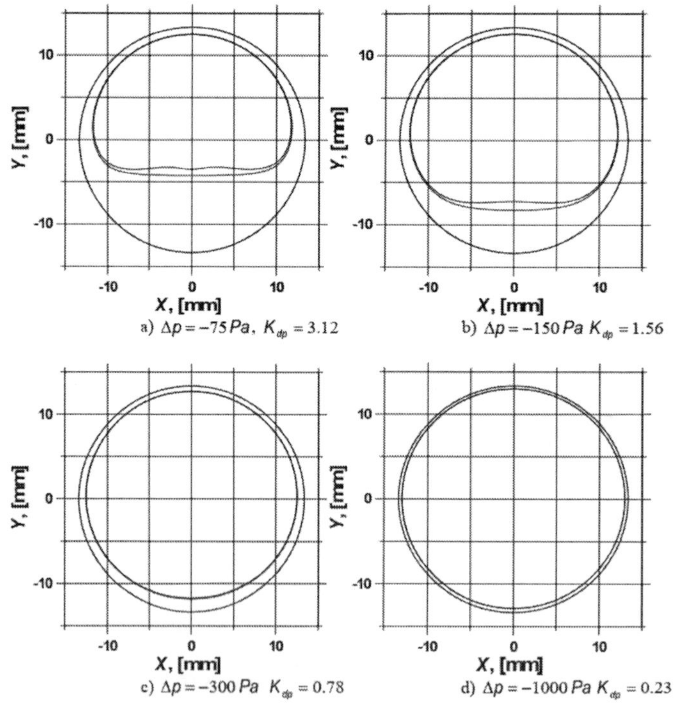

a) $\Delta p = -75\,Pa,\ K_{dp} = 3.12$ b) $\Delta p = -150\,Pa\ K_{dp} = 1.56$

c) $\Delta p = -300\,Pa\ K_{dp} = 0.78$ d) $\Delta p = -1000\,Pa\ K_{dp} = 0.23$

Fig. 3. Meniscus shape at different values of pressure difference liquid – gas.

3 Conclusions

Mathematical model of liquid meniscus shape in cylindrical micro-channel of the separator unit of condensing/separating system has been developed. Moving liquid meniscus in the 10 μm cylindrical micro-channel is used as a liquid lock to recover the liquid obtained by condensation from the separators. The main goal of the liquid locks to prevent penetration of a gas phase in the liquid line at the small flow rate of the condensate and because of pressure fluctuations in the vapor-gas-liquid loop. Calculation of the meniscus shape has been performed for liquid FC-72 at different values of pressure difference gas - liquid and under normal and micro gravity conditions.

The authors gratefully acknowledge the support of this work by the Ministry of Education and Science of the Russian Federation (Agreement 14.616.21.0016, project identifier RFMEFI61614X0016).

References

1. A. Bar-Cohen, P. Wang, ASME. J. Heat Transfer, **134**, 051017 (2012)
2. O.A. Kabov, D.V. Zaitsev, V.V. Cheverda, A. Bar-Cohen, Experimental Thermal and Fluid Science, **35**, 825 (2011)
3. V. Serin, P. Lavieille, M. Miscevic, Microgravity Science and Technology, **21** (1), 103–109, (2009)
4. I.V. Marchuk and O.A. Kabov, Encyclopedia of Two-Phase Heat Transfer and Flow II: Special Topics and Applications, **3**, **5**, 133 (2015)

AUTHOR INDEX

AUTHOR INDEX

EDP Sciences – Web of Conferences
17, Avenue du Hoggar
Parc d'Activité de Courtabouf
BP 112
F-91944 Les Ulis Cedex A
France

ISBN 978-1-5108-5054-5

CURRAN ASSOCIATES INC.
proceedings.com